Breaking Tolerance to Antibody-Mediated Immunotherapy

NOVEL SENSITIZING AGENTS FOR THERAPEUTIC ANTI-EGFR ANTIBODIES

VOLUME 1

Breaking Tolerance to Antibody-Mediated Immunotherapy

Series Editor: Benjamin Bonavida, Ph.D.

Volume 1: Novel Sensitizing Agents for Therapeutic Anti-EGFR Antibodies
Edited by Shi Hu, M.D., Ph.D.

Upcoming Volumes:

Overcoming Cancer Resistant to HER-2 Antibodies
Edited by Benjamin Bonavida, Ph.D.

New Combination Approaches to Enhance Rituximab-Based Lymphoma Therapies
Edited by Benjamin Bonavida, Ph.D.

Resistance to Anti-CD20 Antibodies and Approaches for Their Reversal
Edited by William Chi Shing Cho, Ph.D.

Resistance to Anti-VEGFR1/2 and Mechanisms to Overcome the Resistance
Edited by Bruno Fernandes de Carvalho, Ph.D.

Breaking Tolerance to Antibody-Mediated Immunotherapy

NOVEL SENSITIZING AGENTS FOR THERAPEUTIC ANTI-EGFR ANTIBODIES

VOLUME 1

Edited by

SHI HU, M.D., Ph.D.
Department of Biophysics, College of Basic Medical Sciences, Naval Medical University (Second Military Medical University), Shanghai, China

ELSEVIER

ACADEMIC PRESS
An imprint of Elsevier

Academic Press is an imprint of Elsevier
125 London Wall, London EC2Y 5AS, United Kingdom
525 B Street, Suite 1650, San Diego, CA 92101, United States
50 Hampshire Street, 5th Floor, Cambridge, MA 02139, United States
The Boulevard, Langford Lane, Kidlington, Oxford OX5 1GB, United Kingdom

ISBN: 978-0-12-821584-5

For information on all Academic Press publications
visit our website at https://www.elsevier.com/books-and-journals

Publisher: Stacy Masucci
Acquisitions Editor: Linda Versteeg-Buschman
Editorial Project Manager: Samantha Allard
Production Project Manager: Punithavathy Govindaradjane
Cover Designer: Vicky Pearson

Typeset by STRAIVE, India

Cover Image Insert

The background circuit represents the genetic background of different patients, and the different antibodies represent that the EGFR antibodies have different effects and drug resistance mechanisms in different genetic backgrounds.

Aims and Scope of Series "Breaking Tolerance to Antibody-Mediated Immunotherapy"

The role of the immune system in the eradication of cancer has been the subject of numerous controversial reports. In particular, the role of anticancer antibodies in the prevention and treatment of established cancers in experimental animal models has resulted often in both the protection and the rejection of various experimental tumors; however, the clinical findings in humans were not encouraging and yielded poor responses. Noteworthy, in 1997, the FDA approved the first chimeric anti-CD20 monoclonal antibody, rituximab, or Rituxan, for the treatment of resistant non-Hodgkin lymphoma. The clinical responses achieved were very good for a subset of patients who were treated with a single agent alone, although the clinical response rates were significantly improved with the combination of rituximab and chemotherapy. As a result of the initial successful finding of the therapeutic efficacy of rituximab, in the subsequent years, a large number (>25) of FDA-approved monoclonal antibodies were generated against various cancers, the majority of which were directed at and targeted cancer cells and a few targeted noncancerous cells. However, it was found that not all patients responded to antibody treatment initially and many responding patients developed resistance to further treatments. Therefore, it was important to develop novel strategies to treat the unresponsive patients. Analyses of the various underlying mechanisms of antibody resistance in various cancers have revealed a number of gene products, resistant factors, that can be targeted by various specific interventions, namely, small molecules that can be termed sensitizing agents. The combination of sensitizing agents and antibody therapy resulted, in many cases, in reversal of resistance in various experimental models and also clinically.

The main aim of the new series "Breaking Tolerance to Antibody-Mediated Immunotherapy" is to develop highly focused volumes, each of which will review the role of particular and specific sensitizing agents that reverse the resistance of various therapeutic antibodies in different cancers. In each volume, the chapter reviews will have a detailed discussion on the biochemical, molecular, and underlying genetic mechanisms by which the sensitizing agents mediate their effects when used alone or in combination with a therapeutic antibody.

The scope of the series is to provide scientists and clinicians with updated basic, fundamental, and clinical findings. Such updated information will be helpful for further investigations and the development of more potent and different antibody sensitizing agents for various cancers.

Benjamin Bonavida, Ph.D. (Series Editor)

About the Series Editor

Dr. Benjamin Bonavida, Ph.D. (Series Editor), is currently distinguished research professor at the University of California, Los Angeles (UCLA). He is affiliated with the Department of Microbiology, Immunology and Molecular Genetics, UCLA David Geffen School of Medicine. His research career, thus far, has focused on investigations in the fields of basic immunochemistry and cancer immunobiology. His research investigations have ranged from the biochemical, molecular, and genetic mechanisms of cell-mediated killing and tumor cell resistance to chemoimmuno cytotoxic drugs. The reversal of tumor cell resistance was investigated by the use of various selected sensitizing agents based on molecular mechanisms of resistance. In these investigations, there was the newly characterized dysregulated NF-κB/Snail/YY1/RKIP/PTEN loop in many cancers that was reported to regulate cell survival, proliferation, invasion, metastasis, and resistance. Emphasis was focused on the roles of the tumor suppressor Raf kinase inhibitor protein (RKIP), the tumor promoter Yin Yang 1 (YY1), and the role of nitric oxide as a chemo-immuno-sensitizing factor. Many of the aforementioned studies are centered on the clinical challenging features of cancer patients' failure to respond to both conventional and targeted therapies.

The editor has been active in the organization of regular sequential international mini-conferences that are highly focused on the roles of YY1, RKIP, and nitric oxide in cancer and their potential therapeutic applications. Several books edited or coedited by the editor have been published. In addition, the editor has been the series editor of books (over 23) published by Springer on *Resistance to Anti-Cancer Targeted Therapeutics*. In addition, the editor is presently the series editor of three series published by Elsevier/Academic Press on *Cancer Sensitizing Agents for Chemotherapy*, *Breaking Tolerance to Anti-Cancer Cell-Mediated Immunotherapy*, and *Breaking Tolerance to Antibody-Mediated Immunotherapy*. Lastly, the editor is the editor-in-chief of the journal *Critical Reviews in Oncogenesis*. The editor has published over 500 research publications and reviews in various scientific journals of high impact.

Acknowledgments: The editor wishes to acknowledge the excellent editorial assistance of Ms. Inesa Navasardyan, who has worked diligently for the completion of this volume, by both the editing and formatting the various contributions to this volume. Ms. Navasardyan has recently graduated from UCLA (June 2020) and has also contributed a review chapter in this volume on triple-negative breast cancer.

The editor acknowledges the Department of Microbiology, Immunology and Molecular Genetics and the UCLA David Geffen School of Medicine for their continuous support. The editor also acknowledges the assistance of Mr. Rafael Teixeira, Acquisitions Editor, for Elsevier/Academic Press, and the excellent assistance of Ms. Samantha Allard, Editorial Project Manager, for Elsevier/Academic Press, for their continuous cooperation throughout the development of this book.

Aims and Scope of Volume

The epidermal growth factor receptor (EGFR) signaling pathways play a crucial role in cancer development. The therapeutic application of anti-EGFR antibodies in the prevention and treatment of established cancers in experimental animal models has often resulted in both the protection and the rejection of various experimental tumors; however, the clinical findings in humans were not encouraging and yielded poor responses.

The main aim of the volume *Novel Sensitizing Agents for Therapeutic Anti-EGFR Antibodies* is to present an in-depth description of sensitizers to overcome resistance to anti-EGFR-targeted antibody therapies in cancer. The sensitizers-based therapies that are discussed in this volume include novel engineered antibody drugs or other sensitizers. This volume provides an insight into the landscape of anti-EGFR-based cancer treatment, the challenges of targeted therapy, and a glimpse into the future of antibody therapy.

The scope of the volume is to provide scientists and clinicians with updated findings when they face the problem of anti-EGFR antibody resistance in basic research, pharmaceutical research, and clinical research. The content provides current knowledge to the research community to overcome such resistance and, therefore, will help them in their own research field.

Shi Hu, M.D., Ph.D.

About the Volume Editor

Shi Hu is an Associate Professor of Biomedical Engineering in the Department of Biophysics, Second Military Medical University, China. Professor Hu's research is focused on synthetic immunology and antibody engineering through the development of novel immune therapeutics. In the last decade, he has focused on the development of novel sensitizers for antibody-based therapies (e.g., anti-ERBB antibodies) that underlie the mechanism research of tumor heterogeneity and plasticity. He has authored over 40 scientific publications in peer-reviewed journals and additional reviews and chapters in books on the molecular, biochemical, and cellular aspects of synthetic immunology and antibody engineering.

Preface

Monoclonal antibody-based cancer therapy has been considered one of the most successful therapeutic strategies for hematologic malignancies and solid tumors in the last 2 decades. The epidermal growth factor receptor (EGFR) signaling pathways play a crucial role in cell proliferation, survival, differentiation, regulation as well as cancer development. Hence, anti-EGFR mAbs are among the most promising approaches in cancer therapy, and they are now widely used in clinics worldwide. Monoclonal antibody-based cancer treatment has been established as one of the main therapeutic strategies in the last century. Despite great success in targeted cancer therapy, the resistance to mAb has recently become a significant issue. In a relatively short period, therapeutic sensitizers have entered the research stream of anti-EGFR therapy. However, the successful development of therapeutic sensitizers needs to be established on the precise knowledge of a specific target that drives disease progression, validation of a clinically applicable biomarker, acceptance of a rational endpoint, and understanding of the mechanisms of resistance. Therefore, a thorough publication of therapeutic sensitizers, covering the basic and translational research of such novel agents, will certainly be on the horizon and bodes well for a renaissance of interest in antibody-based immunotherapy, as well as for patients.

The main objective of this book titled *Novel Sensitizing Agents for Therapeutic Anti-EGFR Antibodies* is to provide the integration of basic and clinical studies of novel sensitizers to reverse the resistance of anti-EGFR antibodies, covering both drug discovery and translational research extensively. It offers pertinent basic science information on strategies used for the rational design and discovery of novel sensitizing agents, and, in addition, translational studies involving preclinical and clinical designs and applications of these novel agents. This book covers basic science strategies that are being used in drug discovery and preclinical evaluation focused on EGFR blockage of resistance, as well as clinical trial methodology including clinical pharmacokinetics and imaging to address issues of efficacy evaluation of the new anticancer sensitizers for anti-EGFR drug resistance.

This book consists of 28 review chapters; the highlights of each chapter's contribution are as follows:

Chapter 1 reviews the current mainstream cancer treatment options and related studies targeting EGFR in cancer therapy, and it also provides a comprehensive list of EGFR-targeted cancer therapy drugs. This chapter discusses these newly developed preclinical-stage novel anti-EGFR cancer therapies and medications.

Chapter 2 reviews the biological characteristics of EGFR, its role in human cancer, and the development of anti-EGFR antibody therapies. This chapter further discusses the molecular mechanisms of cetuximab resistance and potential strategies for circumventing such resistance.

Chapter 3 reviews the oligoclonal antibody MM-151, developed to bind to several regions of the EGFR extracellular domain to address resistance and recurrence caused by mutations in the EGFR.

Chapter 4 reviews the design and development of SYM004 (an antibody combination drug synthesized by two monoclonal antibodies against ectopic epitopes of EGFR) in cetuximab-resistant cancer models and highlights the importance of acquired resistance to anti-EGFR therapy caused by MET activation.

Chapter 5 reviews the usage of targeted vector immune prophylaxis to block a variety of receptor tyrosine kinases in the reduction of cancer recurrence rate and the delay of anti-EGFR antibody resistance.

Chapter 6 reviews Pan-HER, a therapeutic drug targeting members of the HER family. Pan-HER is not a novel drug targeting a member of the HER family but a cocktail of six antibodies capable of targeting EGFR, HER2, and HER3. By targeting three targets simultaneously, Pan-HER can effectively prevent compensatory tumor-promoting mechanisms in vivo.

Chapter 7 reviews the development of FL518 (a four-in-one antibody) and DVD-Ig CRTB6 (a tetravalent specific antibody), which can recognize EGFR, HER2, HER3, and vascular endothelial growth factor (VEGF) while simultaneously inhibiting HER-MET crosstalk.

Chapters 8 and 9 review the critical role of the HER3 signaling pathway in the acquired resistance of EGFR inhibitors and describe the combined EGFR and HER3 therapy to overcome acquired resistance to cetuximab.

Chapter 10 reviews the underlying mechanisms for ERBB2-mediated cetuximab resistance. It describes the functional results of simultaneous inhibition of ERBB2 and EGFR in overcoming cetuximab resistance in a series of preclinical models and patients.

Chapter 11 reviews the role of Notch signaling and the expansion of cancer stem cells in accompanying resistance to EGFR-targeted treatment and radiation. It further introduces a dual antibody targeting EGFR and Notch2/3 signaling in potential models in combination with radiotherapy.

Chapter 12 reviews the study that supports cetuximab combined with bromine domain inhibition that provides a novel therapeutic strategy for neck squamous cell carcinoma treatment.

Chapter 13 summarizes the results of Phase IB/II studies that evaluated the efficacy and toxicity profile of the combination of cetuximab, dasatinib, and FOLFOX in metastatic colorectal cancer.

Chapter 14 reviews the development of the antibody JNJ-61186372, a bispecific EGFR-cMet antibody, to inhibit the effects of EGFR mutations and cMet pathway activation.

Chapter 15 emphasizes the hepatocyte growth factor-dependent mesenchymal-epithelial transition and the potential to use Met kinase inhibitor to overcome cetuximab resistance in colorectal cancer.

Chapters 16 and 17 focus on the crosstalk of the VEGF and EGFR pathways and review the antitumor effect of VEGF(R) inhibitors combined with anti-EGFR therapy.

Chapter 18 reviews the use of KRAS-specific siRNA coupled with anti-EGFR antibody cetuximab to treat KRAS-mutated cancers. The siRNA-antibody complex is a promising candidate to address the unmet clinical need in combating KRAS-driven human cancers.

Chapter 19 reviews the activation of MEK involved in primary and acquired resistance to cetuximab and inhibition of EGFR and MEK, which may be a strategy to overcome anti-EGFR resistance in colorectal cancer patients.

Chapter 20 reviews the ability of two MEK inhibitors to respond to EGFR inhibitor resistance in patients with K-ras mutations in both cells and xenotransplantation.

Chapter 21 reviews the role of JIP-4/JNK2 signal bypass in reducing cetuximab tolerance.

This chapter further discusses the resistance mechanism of cetuximab and the intracellular phosphorylation signals in treating head and neck squamous cell carcinomas.

Chapter 22 reviews the development of bispecific antibodies against the EGFR and Notch signal pathways, which could help to overcome the problems caused by the resistance to inhibitors of EGFR and the PI3K-Akt signaling.

Chapter 23 reviews the possibility of targeting metabolic targets to overcome cetuximab resistance. The glutamine transporter solute carrier family 5 (SLC1A5) was identified as a potential therapeutic target to overcome cetuximab resistance and resensitize cancer cells to antitumor therapies in metastatic colorectal cancer.

Chapter 24 describes that targeting EGFR and PYK2/FAK synergistically inhibited basal-like triple-negative breast cancer cell proliferation and attenuated tumor growth in a xenograft model in vitro. Dual targeting of EGFR and PYK2/FAK inhibited complementary critical growth and survival signal pathways activated in tumors.

Chapter 25 reviews autophagy and EGF signaling pathways in drug-resistant tumor cells. The chapter further points out that autophagy inhibitors target autophagosome formation or fusion of autophagosomes and that lysosomes work well in some cancers, such as enhancing sensitivity to EGFR tyrosine kinase inhibitors (EGFR-TKIs).

Chapter 26 describes the activation of the hedgehog signaling pathway to integrate with EGFR signaling to induce tumorigenesis and drug resistance. A phase I clinical trial (NCT01255800) found that the combinatorial therapy of hedgehog signaling and EGFR blockers appears tolerable and showed signs of efficacy.

Chapter 27 reviews the effect of an immune-modified second-generation oligonucleotide on the growth of cetuximab-resistant tumor models. It also suggests the combination of Toll-like receptor 9 agonists, cetuximab, and chemical agent irinotecan in the treatment of cancer.

Chapter 28 describes the molecular bases of the mechanism of monoclonal antibodies directed to the EGFR family and their sensitizing agents capable of reversing tumor resistance. Future therapeutic approaches are suggested to involve the combination of different antineoplastic-targeted agents.

The editor Dr Shi Hu wishes to acknowledge the excellent assistance of Jian Zhao for her help with the various contents of this volume.

Shi Hu, M.D., Ph.D.
Department of Biophysics, College of Basic Medical Sciences, Naval Medical University (Second Military Medical University), Shanghai, China

Contents

Contributors xxv

1. Current status of anti-EGFR agents

Tian Li, Wenyan Fu, Changhai Lei, and Shi Hu

Introduction 2
Targeting EGFR 3
Novel EGFR targeting strategies 7
Conclusions 10
References 10

2. Molecular mechanisms of resistance to the EGFR monoclonal antibody cetuximab

Zetong Ma, Changhai Lei, and Shi Hu

Introduction 14
EGFR biology 14
The role of EGFR in human cancer 16
The function of EGFR-targeted antibodies 16
Cetuximab in the clinic 16
Mechanisms of resistance to EGFR-targeted antibodies 17
EGFR mutations 17
EGFR gene copy number as a predictor of response 17
EGFR ligand state 17
KRAS mutations as a predictor of response 18
BRAF mutation as a predictor of response 18
Mechanisms of resistance against EGFR-targeted antibodies 18
Disorders of EGFR internalization and degradation 19
Oncogenic shift 19
Subcellular localization of EGFR 20
Epithelial cells transform into mesenchymal cells 20
Constitutive activation of EGFR downstream effector molecules 21
Increased expression of HER family growth factors 21
Concluding remarks 22
References 22

3. MM-151 overcomes acquired resistance to cetuximab and panitumumab in colorectal cancers harboring EGFR extracellular domain mutations

Zetong Ma, Changhai Lei, and Shi Hu

Introduction 30
EGFR ECD mutations do not affect ligand–receptor binding but do affect antibody-receptor binding 30
MM-151 can inhibit the growth of LIM1215 cells overexpressing EGFR ECD mutants 31
MM-151 can inhibit the growth of cetuximab-resistant clones in spontaneous models 31
MM-151 reduced the frequency of EGFR-ECD allele mutations 32
Discussion 32
References 33
Further reading 34

4. Sym004 anti-EGFR antibody mixture overcomes resistance to anti-EGFR antibodies in metastatic colorectal cancer

Guangyao Li, Tian Li, Wenyan Fu, Changhai Lei, and Shi Hu

EGFR and CRC 36
SYM004 compared with cetuximab 36
Cell proliferation and induction of apoptosis 36
EGFR-dependent intracellular signaling/MET activation and ERBB2 amplification 37
Tumor xenograft models of human colorectal cancer 37
References 39

5. Broad RTK-targeted therapy overcomes molecular heterogeneity-driven resistance to cetuximab via vectored immunoprophylaxis in colorectal cancer

Guangyao Li, Changhai Lei, and Shi Hu

Study on the related functional signaling network pathway based on the drug resistance of targeted monoclonal antibodies in CRC 42
Phased therapeutic strategy of antibody transformation in vivo based on VIP background 42
In vivo experiments were conducted to characterize the relationship between pathway crosstalk of key gene effectors and drug resistance of targeted monoclonal antibodies 43
Cross design and effect presentation of VIP therapeutic drugs and carriers 44
Final summary and assumption 45
References 45

6. Pan-HER, an antibody mixture, simultaneously targeting EGFR, HER2, and HER3 effectively overcomes resistance to anti-EGFR antibodies

Tian Li, Changhai Lei, and Shi Hu

7. Four-in-one antibodies have superior cancer inhibitory activity against EGFR, HER2, HER3, and VEGF through disruption of HER/MET crosstalk

Zetong Ma, Tian Li, Wenyan Fu, Changhai Lei, and Shi Hu

Introduction 56
Design and characterization of four-in-one antibodies 56
Four-in-one antibodies can effectively inhibit EGFR, HER2, HER3, and VEGFR2 57
The four-in-one antibody has excellent antitumor activity both in vitro and in vivo 57
Only four-in-one antibodies interfere with HER/MET crosstalk 58
MET/HER signaling is critical for resistance to HER-targeted therapy 58
Four-in-one antibodies can overcome resistance to HER therapy 59
Discussion 59
References 60

8. Overcoming acquired resistance to cetuximab by combining EGFR- and HER3-neutralizing monoclonal antibodies

Kewen Qian, Changhai Lei, and Shi Hu

EGFR inhibitor tolerance is related to both EGFR and HER3, which MEHD7945A completely targets 64
MEHD7945A inhibits the growth of two tumor cell lines that are resistant to cetuximab 64
MEHD7945A inhibits the growth of human tumor xenografts that are resistant to EGFR inhibitors 65
MEHD7945A inhibits radiation-induced survival and overcomes cross-resistance to radiation 65
mAb33: A neutralizing anti-HER3 antibody that prevents resistance to an EGF kinase inhibitor 65
Combining osimertinib with mAb33 overcomes resistance by upregulating HER3 and inhibiting the release of the soluble extracellular domain 65
Combination of mAb33, cetuximab, and osimertinib inhibits growth of human tumor xenografts by simultaneously inhibiting other RTKs 66
The drug combination comprising mAb33 delays relapse of a model derived from a patient's metastatic lesion 66
Discussion 67
References 67

9. Dual targeting of EGFR and HER3 with MEHD7945A overcomes acquired resistance to EGFR inhibitors and radiation

Guangyao Li and Shi Hu

HER3 is an essential regulatory target for acquired drug resistance 70
The mechanism of MEHD7945A in overcoming acquired drug resistance through tumor xenografts and cells 71
Study on the mechanism of MEHD7945A in radiation-damaged cells 72
Final thoughts and future directions 73
References 73

10. Targeting ERBB2 overcomes resistance to the anti-EGFR therapeutic antibody cetuximab

Chuqi Wang, Tian Li, Wenyan Fu, and Shi Hu

Introduction 76
ERBB2 amplification mediates cetuximab resistance through Erk1/2 signaling 76
Heregulin activates ERBB2 signaling and mediates resistance to cetuximab without ERBB2 amplification 77
611-CTF, a C-terminal fragment of HER2, is phosphorylated in cetuximab-resistant cells 78
Targeting ERBB2 signaling overcomes resistance to cetuximab: Preclinical studies 79
Dual inhibition of ERBB2 and EGFR in cetuximab-resistant patients: Clinical trials 79
Conclusion 80
References 80

11. CT16: A dual antibody targeting both EGFR and Notch suppresses EGFR blockage and radiation resistance by decreasing tumor-initiating cell frequency

Kewen Qian, Changhai Lei, and Shi Hu

Introduction 84
Combined use of EGFR inhibitors and irradiation can promote EMT and stem cell-like properties in NSCLC cells 84
Enrichment of the ALDH^{+} cell population depends on Notch2/3 receptors 85
Dual blockade of EGFR and Notch2/3 inhibits resistance to EGFR inhibitors in vitro 85
Dual blockade of EGFR and Notch2/3 inhibits resistance to EGFR inhibitors in vivo 86
CT16 enhances the radiosensitivity of cancer cells and reduces the ALDH^{+} cell populations 86
CT16 can reduce the proliferation of tumor cells and shows good antitumor activity 87
Dual blockade of EGFR and Notch2/3 reduces the number of CSCs and delays tumor recurrence after radiotherapy 87
Discussion 88
References 88

12. BET inhibition overcomes receptor tyrosine kinase-mediated cetuximab resistance in HNSCC

Zetong Ma, Tian Li, Wenyan Fu, and Shi Hu

Introduction 92
The HNSCC model of acquired cetuximab resistance relies on surrogate RTK activity 92
Pharmacological inhibition of BET family members overcame cetuximab resistance in vitro 93
Blocking BET can inhibit the expression of RTK in the HNSCC model of cetuximab 94
RTK overexpression mediates resistance to cetuximab and BET inhibition 95
BET blocking prevents the acquisition of cetuximab resistance in the PDX model of HNSCC patients 95
Expression of RTK and BRD4 genes in a clinical cohort of HNSCC 95
Conclusion 96
References 96

13. Dual inhibition of EGFR and c-Src by cetuximab and dasatinib combined with FOLFOX chemotherapy in patients with metastatic colorectal cancer

Kewen Qian, Wenyan Fu, Changhai Lei, and Shi Hu

Combination synergistic chemotherapy of mCRC treatment 100
Phase IB: Safety and efficacy 100
Phase II: The role of KRAS mutation 101
RAS amplification and KRAS mutation 101
IBD and KRAS-amplified mCRC 102
Discussion and future directions 102
References 102

14. A novel bispecific antibody targeting EGFR and cMet is effective against EGFR inhibitor-resistant lung tumors

Tian Li, Changhai Lei, and Shi Hu

15. Use of MET kinase inhibitors to overcome cetuximab resistance in colorectal cancer

Zetong Ma, Wenyan Fu, Changhai Lei, and Shi Hu

Introduction 114
EGFR and MET are coexpressed in a genotypically diverse panel of cetuximab-sensitive CRC cell lines 114
Combined activation of EGFR and MET enhances CRC cell proliferation by augmenting the activation of AKT and MAPK 115
HGF induces resistance to cetuximab through MET activation 115
HGF rescues CRC cells from cetuximab-induced G1 arrest 115
HGF rescued the apoptosis of DiFi cells induced by cetuximab 116
HGF rescue occurs via MET-dependent activation of AKT and MAPK 116
Discussion 116
References 117

16. EGFR- and VEGF(R)-targeted small molecules show synergistic activity in colorectal cancer models refractory to combinations of monoclonal antibodies

Guangyao Li, Tian Li, Wenyan Fu, and Shi Hu

An important oncogenic signaling pathway: Intracellular signal transduction 120
Mechanistic differences and activity comparison between TKIs and mAbs 121
Final thoughts and future directions 122
References 122

17. Antitumor activity of the VEGFR inhibitor ZD6474 in human cancer cells resistant to anti-EGFR therapy

Chuqi Wang, Wenyan Fu, Changhai Lei, and Shi Hu

Introduction 126
ZD6474 efficiently inhibited the growth of EGFR-resistant tumors via inhibition of VEGFR 126
The combination of ZD6474 and cetuximab displayed a synergistic antitumor effect against multiple cancer cell lines 127
Discussion 128
References 129

18. Antibody-mediated delivery of anti-KRAS-siRNA overcomes therapy resistance in colon cancer

Chuqi Wang, Changhai Lei, and Shi Hu

Introduction 132
Generation and characterization of KRAS-siRNA-anti-EGFR antibody complexes 132
In vitro validation of KRAS-siRNA-anti-EGFR antibody complexes: KRAS silencing and tumor inhibitory activity 133
In vivo antitumor effect of the cetuximab-KRAS-esiRNA complex 134
Final thoughts and future directions 134
References 134

19. Primary and acquired resistance of colorectal cancer cells to anti-EGFR antibodies converge on MEK/ERK pathway activation and can be overcome by combined MEK/EGFR inhibition

Tian Li and Shi Hu

20. MEK1/2 inhibitors AS703026 and AZD6244 may be potential therapies for KRAS-mutated colorectal cancer that is resistant to EGFR monoclonal antibody therapy

Guangyao Li, Changhai Lei, and Shi Hu

The status of K-RAS mutations is the key influencing factor in determining cetuximab resistance sensitivity 146
Efficacy in the ERK pathway, proliferation, transformation, and tumor growth in cetuximab-resistant tumor cells caused by K-RAS mutation 147
The development of MEK inhibitors and the discovery of their mechanism of action 147
Final thoughts and future directions 148
References 148

21. EGFR/JIP-4/JNK2 signaling attenuates cetuximab-mediated radiosensitization of squamous cell carcinoma cells

Kewen Qian, Tian Li, Wenyan Fu, Changhai Lei, and Shi Hu

3-Dimensional (3D) extracellular matrix (ECM) for studying how HNSCC cells grow in vivo 152

Growth conditions play an important role in EGFR phosphorylation and signaling 152
Cytotoxicity and radiosensitizing potential of cetuximab 153
Phosphoproteome modifications after cetuximab-mediated EGFR inhibition 153
JNK2 depletion enhances cetuximab-induced radiosensitization 153
JIP-4 connects EGFR to JNK2 154
Discussion 154
References 154

22. EGFR/Notch antagonists enhance the response to inhibitors of the PI3K-Akt pathway by decreasing tumor-initiating cell frequency

Tian Li, Changhai Lei, and Shi Hu

23. Inhibition of SLC1A5 sensitizes colorectal cancer to cetuximab

Zetong Ma, Tian Li, Wenyan Fu, Changhai Lei, and Shi Hu

Introduction 162
SLC1A5 is overexpressed in cancer tissues of cetuximab-resistant CRC patients 162
SLC1A5 promotes the proliferation of CRC both in vitro and in vivo 162
Targeting SLC1A5 significantly enhanced the inhibitory effect of cetuximab on CRC proliferation 163
Targeting SLC1A5 increases ubiquitin–proteasome pathway-mediated EGFR degradation 163
Inhibition of SLC1A5 induces DNA damage in CRC cells 163
Discussion 164
References 165
Further reading 166

24. Targeting of PYK2 synergizes with EGFR antagonists in basal-like TNBC and circumvents HER3-associated resistance via the NEDD4–NDRG1 axis

Tian Li, Changhai Lei, and Shi Hu

25. Overcoming acquired resistance to EGFR-targeted therapy by regulating autophagy

Kewen Qian, Wenyan Fu, and Shi Hu

Overlap between EGFR and autophagy regulatory signaling 174
Autophagy facilitates tumor resistance to targeted biological therapies 174
Targeting autophagy helps solve the problem of resistance to anti-EGFR treatments 175
miRNAs and EVs regulate both autophagy and the response of cancer cells to anti-EGFR treatments 175
Antiautophagic peptides that modulate the effect of anti-EGFR therapy on tumor cells 176
Discussion 176
References 176

26. Hedgehog signaling alters reliance on EGF receptor signaling and mediates anti-EGFR therapeutic resistance in head and neck cancer

Chuqi Wang, Changhai Lei, and Shi Hu

Introduction 180
Crosstalk between EGFR and HhP in HNSCC 181
Combination of cetuximab and IPI-926 in preclinical models 182
Conclusion 183
References 183

27. Novel Toll-like receptor 9 agonist induces epidermal growth factor receptor (EGFR) inhibition and synergistic antitumor activity with EGFR inhibitors

Kewen Qian, Changhai Lei, and Shi Hu

The ability of CpG DNA to affect tumor growth related to EGFR pathways 186
IMO combined with cetuximab inhibits the expression and angiogenesis of signaling proteins and causes potent antitumor activity 186
The in vitro effect of IMO is not obvious 187
Cooperation of IMO and the EGFR tyrosine kinase inhibitor gefitinib has a combined effect 187
IMO has no inhibitory effect on the growth of cetuximab-resistant GEO-CR xenografts but still cooperates with cetuximab 187
Tumor xenografts can achieve eradication under administration of IMO in combination with cetuximab and irinotecan 187
Discussion 188
References 188

28. Targeted cancer therapy: The future of drug combinations

Débora D. Meira, Matheus C. Casotti, Raquel F.R. Braga, Luiz C.G.S. Filho, Alessandra P. Guimarães, Camilly V. Campanharo, Daniel A. Duque, Débora G. Barbosa, Lorena M. Lopes, Victor N.G. Kohls, Lyvia N.R. Alves, Fernanda M. Garcia, Fernando L.H. Vieira, Vitor H. Almeida, Eldamaria V.W. Santos, Raquel S. Reis, and Iúri D. Louro

Introduction 191
Molecular bases of cancer and growth factor pathway 193
Cancer therapy and mechanisms underlying drug action/resistance 195
Therapeutic mAbs: Mechanisms of action and combinations 198
Future expectations regarding the PD-1/PD-L1 pathways 204
A different way of classifying tumor resistance 205
Does endocytosis/EGFR downregulation explain resistance to mAbs? 208
Conclusion and future tendencies 213
References 214

Index 219

Contributors

Vitor H. Almeida Instituto de Bioquímica Médica Leopoldo De Meis, Universidade Federal do Rio de Janeiro, Rio de Janeiro, Brazil

Lyvia N.R. Alves Núcleo de Genética Humana e Molecular, Departamento de Ciências Biológicas, Universidade Federal do Espírito Santo, Espírito Santo, Brazil

Débora G. Barbosa Núcleo de Genética Humana e Molecular, Departamento de Ciências Biológicas, Universidade Federal do Espírito Santo, Espírito Santo, Brazil

Raquel F.R. Braga Núcleo de Genética Humana e Molecular, Departamento de Ciências Biológicas, Universidade Federal do Espírito Santo, Espírito Santo, Brazil

Camilly V. Campanharo Núcleo de Genética Humana e Molecular, Departamento de Ciências Biológicas, Universidade Federal do Espírito Santo, Espírito Santo, Brazil

Matheus C. Casotti Núcleo de Genética Humana e Molecular, Departamento de Ciências Biológicas, Universidade Federal do Espírito Santo, Espírito Santo, Brazil

Daniel A. Duque Núcleo de Genética Humana e Molecular, Departamento de Ciências Biológicas, Universidade Federal do Espírito Santo, Espírito Santo, Brazil

Luiz C.G.S. Filho Núcleo de Genética Humana e Molecular, Departamento de Ciências Biológicas, Universidade Federal do Espírito Santo, Espírito Santo, Brazil

Wenyan Fu Department of Assisted Reproduction, Shanghai Ninth People's Hospital, Shanghai Jiao Tong University School of Medicine, Shanghai, China

Fernanda M. Garcia Núcleo de Genética Humana e Molecular, Departamento de Ciências Biológicas, Universidade Federal do Espírito Santo, Espírito Santo, Brazil

Alessandra P. Guimarães Núcleo de Genética Humana e Molecular, Departamento de Ciências Biológicas, Universidade Federal do Espírito Santo, Espírito Santo, Brazil

Shi Hu Department of Biophysics, College of Basic Medical Sciences, Naval Medical University (Second Military Medical University), Shanghai, China

Victor N.G. Kohls Núcleo de Genética Humana e Molecular, Departamento de Ciências Biológicas, Universidade Federal do Espírito Santo, Espírito Santo, Brazil

Changhai Lei Department of Biophysics, College of Basic Medical Sciences, Naval Medical University (Second Military Medical University), Shanghai, China

Guangyao Li Department of Biophysics, College of Basic Medical Sciences, Naval Medical University (Second Military Medical University), Shanghai, China

Tian Li Department of Biophysics, College of Basic Medical Sciences, Naval Medical University (Second Military Medical University), Shanghai, China

Lorena M. Lopes Núcleo de Genética Humana e Molecular, Departamento de Ciências Biológicas, Universidade Federal do Espírito Santo, Espírito Santo, Brazil

Iúri D. Louro Núcleo de Genética Humana e Molecular, Departamento de Ciências Biológicas, Universidade Federal do Espírito Santo, Espírito Santo, Brazil

Zetong Ma Department of Biophysics, College of Basic Medical Sciences, Naval Medical University (Second Military Medical University), Shanghai, China

Débora D. Meira Núcleo de Genética Humana e Molecular, Departamento de Ciências Biológicas, Universidade Federal do Espírito Santo, Espírito Santo, Brazil

Kewen Qian Department of Biophysics, College of Basic Medical Sciences, Naval Medical University (Second Military Medical University), Shanghai, China

Raquel S. Reis Núcleo de Genética Humana e Molecular, Departamento de Ciências Biológicas, Universidade Federal do Espírito Santo, Espírito Santo, Brazil

Eldamaria V.W. Santos Núcleo de Genética Humana e Molecular, Departamento de Ciências Biológicas, Universidade Federal do Espírito Santo, Espírito Santo, Brazil

Fernando L.H. Vieira Representante da Universidade Federal do Espírito Santo no Comitê de Publicação do Projeto Elsa, Universidade Federal do Espírito Santo, Espírito Santo, Brazil

Chuqi Wang Department of Biophysics, College of Basic Medical Sciences, Naval Medical University (Second Military Medical University), Shanghai, China; Department of Pharmacy, National University of Singapore, Singapore

CHAPTER

1

Current status of anti-EGFR agents

Tian Li[a], Wenyan Fu[b], Changhai Lei[a], and Shi Hu[a]

[a]Department of Biophysics, College of Basic Medical Sciences, Naval Medical University (Second Military Medical University), Shanghai, China [b]Department of Assisted Reproduction, Shanghai Ninth People's Hospital, Shanghai Jiao Tong University School of Medicine, Shanghai, China

Abstract

Epidermal growth factor receptor (EGFR) is one of the most essential and attractive clinical and scientific attention targets in most cancer therapies. It is involved in the pathogenesis and metastasis of related cancers, showing a decisive modulatory role in these pathways. Clinical trials have demonstrated that monoclonal antibodies and tyrosine kinase inhibitors can benefit and receive drug approval by targeting this receptor in cancer therapy. However, therapeutic strategies are minimal and have broad room for improvement. This chapter reviews the current mainstream cancer treatment options and related studies targeting EGFR in cancer therapy. With a brief introduction to the role of EGFR in cancer and cancer therapy, this chapter provides a comprehensive list of EGFR-targeted cancer therapy drugs. Among them, some drugs are approved for clinical use, and some drugs are in preclinical research or have failed to enter clinical use for related reasons. Finally, this chapter discusses newly developed preclinical stage novel anti-EGFR cancer therapies and drugs.

Abbreviations

ADCC	antibody-dependent cell cytotoxicity
ATP	adenosine triphosphate
EGFR	epidermal growth factor receptor
EGFRvIII	human EGFR variant III
KLH	keyhole edge limpet hemocyanin
mAbs	monoclonal antibodies
TKIs	tyrosine kinase inhibitors

Conflict of interest

No potential conflicts of interest were disclosed.

https://doi.org/10.1016/B978-0-12-821584-5.00027-4

Introduction

EGFR, considered a core target in many different types of cancer, is a vital link in the critical signaling pathways for developing many cancers and plays an influential regulatory role [1,2]. Because of the importance of EGFR, related research and clinical trials have attracted the attention of many researchers and enterprises; clinical trials have proven that monoclonal antibodies and tyrosine kinase inhibitors (TKIs) have significant therapeutic effects in EGFR-positive cancers. However, there is still a large room for optimization and progress in drug research and development, and clinical researchers are constantly updating cancer treatment strategies with EGFR as the target (Fig. 1).

Stanley Cohen discovered the epidermal growth factor receptor (EGFR) in 1962, and in 1986, he shared the Nobel Prize with another scientist because of his related work in physiology or medicine. Many other kinds of growth factors and receptors were then discovered. The currently known epidermal growth factor receptor family includes four members, and the researchers named them according to the order of discovery: EGFR, HER2, HER3, and HER4. These receptors have a notable effect on different physiological processes and are also involved in the incidence and progression of many cancers.

Research on EGFR has continued for over 35 years, and it is the most well-studied member of the epidermal growth factor receptor family [3]. EGFR easily binds with its ligands, which usually leads to receptor dimerization, activates the activity of intracellular tyrosine protein kinases, phosphorylates tyrosine residues at the end of the C-terminus, and triggers downstream enzymes, such as Ras, Raf, and PI3K, thus initiating a series of complex and intersecting signal transduction pathways downstream [4,5]. EGFR regulates normal cell growth, differentiation, and apoptosis by mediating these signaling pathways [6,7].

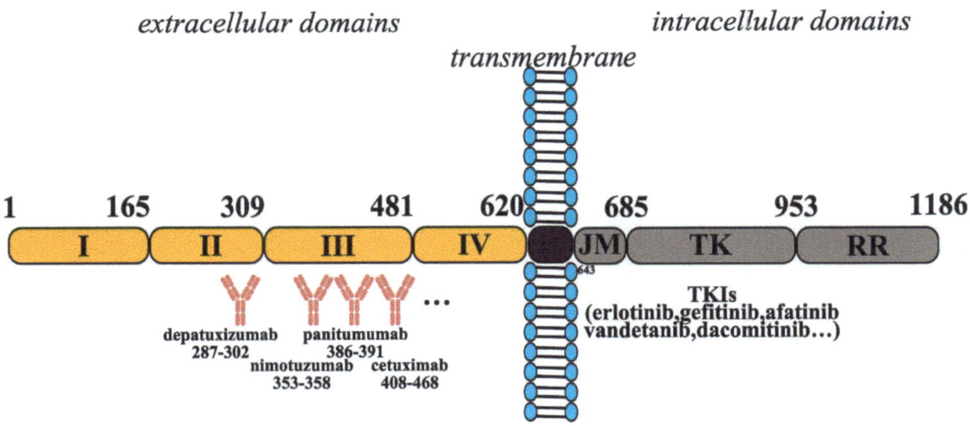

FIG. 1 Schematic diagram of EGFR structure with the extracellular domain (domains I–IV), the transmembrane domain, and the intracellular domain (JM, juxtamembrane; TK, tyrosine kinase domain; and RR, tyrosine-rich regulatory region). Therapeutic antibodies mainly bind to the extracellular domains II and III of EGFR, and TKIs mainly act on the tyrosine kinase domain.

Targeting EGFR

Early researchers found EGFR to be a promising target in cancer treatment and in line with the understanding of biomedical research. Clinically collected patient data found that the receptor is abnormal in many human cancers, such as increased expression levels, abnormal posttranslational modifications, and even sequence-producing mutations. Overexpression of EGFR has been found in the cancer cells of many cancer patients in more than 80% of patients with head and neck cancers, in more than 50% of patients with colorectal cancer, and in non-small cell lung cancer, prostate cancer, or ovarian cancer, which is more than one-third of the detection rate [6–8]. Through the analysis of molecular biology-related experimental methods, it can be confirmed that the expression of EGFR is exceptionally high in cancer cells, and the amount of expression can reach the level of 2 million per cell, which is 20–50 times the number of receptors in normal cells [3,9].

With the help of uncontrolled regulation of signal transduction pathways, cell growth is uncontrolled and deteriorates, and apoptosis of tumor cells is inhibited. The expression levels of EGFR in different tumor tissues are also quite different and are closely related to the degree of tumor differentiation, malignancy, and infiltration [10].

The marked differences in EGFR expression in this disease make targeted anti-EGFR therapy extremely specific. If such therapies are successfully developed, their efficacy may exceed conventional chemotherapy with fewer side effects [11]. There is a discrepancy with that; some research results showed that the treatment effect of targeting EGFR drugs is not correlated with the expression level of the receptors in cancer cells [1]. Targeted EGFR therapy remains a formidable challenge in treating cancer and confirming that it does not recur. Currently approved anti-EGFR drugs for clinical use are mainly divided into tyrosine kinase inhibitors (TKIs) and antibody drugs (Table 1). However, the proportion of complete cures of these drugs is deficient, and only a small number of patients do not develop relevant resistance after a more extended treatment period. Studies have shown that posttranslational modification of EGFR is the leading cause of drug resistance; different acquired modifications have been proven to affect the treatment efficacy of targeted drugs. From another perspective,

TABLE 1 Anti-EGFR drugs that have been approved.

Anti-EGFR antibodies that have been approved			
Agent	**Indications**	**First approval**	**Company**
Cetuximab	Metastatic colorectal cancer HNSCC	2004	Eli Lilly and Company, Merck KGaA, Bristol-Myers Squibb Company
Panitumumab	Metastatic colorectal cancer	2006	Amgen Inc., Takeda Pharmaceutical Company Ltd
Necitumumab	NSCLC	2015	Eli Lilly and Company
Nimotuzumab	HNSCC	2006	InnoMab PTE. Ltd., Biocon Ltd.
Amivantamab	mNSCLC	2021	Janssen Biotech Inc.

Continued

TABLE 1 Anti-EGFR drugs that have been approved—Cont'd

Anti-EGFR small molecule inhibitors that have been approved			
Agent	**Indications**	**First approval**	**Company**
Afatinib	NSCLC	2013	Boehringer Ingelheim GmbH
Erlotinib	NSCLC Metastatic pancreatic cancer	2004	Astellas Pharma Inc., F. Hoffmann-La Roche Ltd.
Furmonertinib	NSCLC	2021	Allist Pharmaceuticals
Gefitinib	NSCLC	2002	AstraZeneca Plc
Icotinib	NSCLC	2011	Betta Pharmaceuticals Co. Ltd.
Lapatinib	HER2-positive breast cancer HER2-positive metastatic breast cancer	2007	Novartis AG
Olmutinib	NSCLC	2016	Hanmi Pharmaceutical Co Ltd.
Osimertinib	NSCLC	2015	AstraZeneca Plc
Vandetanib	Thyroid cancer	2011	AstraZeneca Plc

researchers speculate that the modifications can be positive factors for targeted therapy research; they can also be used as markers for the development of drug resistance [12].

Colorectal cancer patients who received cetuximab for several months were at risk of developing resistance to the initial treatment, and the researchers found that this was mainly related to mutations in the extracellular segment of EGFR. More than six mutations have been found to be associated with treatment resistance. According to the literature, the most common cause is the mutation of EGFR at S492R [13]. This mutation only affects the therapeutic effect of cetuximab but does not interfere with the binding of EGFR to panitumumab.

EGFRvIII is the most representative EGFR activating modification in clinical studies, and amino acids 6–273 encoded by exons 2–7 are missing. The mutant retains the most activity of the relevant link in the signaling pathway in a ligand-independent mechanism [14]. Patients with high expression of this mutation often mean that long-term treatment effects are expected to be poor [1]. Most notably, in response to treatment in glioma patients, the mutation could affect the ligand-binding mechanism, rendering antibodies (cetuximab or panitumumab) that function through that mechanism ineffective. Other EGFR mutations include missing exon 19 and substitutions at the L858R site. These relevant mutation positions may be a key marker of cancer cell susceptibility to first-generation TKIs targeting EGFR. First-generation TKIs such as gefitinib and erlotinib require receptors in the active state to function. In addition, these modifications also affect the resistance of cancer cells to lapatinib, which researchers speculate can only bind to the inactive EGFR kinase domain [15].

Long-term targeted therapy induces cancer cells to become highly differentiated and increases the proportion of cancer cells carrying other modifications that ultimately lead to tumor insensitivity to TKIs and thus drug resistance. T790M substitution of EGFR, a relatively common acquired EGFR mutation, occurs in more than half of patients with advanced TKIs treated with TKIs [16]. Some modifications associated with TKIs, such as T854A substitution, lead to a decrease in the receptor's binding stability and increase in the flexibility of the receptor structure [17].

Clinical samples can also reveal various other EGFR modifications whose effects have not been confirmed, such as EGFRvI, EGFRII, EGFRIV, EGFRV, and EGFR. TDM/2–7, EGFR. TDM/18–25, or EGFR. TDM/18–26, including gene sequence deletion, sequence repeat region, or kinase regulation signal defect [1]. Nevertheless, due to the slightly lower importance, there is currently a lack of further research and data support.

Researchers proposed the concept of treating cancer by targeting EGFR receptors as early as 35 years ago. Since conception, researchers have implemented and evaluated several strategies to affect the EGFR receptor through various approaches, such as antibody drugs (mAbs), TKIs, immunotherapy, and antisense oligonucleotide drugs [18]. Unfortunately, only therapeutic strategies with monoclonal antibodies and TKIs have benefited clinical therapeutics.

Specific mAbs can bind to the relevant extracellular segment of the receptor. The binding capacity between this antibody and the receptor is firm. It can block the binding of ligands to EGFR, thereby blocking the activation of EGFR by ligands, while promoting the endocytosis mechanism of EGFR and reducing EGFR on the cell membrane surface IgG1 antibody can also induce antibody-dependent cytotoxicity (ADCC) [19]. In addition, since the antibody binds to the extracellular region of EGFR, it is effective against both wild-type and EGFR with intracellular tyrosine kinase activity domain variation. In contrast to TKIs targeting the EGFR mechanism, monoclonal antibodies typically recruit immune cells to deal with the abnormal condition [9]. Monoclonal antibodies targeting EGFR receptors are important achievements in the early realization of drug treatment of cancer. This therapeutic idea has been highly successful, and many improved technologies have been derived. The original targeting EGFR antibody is directly bound to the loop located at domain III of EGFR. This finding provided a solid rationale for researching the following monoclonal antibodies for therapeutic purposes that also bind to domain III. These antibodies work by blocking ligand binding and preventing EGFR activation [20].

Cetuximab is the first FDA-approved EGFR mAb. It is a chimeric antibody of type IgG1 that binds to domain III of EGFR. Researchers found that this region may contain the vast majority of EGF-binding interfaces [21,22]. Panitumumab is a monoclonal antibody of type IgG2 and has been humanized. It can compete with endogenous ligands for binding to the extracellular domain of EGFR with high affinity and can also induce ADCC to kill tumor cells [23]. The specific binding region is the 386–391 position. In March 2006, the FDA approved cetuximab to treat unresectable head and neck squamous cell carcinoma, the first approved head and neck cancer treatment. Studies have shown that the efficacy or overall survival of cetuximab is related to whether the K-ras gene is mutated. In August 2008, EMEA approved cetuximab combined with chemotherapy as a first-line treatment for K-ras wild-type and EGFR-positive metastatic colorectal cancer patients. The mechanism of action of panitumumab is similar to that of cetuximab. The epitopes of the two antibodies bind to different epitopes, but both have an ideal therapeutic effect on colorectal cancer. They provide more options for patients with different mutations in the EGFR-encoding gene [22]. Usually, because panitumumab has no murine protein, it has a longer half-life, higher receptor affinity, and better immune tolerance.

The humanized IgG1 antibody nimotuzumab targets the different regions exposed outside the cell membrane in the same domain. Compared with other anti-EGFR monoclonal antibodies, nimotuzumab combined with chemoradiotherapy can significantly improve the antitumor effect and has higher safety and tolerability [24].

Although the effects of binding to EGFR domain III appear to be exceptionally therapeutically likely, not all therapeutic agents of similar design have been validated in clinical trials. The development of monoclonal antibodies such as matuzumab has failed [25,26].

Necitumumab belongs to the second generation of antibody drugs to treat EGFR, a recombinant human IgG1 antibody. According to clinical data feedback, survival can be improved when necitumumab and chemotherapy are combined. These patients with squamous cell carcinoma had better overall and progression-free survival than chemotherapy alone [27,28]. It was approved by the FDA in 2015 for the therapeutic schedule of metastatic squamous non-small cell lung cancer in combination with gemcitabine and cisplatin. According to the feedback from previous monoclonal antibody treatment data, the K-ras gene mutation status is an important marker for treating colorectal cancer with cetuximab and panitumumab [29]. The temporary lack of effective efficacy prediction biomarkers may also be one of the main obstacles to the clinical application of necitumumab.

Another group of widely recognized targeting EGFR drugs are small molecules. Small molecule drugs exhibit high specificity and effective cell growth inhibition in cancer treatment. Most drugs that have been approved for clinical application or are in clinical trials are TKIs for the EGFR catalytic domain, and such inhibitory effects can be classified as irreversible or reversible [30]. The primary mechanism of action is that small molecules bind to the EGFR intracellular tyrosine kinase domain, which can competitively inhibit the binding of tyrosine kinase adenosine triphosphate (ATP) to the EGFR intracellular tyrosine kinase domain, thereby affecting tyrosine residue phosphorylation and inhibiting signal transduction downstream of EGFR [31,32]. The EGFR gene includes 28 exons, and exons 18–21 are the coding region of the ATP-binding site of the EGFR gene. Differences in inhibition patterns are the main differences between TKI drugs, and their specificity is generally close. The development of TKIs targeting EGFR has gone through four generations.

Erlotinib and gefitinib are first-generation inhibitors. These two TKIs bind reversibly to the target and are positive in the therapeutic schedule of cancers containing EGFR mutations [16,33]. Both erlotinib and gefitinib are used to treat locally advanced or metastatic non-small cell lung cancer. EGFR is overexpressed in 20%–80% of NSCLC, and the deletion of exon 19 and the point mutation of exon 21 (L858R) are considered drivers of NSCLC pathogenesis [16,33]. It has been confirmed that the efficacy of the first generation of TKI drugs is related to the presence or absence of mutations in EGFR, which is the best indicator of its efficacy. Acquired resistance develops after 9–14 months of continuous use [34]. A mutation in the 790-position threonine residue of exon 20 was replaced by methionine (T790 M) [34,35]. The T790 M mutation led to the formation of a steric hindrance effect, which affected the binding of the anti-EGFR TKI to the previously functioning target, thereby improving the affinity of ATP and EGFR, restoring kinase activity, and restarting the downstream signaling pathway [36].

The second generation of TKIs targeting EGFR are irreversible molecules and were originally designed to address the problem of acquired resistance in the first generation of anti-EGFR TKI drugs. Second-generation anti-EGFR TKIs can be covalently bound to EGFR. However, from the actual clinical effect analysis, most of the second-generation anti-EGFR TKIs are not as effective as expected to prevent the increased affinity between EGFR and ATP caused by the T790 M mutation, and side effects (including rash and gastrointestinal toxicity) severely limit their widespread application in the clinic [37]. Only afatinib has been approved for the therapeutic schedule in non-small cell lung cancer [38,39]. The clinical status that cannot be ignored is that three out of five patients have the T790 M mutation in EGFR, which causes patients to become resistant to most of the available EGFR TKIs in clinical treatment [16].

To overcome this problem, third-generation inhibitors have emerged that have good selectivity for EGFR double mutations and can selectively covalently bind to Cys797 to overcome the severe clinical side effects of second-generation anti-EGFR TKIs. Third-generation inhibitors, including osimertinib, rociletinib, olmutinib, and naquotinib, have been clinically studied or temporarily halted in some research [16]. Among them, the FDA approved the listing of osimertinib in November 2015 for the therapeutic schedule of patients with non-small cell lung cancer who have the T790 M mutation in EGFR [40]. Although third-generation TKIs targeting EGFR have shown significant clinical efficacy, the problem of drug resistance will soon appear. Studies have shown that mutations at the C797S site are found in osimertinib-related acquired resistant lung cancer samples, and this mutation blocks osimertinib binding to EGFR. At the same time, increased expression levels of HER2 or c-MET also occurred in drug-resistant patients, allowing tumor cells to bypass signaling pathways dependent on EGFR to avoid the inhibitory effect caused by the drug [40].

A few fourth-generation inhibitors are also under development, and EAI045 is the first reported fourth-generation anti-EGFR TKI that effectively inhibits tumor cell proliferation in various models. EAI045 may need to be combined with cetuximab to address EGFR dimerization [41]. According to news and literature reports, fourth-generation anti-EGFR TKI drugs developed by different teams, such as ES-072 and TQB3804, have also been approved for clinical trials [42,43].

To further optimize the therapeutic effect of anti-EGFR TKIs, drugs or multistrategy combination therapy is undergoing clinical trials. Clinical studies suggest that responses have better efficacy in terms of survival in NSCLC with EGFR mutations using a variety of gefitinib and carboplatin/pemetrexed combinations [44]. The clinical combination of erlotinib and anti-VEGF antibody drug showed a twofold improvement in survival in patients with NSCLC with EGFR mutations, and the synergistic effect was pronounced [45]. This combination is approved by the CHMP and approved in the European Union for the first-line therapeutic schedule of unresectable advanced, metastatic, or recurrent NSCLC adults with EGFR-activated mutations [46].

Novel EGFR targeting strategies

In addition to the more mature treatment options of monoclonal antibodies and TKIs, researchers still develop various other treatment options according to the drug resistance of cancer, such as immunotherapy, bispecific antibodies, and antibody-drug conjugates [19]. Although the immunotherapy method is still in the clinical trial stage, it has attracted the attention of researchers and clinicians and is considered to have broad prospects. Rindopepimut immunotherapy is a vaccine capable of fighting cancer, which was developed for application in treating recurrent glioblastoma [47], such as conjugation of a specific peptide of human EGFR variant III (EGFRvIII) to keyhole edge limpet hemocyanin (KLH), followed by intradermal injection of the drug candidate, a nonspecific immunomodulator with potential antitumor activity. Immunization with rindopepimut induces cytotoxic T lymphocyte immune responses against EGFRvIII-expressing tumor cells. EGFRvIII consists of a mutant caused by a deletion of amino acids in the extracellular domain of EGFR and is the most common EGFR mutant in tumours [48]. This mutant was highly expressed in GBM tumor cells and not in normal cells, with tumor specificity and immunogenicity.

According to existing reports, all antibody drugs (mAbs) tested or marketed by pharmaceutical companies that target EGFR all recognize a specific epitope of domain III of the extracellular part of EGFR and then block signal pathway transduction to play a role in the proliferation of cancer cells by blocking ligand binding. This phenomenon opens a new door to developing new drugs targeting other functional domains of EGFR. Depatuxizumab is a humanized monoclonal antibody that can target epitopes located within domain II of the extracellular part of EGFR (287–302). It can target the forms of EGFR and EGFRvIII missing variants. After EGFR binds to depatuxizumab, the process of receptor dimerization is blocked by interference at the interface [49,50]. Through clinical data analysis, depatuxizumab showed higher cancer selectivity and safety than cetuximab [51]. These further warranty trials may be due to the selective function of depatuxizumab only for activating mutant receptors present or constituting activity on cancer cells [52]. Further research data may prove whether it is a reliable strategy for developing new drugs targeting EGFR.

Another way to optimize targeting EGFR antibody drugs is to expand the objects they identify to other specific cancer receptors. The development may be precious when the inactivation of different receptors has a joint action. Another development strategy is to develop multispecific antibody drugs.

Duligotuzumab is a bispecific antibody under development that can simultaneously bind to EGFR and HER3 [53]. At present, duligotuzumab is in phase II clinical trial stage, and preliminary results show that it is superior to the single-target EGFR antibody cetuximab in efficacy. Some studies have shown that duligotuzumab can overcome some of the resistance caused by cetuximab. The phosphorylation of EGFR and HER3 activates the downstream Ras/MAPK and PI3K/AKT signaling pathways, promoting cell growth and proliferation. Antibody drugs that can only bind to one site against EGFR or HER3 cannot thoroughly inhibit the proliferation signaling pathway. Combining the two specific antibodies against EGFR and HER3 can completely inhibit the Ras/MAPK and PI3K/AKT signaling systems and improve the therapeutic efficacy of cancer [54]. The two antibody molecules have a good affinity for EGFR and HER3. The existence of the Fc region enables this bispecific antibody to have an immune effect similar to normal human IgG at the same time.

LY3164530 is a reconstructive bispecific antibody drug by fusing EGFR antibody single-chain variable fragment to the N-terminus of emibetuzumab (targeting c-MET monoclonal antibody) heavy chain [55]. LY3164530 can bind to c-MET and EGFR and promote their endocytosis without activating activity. In the NSCLC xenograft model, the drug showed better antitumor efficacy than the combination regimen of emibetuzumab and cetuximab. Another EGFR/c-MET bispecific antibody drug (amivantamab) has been developed, which directly causes the killing effect of EGFR-mutant lung cancer [55]. In America, it has received approval for treating adult patients with locally advanced or metastatic NSCLC harboring EGFR exon 20 insertion mutations [56]. AFM21 CD16A is also an anticancer drug worth introducing. It is a bispecific tetravalent antibody targeting both EGFRvIII and CD16A locations. The engineered antibody can well recognize EGFRvIII, a highly tumor-specific receptor variant with a high binding force to CD16A on NK cells, thereby guiding NK cells to destroy cancer cells that the antibody targets at the same time. Researchers have also developed the drug AFM21 CD3, which has similar functions to AFM21 CD16A. It is a bispecific tetravalent antibody that targets both EGFRvIII and CD3. The antibody has high binding power to CD3 on T cells based on recognizing EGFRvIII and can similarly guide T cells to destroy cancer cells.

Researchers have also included EGFR and IGF-1 receptors or c-MET tyrosine kinases as therapeutic targets [57]. Antyra Inc. developed B-peptimetics and conducted related research, and the experimental results showed that the drug is a new generation drug that can inhibit the activity of EGFR and IGFR-1 by binding. Bristol-Myers Squibb has developed a bispecific adnectin, El-Tandem, that binds to EGFR and IGFR-1 [57]. Preclinical studies have shown promising results related to reduced tumor growth, reduced microvascular density, and promotion of tumor structure normalization.

Multiantibody mix therapy is also a potential therapeutic strategy. The researchers mixed three human monoclonal antibodies to form the MM151 formulation, which binds to EGFR in different epitopes without overlap. The drug is already in phase I clinical trials, and some data lay the molecular basis for the use of MM151 in patients who have developed resistance to earlier-developed mAbs [58]. In addition, there are many promising drug mixture applications, such as futuximab (Sym-004), a mixture of two synergistic full-length antibodies expected to overcome drug resistance and achieve further therapeutic effects [59]. The antibody binds to two separate nonoverlapping epitopes in EGFR ectodomain III, preventing ligand binding. Mixtures targeting more than one cancer-related receptor simultaneously may also be effective, based on previous experience with combination therapy. The researchers then developed a cocktail of humanized full-length mAbs that simultaneously targeted three receptors, EGFR, ErbB2, and ErbB3, and named it Sym-013. The results exhibit significant synergy compared with single receptor-targeting antibody formulations for therapeutic benefit [59].

Nanoantibodies as cutting-edge technologies have also been applied to the development of therapeutic strategies to target EGFR, the smallest antigen-binding fragments of antibody heavy chains [60], and affibody, a novel affinity ligand consisting of 58 amino acid residues with a relative molecular mass of approximately 6.5×10^3, derived from a small protein structure of the Z domain of protein A [61], making it possible to develop small molecules to bind to different targets. This biomolecule exhibits low immunogenicity, rapid clearance, and high targeting specificity. Polyvalent antibodies consisting of nanoantibodies and affinity moieties targeting EGFR and ErbB2 showed lower cytotoxicity and optimized antitumor activity than standard therapeutic mAbs. The more well-known ones are cetuximab and trastuzumab [62].

Based on research on TKIs, many molecular therapeutic regimens that act as EGF antagonists have also been discovered. Naphthoquinone-derived molecules such as shikonin and FR18 can inhibit the proliferation of cancer cells and induce apoptosis. These effects are closely related to the inhibition of the EGFR pathway [63]. The experimental results confirm that FR18 can block the binding of EGF to EGFR and that the drug does not inhibit tyrosine kinase activity [64].

A more direct study than the development of new technologies targeting specific signaling pathway components is to regulate individual signaling pathways in cell proliferation at the gene expression level. Such strategies are also considered interesting for anti-EGFR therapy in cancer. Antisense oligonucleotides or RNA interference techniques have been used to reduce the EGFR-activated growth and proliferation signaling pathways [18,65–67]. Two small RNA interference treatment regimens have reported moderately positive results in the early stages of development [68].

Conclusions

While the benefits of existing anti-EGFR therapies are indisputable, many steps should be taken to improve efficacy and safety and avoid resistance. Cancer treatment strategies targeting EGFR and other ErbB family receptors have produced many positive clinical trial results and have significant advantages in various aspects. This makes this type of research more scientifically significant and attracts relevant pharmaceutical companies and funds to invest in research on relevant targets. Future basic research is crucial to help us understand the role of EGFR in cancer development and progression, and adequate research can more accurately predict the response to EGFR inhibition.

References

[1] Normanno N, et al. Epidermal growth factor receptor (EGFR) signaling in cancer. Gene 2006;366:2–16.
[2] Holbro T, Civenni G, Hynes NE. The ErbB receptors and their role in cancer progression. Exp Cell Res 2003;284:99–110.
[3] Kamath S, Buolamwini JK. Targeting EGFR and HER-2 receptor tyrosine kinases for cancer drug discovery and development. Med Res Rev 2006;26:569–94.
[4] Lemmon MA, Schlessinger J. Cell signaling by receptor tyrosine kinases. Cell 2010;141:1117–34.
[5] Ocaña A, Pandiella A. Targeting HER receptors in cancer. Curr Pharm Des 2013;19:808–17.
[6] Yewale C, Baradia D, Vhora I, Patil S, Misra A. Epidermal growth factor receptor targeting in cancer: a review of trends and strategies. Biomaterials 2013;34:8690–707.
[7] Mamot C, Rochlitz C. Targeting the epidermal growth factor receptor (EGFR)—a new therapeutic option in oncology? Swiss Med Wkly 2006;136:4–12.
[8] Normanno N, Bianco C, De Luca A, Maiello MR, Salomon DS. Target-based agents against ErbB receptors and their ligands: a novel approach to cancer treatment. Endocr Relat Cancer 2003;10:1–21.
[9] Bianco R, Gelardi T, Damiano V, Ciardiello F, Tortora G. Rational bases for the development of EGFR inhibitors for cancer treatment. Int J Biochem Cell Biol 2007;39:1416–31.
[10] Baselga J. Why the epidermal growth factor receptor? The rationale for cancer therapy. Oncologist 2002;7(Suppl. 4):2–8.
[11] Huang M, Shen A, Ding J, Geng M. Molecularly targeted cancer therapy: some lessons from the past decade. Trends Pharmacol Sci 2014;35:41–50.
[12] Balak MN, et al. Novel D761Y and common secondary T790M mutations in epidermal growth factor receptor-mutant lung adenocarcinomas with acquired resistance to kinase inhibitors. Clin Cancer Res 2006;12:6494–501.
[13] Montagut C, et al. Identification of a mutation in the extracellular domain of the epidermal growth factor receptor conferring cetuximab resistance in colorectal cancer. Nat Med 2012;18:221–3.
[14] Wikstrand CJ, Reist CJ, Archer GE, Zalutsky MR, Bigner DD. The class III variant of the epidermal growth factor receptor (EGFRvIII): characterization and utilization as an immunotherapeutic target. J Neurovirol 1998;4:148–58.
[15] Sogabe S, et al. Structure-based approach for the discovery of pyrrolo[3,2-d]pyrimidine-based EGFR T790M/L858R mutant inhibitors. ACS Med Chem Lett 2013;4:201–5.
[16] Wang S, Cang S, Liu D. Third-generation inhibitors targeting EGFR T790M mutation in advanced non-small cell lung cancer. J Hematol Oncol 2016;9:34.
[17] Goyal S, Jamal S, Shanker A, Grover A. Structural investigations of T854A mutation in EGFR and identification of novel inhibitors using structure activity relationships. BMC Genomics 2015;16(Suppl. 5):S8.

[18] de Bono JS, Rowinsky EK. The ErbB receptor family: a therapeutic target for cancer. Trends Mol Med 2002;8: S19–26.

[19] Mendelsohn J, Baselga J. Status of epidermal growth factor receptor antagonists in the biology and treatment of cancer. J Clin Oncol Off J Am Soc Clin Oncol 2003;21:2787–99.

[20] Wu DG, et al. Human epidermal growth factor (EGF) receptor sequence recognized by EGF competitive monoclonal antibodies. Evidence for the localization of the EGF-binding site. J Biol Chem 1989;264:17469–75.

[21] Li S, Schmitz KR, Jeffrey PD, Wiltzius JJ, Kussie P, Ferguson KM. Structural basis for inhibition of the epidermal growth factor receptor by cetuximab. Cancer Cell 2005;7:301–11.

[22] Voigt M, et al. Functional dissection of the epidermal growth factor receptor epitopes targeted by panitumumab and cetuximab. Neoplasia (New York, NY) 2012;14:1023–31.

[23] Wu M, Rivkin A, Pham T. Panitumumab: human monoclonal antibody against epidermal growth factor receptors for the treatment of metastatic colorectal cancer. Clin Ther 2008;30:14–30.

[24] Ramakrishnan MS, et al. Nimotuzumab, a promising therapeutic monoclonal for treatment of tumors of epithelial origin. MAbs 2009;1:41–8.

[25] Kamat V, et al. Enhanced EGFR inhibition and distinct epitope recognition by EGFR antagonistic mAbs C225 and 425. Cancer Biol Ther 2008;7:726–33.

[26] Schmiedel J, Blaukat A, Li S, Knöchel T, Ferguson KM. Matuzumab binding to EGFR prevents the conformational rearrangement required for dimerization. Cancer Cell 2008;13:365–73.

[27] Thatcher N, et al. Necitumumab plus gemcitabine and cisplatin versus gemcitabine and cisplatin alone as first-line therapy in patients with stage IV squamous non-small-cell lung cancer (SQUIRE): an open-label, randomised, controlled phase 3 trial. Lancet Oncol 2015;16:763–74.

[28] Ciuleanu T, et al. Efficacy and safety of necitumumab continuation therapy in the phase III SQUIRE study of patients with stage IV squamous non-small-cell lung cancer. Clin Lung Cancer 2018;19:130–138.e132.

[29] Makki D, Haddad BZ, Mahmood Z, Shahid MS, Pathak S, Garnham I. Efficacy of corticosteroid injection versus size of plantar interdigital neuroma. Foot Ankle Int 2012;33:722–6.

[30] Patel TB, Bertics PJ. Epidermal Growth Factor: Methods and Protocols. Springer Science & Business Media; 2006.

[31] Jiang N, Saba NF, Chen ZG. Advances in targeting HER3 as an anticancer therapy. Chemother Res Pract 2012;2012, 817304.

[32] Alaoui-Jamali MA, Morand GB, da Silva SD. ErbB polymorphisms: insights and implications for response to targeted cancer therapeutics. Front Genet 2015;6:17.

[33] Liao BC, Lin CC, Yang JC. Second and third-generation epidermal growth factor receptor tyrosine kinase inhibitors in advanced nonsmall cell lung cancer. Curr Opin Oncol 2015;27:94–101.

[34] Pao W, et al. Acquired resistance of lung adenocarcinomas to gefitinib or erlotinib is associated with a second mutation in the EGFR kinase domain. PLoS Med 2005;2, e73.

[35] Lu X, Yu L, Zhang Z, Ren X, Smaill JB, Ding K. Targeting EGFR(L858R/T790M) and EGFR(L858R/T790M/C797S) resistance mutations in NSCLC: current developments in medicinal chemistry. Med Res Rev 2018;38:1550–81.

[36] Yun CH, et al. The T790M mutation in EGFR kinase causes drug resistance by increasing the affinity for ATP. Proc Natl Acad Sci U S A 2008;105:2070–5.

[37] Sos ML, et al. Chemogenomic profiling provides insights into the limited activity of irreversible EGFR inhibitors in tumor cells expressing the T790M EGFR resistance mutation. Cancer Res 2010;70:868–74.

[38] Li D, et al. BIBW2992, an irreversible EGFR/HER2 inhibitor highly effective in preclinical lung cancer models. Oncogene 2008;27:4702–11.

[39] Harvey RD, Adams VR, Beardslee T, Medina P. Afatinib for the treatment of EGFR mutation-positive NSCLC: a review of clinical findings. J Oncol Pharm Pract Off Publ Int Soc Oncol Pharm Pract 2020;26:1461–74.

[40] Greig SL. Osimertinib: first global approval. Drugs 2016;76:263–73.

[41] Wang S, Song Y, Liu D. EAI045: the fourth-generation EGFR inhibitor overcoming T790M and C797S resistance. Cancer Lett 2017;385:51–4.

[42] Zheng J, et al. First-in-human phase 1 study of ES-072, an oral mutant-selective EGFR T790M inhibitor, non-small-cell lung cancer. Clin Lung Cancer 2020;21:509–519.e501.

[43] Dong RF, et al. EGFR mutation mediates resistance to EGFR tyrosine kinase inhibitors in NSCLC: from molecular mechanisms to clinical research. Pharmacol Res 2021;167, 105583.

[44] Sugawara S, et al. Randomized phase II study of concurrent versus sequential alternating gefitinib and chemotherapy in previously untreated non-small cell lung cancer with sensitive EGFR mutations: NEJ005/TCOG0902. Ann Oncol Off J Eur Soc Med Oncol 2015;26:888–94.

[45] Seto T, et al. Erlotinib alone or with bevacizumab as first-line therapy in patients with advanced non-squamous non-small-cell lung cancer harbouring EGFR mutations (JO25567): an open-label, randomised, multicentre, phase 2 study. Lancet Oncol 2014;15:1236–44.
[46] T-Cells C, Bento F. Roche receives EU approval of avastin in combination with Tarceva for patients with a specific type of advanced lung cancer. Financial 2016.
[47] Giotta Lucifero A, Luzzi SJBS. Against the resilience of high-grade gliomas: the immunotherapeutic approach (part I). Brain Sci 2021;11:386.
[48] Mass RD. The HER receptor family: a rich target for therapeutic development. Vol. 58; 2004. p. 932–40.
[49] Johns TG, et al. Identification of the epitope for the epidermal growth factor receptor-specific monoclonal antibody 806 reveals that it preferentially recognizes an untethered form of the receptor. J Biol Chem 2004;279: 30375–84.
[50] Sivasubramanian A, Chao G, Pressler HM, Wittrup KD, Gray JJ. Structural model of the mAb 806-EGFR complex using computational docking followed by computational and experimental mutagenesis. Structure (London, England: 1993) 2006;Vol. 14:401–14.
[51] Scott AM, et al. A phase I clinical trial with monoclonal antibody ch806 targeting transitional state and mutant epidermal growth factor receptors. Proc Natl Acad Sci U S A 2007;104:4071–6.
[52] Reilly EB, et al. Characterization of ABT-806, a humanized tumor-specific anti-EGFR monoclonal antibody. Mol Cancer Ther 2015;14:1141–51.
[53] McKnight BN, et al. Imaging EGFR and HER3 through (89)Zr-labeled MEHD7945A (Duligotuzumab). Sci Rep 2018;8:9043.
[54] Schaefer G, et al. A two-in-one antibody against HER3 and EGFR has superior inhibitory activity compared with monospecific antibodies. Cancer Cell 2011;20:472–86.
[55] Patnaik A, et al. A phase I study of LY3164530, a bispecific antibody targeting MET and EGFR, in patients with advanced or metastatic cancer. Cancer Chemother Pharmacol 2018;82:407–18.
[56] Syed YY. Amivantamab: first approval. Drugs 2021;81:1349–53.
[57] Yeo CD, et al. Expression of insulin-like growth factor 1 receptor (IGF-1R) predicts poor responses to epidermal growth factor receptor (EGFR) tyrosine kinase inhibitors in non-small cell lung cancer patients harboring activating EGFR mutations. Lung Cancer (Amsterdam, Netherlands) 2015;87:311–7.
[58] Napolitano S, et al. Antitumor efficacy of triple monoclonal antibody inhibition of epidermal growth factor receptor (EGFR) with MM151 in EGFR-dependent and in cetuximab-resistant human colorectal cancer cells. Oncotarget 2017;8:82773–83.
[59] Jones S, et al. Targeting of EGFR by a combination of antibodies mediates unconventional EGFR trafficking and degradation. Sci Rep 2020;10:663.
[60] Muyldermans S. Single domain camel antibodies: current status. J Biotechnol 2001;74:277–302.
[61] Nygren PA. Alternative binding proteins: affibody binding proteins developed from a small three-helix bundle scaffold. FEBS J 2008;275:2668–76.
[62] Ding L, et al. Small sized EGFR1 and HER2 specific bifunctional antibody for targeted cancer therapy. Theranostics 2015;5:378–98.
[63] Tian R, Li Y, Gao M. Shikonin causes cell-cycle arrest and induces apoptosis by regulating the EGFR-NF-κB signalling pathway in human epidermoid carcinoma A431 cells. Biosci Rep 2015;35.
[64] Calonghi N, et al. A new EGFR inhibitor induces apoptosis in colon cancer cells. Biochem Biophys Res Commun 2007;354:409–13.
[65] Padfield E, Ellis HP, Kurian KM. Current therapeutic advances targeting EGFR and EGFRvIII in glioblastoma. Front Oncol 2015;5:5.
[66] Azizi E, et al. Release profile and stability evaluation of optimized chitosan/alginate nanoparticles as EGFR antisense vector. Int J Nanomedicine 2010;5:455–61.
[67] Lu Y, et al. siRNA delivered by EGFR-specific scFv sensitizes EGFR-TKI-resistant human lung cancer cells. Biomaterials 2016;76:196–207.
[68] Kwapiszewski R, Pawlak SD, Adamkiewicz K. Anti-EGFR agents: current status, forecasts and future directions. Target Oncol 2016;11:739–52.

CHAPTER

2

Molecular mechanisms of resistance to the EGFR monoclonal antibody cetuximab

Zetong Ma, Changhai Lei, and Shi Hu

Department of Biophysics, College of Basic Medical Sciences, Naval Medical University (Second Military Medical University), Shanghai, China

Abstract

Human epithelial carcinomas, such as breast cancer, nonsmall cell lung cancer (NSCLC), glioblastoma, and colorectal cancer (CRC), are often associated with increased activation caused by EGFR protein overexpression, gene amplification, or mutation. To date, several small-molecule tyrosine kinase inhibitors (TKIs) and monoclonal antibodies targeting EGFR have been approved by the FDA for use in clinical application. However, the mechanisms of inherent resistance and acquired resistance to cetuximab have been widely studied. This chapter reviews the biological characteristics of EGFR, its role in human cancer, and the development of anti-EGFR antibody therapies. We further discuss the molecular mechanisms of cetuximab resistance and potential strategies for circumventing such resistance.

Abbreviations

CLIA	clinical laboratory improvement amendment
CRC	colorectal cancer
ECD	extracellular domain
EGFR GCN	EGFR gene copy number
HB-EGF	heparin-binding EGF
HNSCC	head and neck squamous cell carcinoma
NLS	near nuclear localization signal domain
NSCLC	nonsmall cell lung cancer
PFS	progression-free survival
RTK	receptor tyrosine kinases
TKD	tyrosine kinase domain
TKIs	tyrosine kinase inhibitors

https://doi.org/10.1016/B978-0-12-821584-5.00007-9

Conflict of interest

No potential conflicts of interest were disclosed.

Introduction

EGFR was discovered and identified by Carpenter and colleagues [1] approximately 50 years ago and isolated, cloned, and sequenced from normal placental cells and A431 tumor cells by other researchers in 1984 [2]. EGFR has been identified as a receptor protein tyrosine kinase (RTK) [3,4]. Protein modification through tyrosine residue phosphorylation has been recognized as a critical step in tumorigenesis [5,6]. Some reports have described that EGFR signal transduction disorder plays a crucial role in human cancer. EGFR antibodies and small-molecule tyrosine kinase inhibitor development make many patients benefit from this targeted therapy, but there is still resistance to targeted therapy. The understanding of EGFR inhibitor endogenous and secondary drug resistance molecular mechanisms is fundamental. This chapter reviews the biological characteristics of EGFR and its role in human cancer. The development and experience of EGFR-targeted antibody therapies, known mechanisms of resistance, and possible optimization strategies are discussed (Fig. 1).

EGFR biology

EGFR belongs to the EGF receptor tyrosine kinase (RTK) family, including EGFR (ErbB1/HER1), HER2/NEU (ErbB2), HER3 (ErbB3), and HER4 (ErbB4) [7]. EGFR consists of an extracellular ligand-binding domain (domain I-IV), a transmembrane domain, a near nuclear localization signal domain (NLS), a cytoplasmic tyrosine kinase domain (TKD), and a C-terminal tail. The EGFR ligand precursor is cleaved on the cell surface by ADAM protease and binds via autocrine or paracrine to leucine-rich repeats in the extracellular domains I and III of EGFR, triggering a conformational change in the receptor that exposes the dimeric ring (domain II), leading to homologous or heterodimerization with other members of the HER family and activation of EGFR kinase [8,9]. EGFR kinase activation leads to autophosphorylation and transphosphorylation of the c-terminal tail of the cytoplasm. HER3 can also be phosphorylated by heterodimerization, lacking intrinsic kinase activity [10,11].

EGFR activation stimulates several complex intracellular signaling pathways. Three major signaling pathways of EGFR include the RAS/RAF/MEK/ERK, PI3K/AKT, and PLCγ/PKC axes and SRC tyrosine kinase and STAT activation. These signaling pathways are closely regulated by ligands, heterodimer composition, and phosphotyrosine-binding protein availability, leading to complex cellular processes including proliferation, apoptosis, cell survival, and cell migration [12].

EGFR can be detected in cancer cell nuclei, primary tumor specimens, and other proliferative tissues [13–17], suggesting that nuclear EGFR may also be involved in cancer progression. In breast cancer, oropharyngeal squamous cell carcinoma, and ovarian cancer, increased nEGFR localization suggests adverse clinical outcomes [18–20]. The mechanisms of EGFR transport to the nucleus include ligand binding, dimerization, activation, and

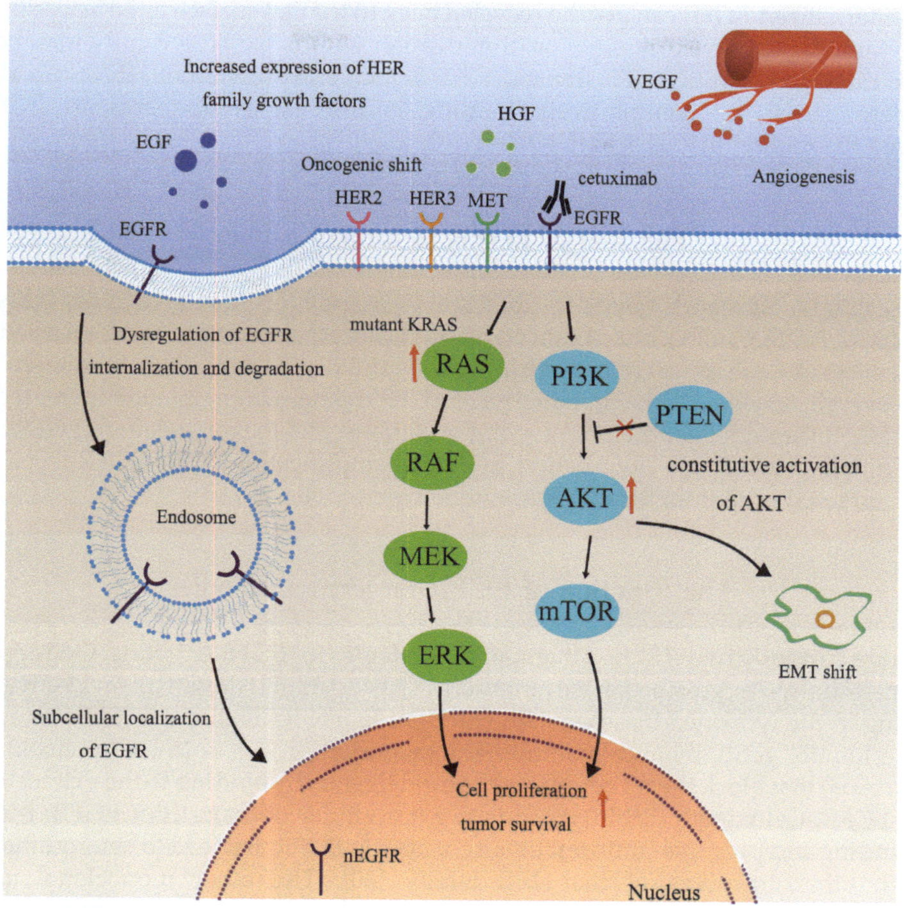

FIG. 1 Mechanisms of resistance to EGFR-targeted antibodies. EGFR-blocking antibodies increased the expression of VEGF in resistant cells, which is a potent stimulator of neovascularization. Cetuximab-resistant cells have dysregulated EGFR internalization and degradation and have increased expression levels of HER family growth factors and EGFR, HER2, HER3, and MET. EGFR downstream effector molecules such as AKT are constitutively activated due to the degradation of PTEN, which induces resistance to cetuximab. The expression of E-cadherin decreased and the expression of vimentin increased in cetuximab-resistant cells, suggesting that epithelial-mesenchymal transformation is another mechanism by which tumor cells resist EGFR blockade.

internalization [21]. EGFR acts as a tyrosine kinase and phosphorylates and stabilizes CNA, thereby promoting the proliferation of cancer cells [22]. In addition, Src family kinases (SFKs) mediate radiation-induced nuclear transport of EGFR [23]. Cetuximab also causes translocation of EGFR to the nucleus [24]. These findings suggest that EGF ligands, radiotherapy, and cetuximab contribute to EGFR accumulation in the nucleus.

Attenuation of the EGFR signaling pathway is essential to EGFR. EGFR decay involves several processes, including receptor ubiquitination, dephosphorylation, ligand pathway depletion, receptor transport to lysosomes, and subsequent destruction [25]. Epidermal-associated EGFR is rapidly internalized and degraded mainly through clathrin-mediated endocytosis

[26–30]. Internalized EGFR can also be recycled back to the cell surface or be active inside the cell [25]. Mutations, overexpression, or dimerization with other HER family members may influence the regulation of EGFR transport [31–37]. Overexpression of HER2 can also alter EGFR degradation by competitively binding endosomal components to EGFR and preventing its binding to EGFR [38].

The role of EGFR in human cancer

A large number of reports have described that dysregulation of EGFR activity is associated with HNSCC, NSCLC, CRC, breast cancer, bladder cancer, pancreatic cancer, ovarian cancer, cervical cancer, gastric cancer, endometrial cancer, and brain cancer, and ultimately leads to reduced overall survival [2,39–55]. Patterns of EGFR dysregulation include increased stimulation of EGFR through ligand binding, point mutations, EGFR exon deletion (EGFRvIII) mutations, impaired EGFR downregulation, and gene amplification of the EGFR locus [56–59]. Effective targeted inhibitors for EGFR are urgently needed.

The function of EGFR-targeted antibodies

There are currently two effective therapeutic strategies for EGFR targeting. One approach is to use small-molecule tyrosine kinase inhibitors (TKIs) to inhibit downstream EGFR signaling by binding to the tyrosine kinase domain (TKD) of EGFR. TKIs that the FDA has approved include erlotinib, gefitinib, and lapatinib. The second approach is to use monoclonal antibodies (mAbs) that block the recognition of natural ligands by binding to the extracellular domain of EGFR. Cetuximab (Erbitux) binds to extracellular domain III of EGFR. It prevents ligand binding and prevents conformational changes in EGFR that lead to heterodimer formation with other members of the HER family [60]. The other monoclonal antibody, panitumumab (Vectibix), is similar to cetuximab, except that it is fully humanized [61]. Targeted therapy with monoclonal antibodies effectively blocks EGFR ligand binding, and receptor dimerization promotes EGFR phosphorylation and internalization [62], ultimately reducing cell proliferation in various cancer models [62,63]. Cetuximab induces arrest of the G1 phase of the cell cycle and increases induction of P27KIP1 levels, which in turn leads to increased dimerization of the P27KIP1-CDK2 complex, ultimately preventing its withdrawal from G1 [64–66]. Various studies have shown that cetuximab treatment can also inhibit tumor survival through various mechanisms, including downregulation of angiogenic factors and matrix metalloproteinases responsible for cell adhesion to reduce cancer cell metastasis, upregulation of Bax and other pro-apoptotic factors, downregulation of Bcl2 and activation of caspases, and induction of ADCC in vivo by recruiting immune cells to tumor cells [67–72].

Cetuximab in the clinic

In 1997–98, a phase I trial provided the first clinical evidence that the addition of cetuximab to radiotherapy improves tumor response and disease control in HNSCC patients [73]. A phase III trial confirmed an overall survival advantage of 10% for patients with advanced

HNSCC who received cetuximab combination therapy compared with radiotherapy alone [74]. The results of a phase II clinical trial of cetuximab in combination with platinum showed that the addition of cetuximab to first-line cisplatin and vinorelbine improved the overall survival. Cetuximab combined with carboplatin/taxane improved the overall response rates, but progression-free survival (PFS) was not significantly improved [75–77].

Mechanisms of resistance to EGFR-targeted antibodies

A number of studies have found that markers in a patient's tumor can predict an individual's response to cetuximab therapy and positive clinical outcomes. The EGFR expression level is a predictive biomarker of the response to cetuximab treatment. Breast cancer patients with high HER2 expression are more likely to benefit from trastuzumab. However, the correlation between EGFR expression levels and clinical response to EGFR inhibitor therapy is not fully understood. Chung et al. confirmed that some CRC patients with unmeasurable EGFR benefit from cetuximab [78]. These studies suggest that EGFR expression levels do not accurately predict the cetuximab response.

EGFR mutations

In 2004, mutations in the EGFR tyrosine kinase domain were found in a subset of patients with nonsmall cell lung cancer resistant to erlotinib and gefitinib [79–81]. Pao et al. reported a secondary mutation in exon 20 of EGFR in patients with acquired resistance to gefitinib or erlotinib, resulting in methionine replacement of THR in the kinase domain 790 (T790M) [82]. Other mutations of EGFR include G465R and G465E, S468R and S492R, which are located in extracellular domains (ECDs) at the EGFR-mab interaction interface, preventing the mAb from binding [83–85].

EGFR gene copy number as a predictor of response

A large CRC clinical cohort has demonstrated the association between increased EGFR gene copy number (EGFR GCN) and cetuximab clinical response [86–89]. It is important to note that an increase in EGFR gene copy number does not necessarily increase EGFR expression [86,90]. A total of 30%–60% of NSCLC patients have increased EGFR gene copy number and benefit from cetuximab plus chemotherapy treatment [91]. In 2013, the meta-analysis showed that in patients treated with anti-EGFR, increased EGFR GCN appeared to be associated with improved survival outcomes, and the effect on survival appeared to be related to the treatment line the patient received [92].

EGFR ligand state

Increased EGFR ligand AR and EPR expression levels may drive tumor proliferation through the autocrine/paracrine mode and influence the effect of CRC on cetuximab treatment, which can be used as a predictor of response [93]. Increased EGFR ligand mRNA expression in tumor cells is associated with improved disease control and progression-free survival. These results suggest that the EGFR ligand expression level may influence the therapeutic effect of cetuximab.

KRAS mutations as a predictor of response

Mutation status of the KRAS (a small GTPase) gene is one of the predictive biomarkers most closely associated with cetuximab response. Activation of EGFR will cause KRAS to bind to GTP, resulting in activation of downstream signaling pathways such as RAF/MEK/ERK pathways. Mutations in KRAS codons 12 or 13 lead to the accumulation of an active GTP-bound conformation, resulting in sustained activation of downstream signaling pathways, which may predict cetuximab resistance [94,95]. Lievre et al. found that KRAS mutations were present in 43% of tumors in metastatic CRC patients treated with cetuximab and were significantly associated with cetuximab resistance [96], which was consistent with Di Fiore et al.'s findings in mCRC patients treated with cetuximab in combination with chemotherapy [97]. A larger study [98] shows that wild-type KRAS was a strong predictor of significantly increased overall survival in patients receiving cetuximab-refractory mCRC. Patients with KRAS wild-type tumors can benefit from cetuximab plus FOLFIRI first-line therapy [99]. ASCO's 2009 interim clinical opinion states that all mCRC patients should be tested for KRAS mutations by the Clinical Laboratory Improvement Amendment (CLIA) before anti-EGFR antibody therapy. Patients should not be treated with anti-EGFR antibodies if KRAS codon 12 or 13 is mutated [100]. However, some patients with KRAS mutation still respond to anti-EGFR antibodies [87,101–103]. How these individuals are distinguished from other patients with KRAS mutations is worthy of further investigation.

BRAF mutation as a predictor of response

BRAF is a serine–threonine kinase that belongs to the RAF protein kinase family. BRAF is activated upon GTP binding to KRAS, which activates the MEK pathway [104–106]. Preliminary studies have shown that BRAF mutations impair the EGFR antibody response. CRC patients carrying wild-type BRAF are clinically responsive to cetuximab, while valine 600 (V600E) mutations are present in approximately 14% of nonreactive tumors [107]. Some reports suggest that BRAF gene mutations are relatively rare events in lung cancer and HNSCC [108–113].

Mechanisms of resistance against EGFR-targeted antibodies

Angiogenesis

Some studies have demonstrated that angiogenesis inhibition approves the effect of cetuximab [114–116]. VEGF is a powerful stimulator of neovascularization [117]. In 2001,

Viloria-Petit et al. confirmed that increased VEGF production in tumors is resistant to EGFR-blocking antibodies [118]. Researchers treated A431 cells with three anti-EGFR antibodies, including cetuximab, hR3, or mR3, to obtain drug-resistant mutant cell lines (no cetuximab-resistant cell lines were established) and found that EGFR levels in the mutant cells were similar to those in the parent cell lines, while VEGF mRNA and protein expression levels were more than twice as high as those in the parent cells. When mice were inoculated with the parental cell line A431, which overexpressed VEGF, tumor vascularization increased, and anti-EGFR antibodies showed resistance. Another combination of a VEGF inhibitor and cetuximab shows better antitumor efficacy in inhibiting tumor growth and inducing apoptosis in gastric cancer-transplanted tumors [119]. Ciardiello et al. performed xenotransplantation of human GEO colon tumors using cetuximab or ZD1839 (gefitinib) for 4 weeks, during which time all tumors were found to have completely retreated [120]. The mice that stopped treatment after complete regression had tumor recurrence within 3–5 weeks, while those that continued treatment had tumor recurrence and growth after 11–12 weeks. The sensitivity of the recurrent tumor to cetuximab was reduced. Two drug-resistant tumor cell lines were established in vitro, and western blot analysis showed that the EGFR expression level remained unchanged. The expression of COX-2, phosphorylated MAPK, and VEGF increased 5–10-fold, and these resistant variants secreted more VEGF in the surrounding mediators. Researchers further treated xenograft tumors and drug-resistant cell lines with the dual VEGFR-EGFR tyrosine kinase inhibitor and found that xenograft tumors did not recur after treatment and that chemoresistant variants were also sensitive to ZD6474. Bianco et al. established in vivo drug-resistant strains of different cancer cell lines, including breast, colon, and prostate cancer, and found that ZD6474 significantly reduced the phosphorylation levels of VEGFR-1- and VEGFR-2-resistant cell lines, which have VEGFR-1 overexpression at the mRNA and protein levels [121].

Disorders of EGFR internalization and degradation

Lu et al. treated the colon cancer cell line DiFi with cetuximab to obtain the drug-resistant cell line DiFi5 [122]. Compared with the parent cells, the resistant cells showed EGFR polyubiquitination, and the epidermal growth factor receptor and total level on the cell surface decreased. In addition, increased expression of the ubiquitin E3 ligase Cbl in cetuximab-resistant cells is consistent with high levels of EGFR ubiquitination. The decrease in EGFR stability in this drug resistance model requires further study, and the researchers hypothesize that other receptors/pathways are involved in cell growth and survival.

Oncogenic shift

In NSCLC and HNSCC models, drug-resistant cell lines were continuously treated with cetuximab, and phosphorylated RTK levels were compared in cetuximab-resistant and cetuximab-sensitive parental cell lines. In both models of cetuximab resistance, compared with parental controls, EGFR HER2 and HER3 protein levels and MET receptor upregulation immunoprecipitation in drug-resistant clones showed increased homeostasis of EGFR binding to HER2, HER3, and cMET in drug-resistant lines. These data suggest that activation of various RTKs and heterodimerization with EGFR may be a mechanism of tumor cell resistance to cetuximab antagonists [123].

Subcellular localization of EGFR

EGFR is distributed in subcellular compartments including endosomal, mitochondria, and nucleus [124]. nEGFR has been used as a prognostic factor in various tumors, including brain cancer, breast cancer, bladder cancer, ovarian cancer, NSCLC, and HNSCC [15,16,20,21,125–128]. In 2009, Li et al. demonstrated that increased EGFR levels in drug-resistant clones were consistent with increased levels of genes regulating EGFR expression, such as cyclin D1, B-myb, and PCNA. In addition, SFK expression increased in cetuximab-resistant cells [20]. The use of the PAN-SFK inhibitor dasatinib resulted in a decrease in nEGFR expression and a significant increase in surface-associated EGFR expression. These data suggest that SFK is necessary to promote nuclear translocation of EGFR and contribute to developing tumor resistance to cetuximab. The combination of cetuximab with SFK inhibitors may have some clinical benefit [129]. In another study, mammary-derived growth inhibitor (MDGI), an intracellular fatty acid-binding protein expressed in 40% of breast cancers and 85% of lung cancers, made breast and lung cancer cell models resistant to cetuximab by stimulating the production of nEGFR [130–133]. In cell culture models, MDGI-transfected cells have reduced EGFR levels on the surface and increased intracellular biotin-labeled EGFR. Increased intracellular transport of EGFR appears to be independent of increased activity of other HER family members or alveolar proteins. In MDGI-transfected cells, EGFR remains intracellularly phosphorylated. In a 3D matrix model, GFP-transfected control cells remained sensitive to cetuximab, while MDGI-transfected lung cancer cells were resistant to cetuximab and grew much faster than GFP-controlled tumors. In conclusion, this study identifies MDGI as a determinant of the intracellular localization of EGFR and provides a theoretical basis for considering the MDGI expression level as a biomarker for cetuximab sensitivity.

Epithelial cells transform into mesenchymal cells

EMT is characterized by loss of adhesion connexins, for example, the acquisition of cytoskeletal filaments, and increased cell motility and aggressiveness, and is considered a hallmark of cancer [134–136]. These features have been used to identify epithelial and mesenchymal cells. Fuchs et al. tested E-cadherin and Vimentin expression levels in 12 human hepatoma cell lines and identified them as epithelial cells or mesenchymal cells. Mesenchymal cell lines resisted anti-EGFR therapy, while epithelial cell lines were sensitive to EGFR inhibitors. Increased AKT, STAT3, and integrin-linked kinase (ILK) activation were observed in mesenchymal cell lines. The expression of e-cadherin increased in ILK-inactivated mutants, indicating a reversal of EMT. This transition is associated with a decrease in the AKT signaling pathway, resulting in a recovery of cell sensitivity to EGFR inhibitors [137]. In 2010, researchers found that mesenchymal squamous cell carcinoma was resistant to cetuximab [138]. The cell types were classified into low-expression e-cadherin subsets

(E-cadherin/Low) and high-expression e-cadherin subsets (E-cadherin/high). The expression of multiple EMT-related genes in the E-cadherin/low subpopulation increases, the cell expression is lower, and it is also inhibited during the G0 phase of the cell cycle, which means the cell replacement rate is lower. The E-cadherin/low subgroup EGFR decreased, and the ligand EGF failed to stimulate downstream signaling pathways, suggesting that the regulation of AKT and ERK activation is independent of EGF activation of EGFR. In both in vitro and in vivo models, increased vimentin expression was observed when xenografts in the e-cadherin/high subgroup and low subgroup were treated with cetuximab for 12 days. Meanwhile, the E-cadherin/low subgroup remained resistant to cetuximab. Therefore, tumor cells may develop resistance to cetuximab by increasing the proportion of mesenchymal cells through EMT.

Constitutive activation of EGFR downstream effector molecules

Degradation of the PTEN protein and consequent constitutive activation of AKT was observed in an NSCLC cell line (NCI-HCC827) resistant to cetuximab [139]. PTEN degradation and AKT activation reduction were observed in drug-resistant cells after treatment with the proteome inhibitor MG-132. These results suggest that downregulation of PTEN expression can lead to decreased sensitivity of cells to cetuximab. Another study showed that the intracellular kinase Src activates the PI3K/AKT pathway in cetuximab-resistant NSCLC cells [129]. The combination of cetuximab and dasatinib (a pan-SFK inhibitor) can more effectively inhibit the activation of the downstream AKT signaling pathway and ultimately hinder the proliferation of drug-resistant clones. For mCRC cell lines carrying KRAS, the combination of cetuximab and dasatinib also significantly inhibited cell proliferation, with decreased activity of the MAPK, STAT, or β-catenin pathways observed in the cells. In mouse KRAS mutant xenograft tumors, cetuximab combined with dasatinib from the onset of tumors has been found to have significant growth inhibition. Overall, in KRAS mutant cell lines, SFK blockade reduces vital signaling pathways critical for cell proliferation and survival [140]. These data provide a theoretical basis for the clinical use of cetuximab and dasatinib in combination with KRAS mutant CRC patients.

Increased expression of HER family growth factors

Hatakeyama et al. obtained a cetuximab-resistant clone (1Cc8) from the HNSCC UMSCC-1-cell line after long-term cetuximab treatment. A heparin-binding EGF (HB-EGF) increase was observed in 1Cc8 cells, and inhibition of HB-EGF expression resensitized these cells to cetuximab treatment, suggesting that dysregulation of ligand expression contributes to cetuximab resistance to some degree. MiR-212 was 27 times lower in 1Cc8 cells than in the cetuximab-sensitive cell line, and miR-212 mimics could resensitize 1Cc8 cells to cetuximab. Furthermore, HB-EGF, TGF-α, and AR expression in recurrent tumors was higher. In general, increased HB-EGF expression and downregulation of specific miRs are one of the ways tumor cells escape cetuximab therapy [141].

Concluding remarks

The discovery of EGFR and the success of targeted therapies have benefited countless patients, yet primary and secondary resistance still limits the use of these therapies. Understanding the resistance mechanisms will also help further improve this therapy, with crucial areas remaining, such as identifying effect-related biomarkers and key signaling pathways that enhance cetuximab activity or are associated with cetuximab resistance. A better understanding of EGFR and cetuximab will pave the way for other antibody therapies to target tumors in the future.

References

[1] Carpenter G, Lembach KJ, Morrison MM, Cohen S. Characterization of the binding of 125-I-labeled epidermal growth factor to human fibroblasts. J Biol Chem 1975;250:4297–304.
[2] Ullrich A, et al. Human epidermal growth factor receptor cDNA sequence and aberrant expression of the amplified gene in A431 epidermoid carcinoma cells. Nature 1984;309:418–25.
[3] Riedel H, Dull TJ, Schlessinger J, Ullrich A. A chimaeric receptor allows insulin to stimulate tyrosine kinase activity of epidermal growth factor receptor. Nature 1986;324:68–70.
[4] Schlessinger J. Signal transduction by allosteric receptor oligomerization. Trends Biochem Sci 1988;13:443–7.
[5] Eckhart W, Hutchinson MA, Hunter T. An activity phosphorylating tyrosine in polyoma T antigen immunoprecipitates. Cell 1979;18:925–33.
[6] Hunter T, Sefton BM. Transforming gene product of Rous sarcoma virus phosphorylates tyrosine. Proc Natl Acad Sci U S A 1980;77:1311–5.
[7] Roskoski R. The ErbB/HER family of protein-tyrosine kinases and cancer. Pharmacol Res 2014;79:34–74.
[8] Singh AB, Harris RC. Autocrine, paracrine and juxtacrine signaling by EGFR ligands. Cell Signal 2005;17: 1183–93.
[9] Baselga J, Swain SM. Novel anticancer targets: revisiting ERBB2 and discovering ERBB3. Nat Rev Cancer 2009;9:463–75.
[10] Wallasch C, Weiss FU, Niederfellner G, Jallal B, Issing W, Ullrich A. Heregulin-dependent regulation of HER2/neu oncogenic signaling by heterodimerization with HER3. EMBO J 1995;14:4267–75.
[11] Guy PM, Platko JV, Cantley LC, Cerione RA, Carraway KL. Insect cell-expressed p180erbB3 possesses an impaired tyrosine kinase activity. Proc Natl Acad Sci U S A 1994;91:8132–6.
[12] Sebolt-Leopold JS, Herrera R. Targeting the mitogen-activated protein kinase cascade to treat cancer. Nat Rev Cancer 2004;4:937–47.
[13] Marti U, et al. Localization of epidermal growth factor receptor in hepatocyte nuclei. Hepatology 1991;13:15–20.
[14] Cao H, Lei ZM, Bian L, Rao CV. Functional nuclear epidermal growth factor receptors in human choriocarcinoma JEG-3 cells and normal human placenta. Endocrinology 1995;136:3163–72.
[15] Lin SY, et al. Nuclear localization of EGF receptor and its potential new role as a transcription factor. Nat Cell Biol 2001;3:802–8.
[16] from traditional signal transduction to direct nuclear translocalization, editor. Lo H—W, Hsu S—C, Hung M-C. EGFR signaling pathway in breast cancers. Breast Cancer Res Treat 2006;95:211–8.
[17] Lo HW, Hung MC. Nuclear EGFR signaling network in cancers: linking EGFR pathway to cell cycle progression, nitric oxide pathway and patient survival. Br J Cancer 2006;94:184–8.
[18] Lo HW, Xia W, Wei Y, Ali-Seyed M, Huang SF, Hung M-C. Novel prognostic value of nuclear epidermal growth factor receptor in breast cancer. Cancer Res 2005;65:338–48.
[19] Psyrri A, et al. Quantitative determination of nuclear and cytoplasmic epidermal growth factor receptor expression in oropharyngeal squamous cell cancer by using automated quantitative analysis. Clin Cancer Res 2005;11:5856–62.
[20] Xia W, et al. Nuclear expression of epidermal growth factor receptor is a novel prognostic value in patients with ovarian cancer. Mol Carcinog 2009;48:610–7.

[21] Lo HW, Ali-Seyed M, Wu Y, Bartholomeusz G, Hsu SC, Hung M-C. Nuclear-cytoplasmic transport of EGFR involves receptor endocytosis, importin beta1 and CRM1. J Cell Biochem 2006;98:1570–83.
[22] Wang S-C, et al. Tyrosine phosphorylation controls PCNA function through protein stability. Nat Cell Biol 2006;8:1359–68.
[23] Dittmann K, Mayer C, Kehlbach R, Rodemann HP. Radiation-induced caveolin-1 associated EGFR internalization is linked with nuclear EGFR transport and activation of DNA-PK. Mol Cancer 2008;7:69.
[24] Liao H-J, Carpenter G. Cetuximab/C225-induced intracellular trafficking of epidermal growth factor receptor. Cancer Res 2009;69:6179–83.
[25] Citri A, Yarden Y. EGF-ERBB signaling: towards the systems level. Nat Rev Mol Cell Biol 2006;7:505–16.
[26] Haglund K, Sigismund S, Polo S, Szymkiewicz I, Di Fiore PP, Dikic I. Multiple monoubiquitination of RTKs is sufficient for their endocytosis and degradation. Nat Cell Biol 2003;5:461–6.
[27] Mineo C, Gill GN, Anderson RG. Regulated migration of epidermal growth factor receptor from caveolae. J Biol Chem 1999;274:30636–43.
[28] Vieira AV, Lamaze C, Schmid SL. Control of EGF receptor signaling by clathrin-mediated endocytosis. Science 1996;274:2086–9.
[29] Yamabhai M, Anderson RGW. Second cysteine-rich region of epidermal growth factor receptor contains targeting information for caveolae/rafts. J Biol Chem 2002;277:24843–6.
[30] Soubeyran P, Kowanetz K, Szymkiewicz I, Langdon WY, Dikic I. Cbl-CIN85-endophilin complex mediates ligand-induced downregulation of EGF receptors. Nature 2002;416:183–7.
[31] Bache KG, Slagsvold T, Stenmark H. Defective downregulation of receptor tyrosine kinases in cancer. EMBO J 2004;23:2707–12.
[32] Hendriks BS, Opresko LK, Wiley HS, Lauffenburger D. Coregulation of epidermal growth factor receptor/human epidermal growth factor receptor 2 (HER2) levels and locations: quantitative analysis of HER2 overexpression effects. Cancer Res 2003;63:1130–7.
[33] Sorkin A, Waters CM. Endocytosis of growth factor receptors. Bioessays 1993;15:375–82.
[34] Baulida J, Kraus MH, Alimandi M, Di Fiore PP, Carpenter G. All ErbB receptors other than the epidermal growth factor receptor are endocytosis impaired. J Biol Chem 1996;271:5251–7.
[35] Worthylake R, Opresko LK, Wiley HS. ErbB-2 amplification inhibits down-regulation and induces constitutive activation of both ErbB-2 and epidermal growth factor receptors. J Biol Chem 1999;274:8865–74.
[36] Wang Z, Zhang L, Yeung TK, Chen X. Endocytosis deficiency of epidermal growth factor (EGF) receptor-ErbB2 heterodimers in response to EGF stimulation. Mol Biol Cell 1999;10:1621–36.
[37] Chung BM, et al. Aberrant trafficking of NSCLC-associated EGFR mutants through the endocytic recycling pathway promotes interaction with Src. BMC Cell Biol 2009;10:84.
[38] Hendriks BS, Wiley HS, Lauffenburger D. HER2-mediated effects on EGFR endosomal sorting: analysis of biophysical mechanisms. Biophys J 2003;85:2732–45.
[39] Weichselbaum RR, et al. Epidermal growth factor receptor gene amplification and expression in head and neck cancer cell lines. Head Neck 1989;11:437–42.
[40] Veale D, Ashcroft T, Marsh C, Gibson GJ, Harris AL. Epidermal growth factor receptors in non-small cell lung cancer. Br J Cancer 1987;55:513–6.
[41] Libermann TA, Razon N, Bartal AD, Yarden Y, Schlessinger J, Soreq H. Expression of epidermal growth factor receptors in human brain tumors. Cancer Res 1984;44:753–60.
[42] Libermann TA, et al. Amplification, enhanced expression and possible rearrangement of EGF receptor gene in primary human brain tumors of glial origin. Nature 1985;313:144–7.
[43] Libermann TA, et al. Amplification and overexpression of the EGF receptor gene in primary human glioblastomas. J Cell Sci Suppl 1985;3:161–72.
[44] Maurizi M, et al. Prognostic significance of epidermal growth factor receptor in laryngeal squamous cell carcinoma. Br J Cancer 1996;74:1253–7.
[45] Fischer-Colbrie J, et al. EGFR and steroid receptors in ovarian carcinoma: comparison with prognostic parameters and outcome of patients. Anticancer Res 1997;17:613–9.
[46] Kersemaekers AM, et al. Oncogene alterations in carcinomas of the uterine cervix: overexpression of the epidermal growth factor receptor is associated with poor prognosis. Clinical cancer research: an official journal of the American Association for Cancer Research 1999;5:577–86.
[47] Mellon K, Wright C, Kelly P, Horne CH, Neal DE. Long-term outcome related to epidermal growth factor receptor status in bladder cancer. J Urol 1995;153:919–25.

[48] Inada S, Koto T, Futami K, Arima S, Iwashita A. Evaluation of malignancy and the prognosis of esophageal cancer based on an immunohistochemical study (p53, E-cadherin, epidermal growth factor receptor). Surg Today 1999;29:493–503.
[49] Watanabe K, Tachibana O, Sata K, Yonekawa Y, Kleihues P, Ohgaki H. Overexpression of the EGF receptor and p53 mutations are mutually exclusive in the evolution of primary and secondary glioblastomas. Brain Pathol 1996;6.
[50] Nicholson RI, Gee JM, Harper ME. EGFR and cancer prognosis. Eur J Cancer 2001;37(Suppl 4):S9–15.
[51] Nielsen JS, Jakobsen E, Hølund B, Bertelsen K, Jakobsen A. Prognostic significance of p53, Her-2, and EGFR overexpression in borderline and epithelial ovarian cancer. Int J Gynecol Cancer 2004;14:1086–96.
[52] Santini J, et al. Characterization, quantification, and potential clinical value of the epidermal growth factor receptor in head and neck squamous cell carcinomas. Head Neck 1991;13:132–9.
[53] Yasui W, et al. Expression of epidermal growth factor receptor in human gastric and colonic carcinomas. Cancer Res 1988;48:137–41.
[54] Piyathilake CJ, et al. Differential expression of growth factors in squamous cell carcinoma and precancerous lesions of the lung. Clin Cancer Res 2002;8:734–44.
[55] Scambia G, et al. Significance of epidermal growth factor receptor expression in primary human endometrial cancer. Int J Cancer 1994;56:26–30.
[56] Humphrey PA, et al. Anti-synthetic peptide antibody reacting at the fusion junction of deletion-mutant epidermal growth factor receptors in human glioblastoma. Proc Natl Acad Sci U S A 1990;87:4207–11.
[57] Sizeland AM, Burgess AW. Anti-sense transforming growth factor alpha oligonucleotides inhibit autocrine stimulated proliferation of a colon carcinoma cell line. Mol Biol Cell 1992;3:1235–43.
[58] Malden LT, Novak U, Kaye AH, Burgess AW. Selective amplification of the cytoplasmic domain of the epidermal growth factor receptor gene in glioblastoma multiforme. Cancer Res 1988;48:2711–4.
[59] Peschard P, Park M. Escape from Cbl-mediated downregulation: a recurrent theme for oncogenic deregulation of receptor tyrosine kinases. Cancer Cell 2003;3:519–23.
[60] Li S, Schmitz KR, Jeffrey PD, Wiltzius JJW, Kussie P, Ferguson KM. Structural basis for inhibition of the epidermal growth factor receptor by cetuximab. Cancer Cell 2005;7:301–11.
[61] Yang XD, Jia XC, Corvalan JR, Wang P, Davis CG. Development of ABX-EGF, a fully human anti-EGF receptor monoclonal antibody, for cancer therapy. Crit Rev Oncol Hematol 2001;38:17–23.
[62] Sunada H, Magun BE, Mendelsohn J, MacLeod CL. Monoclonal antibody against epidermal growth factor receptor is internalized without stimulating receptor phosphorylation. Proc Natl Acad Sci U S A 1986;83:3825–9.
[63] Goldstein NI, Prewett M, Zuklys K, Rockwell P, Mendelsohn J. Biological efficacy of a chimeric antibody to the epidermal growth factor receptor in a human tumor xenograft model. Clin Cancer Res 1995;1:1311–8.
[64] Huang SM, Bock JM, Harari PM. Epidermal growth factor receptor blockade with C225 modulates proliferation, apoptosis, and radiosensitivity in squamous cell carcinomas of the head and neck. Cancer Res 1999;59:1935–40.
[65] Peng D, Fan Z, Lu Y, DeBlasio T, Scher H, Mendelsohn J. Anti-epidermal growth factor receptor monoclonal antibody 225 up-regulates p27KIP1 and induces G1 arrest in prostatic cancer cell line DU145. Cancer Res 1996;56:3666–9.
[66] Wu X, et al. Involvement of p27KIP1 in G1 arrest mediated by an anti-epidermal growth factor receptor monoclonal antibody. Oncogene 1996;12:1397–403.
[67] Tortora G, et al. Cooperative inhibitory effect of novel mixed backbone oligonucleotide targeting protein kinase A in combination with docetaxel and anti-epidermal growth factor-receptor antibody on human breast cancer cell growth. Clin Cancer Res 1999;5:875–81.
[68] Liu B, Fang M, Schmidt M, Lu Y, Mendelsohn J, Fan Z. Induction of apoptosis and activation of the caspase cascade by anti-EGF receptor monoclonal antibodies in DiFi human colon cancer cells do not involve the c-jun N-terminal kinase activity. Br J Cancer 2000;82:1991–9.
[69] Mandal M, Adam L, Mendelsohn J, Kumar R. Nuclear targeting of Bax during apoptosis in human colorectal cancer cells. Oncogene 1998;17.
[70] Wu X, Fan Z, Masui H, Rosen N, Mendelsohn J. Apoptosis induced by an anti-epidermal growth factor receptor monoclonal antibody in a human colorectal carcinoma cell line and its delay by insulin. J Clin Invest 1995;95:1897–905.
[71] Kurai J, et al. Antibody-dependent cellular cytotoxicity mediated by cetuximab against lung cancer cell lines. Clin Cancer Res 2007;13:1552–61.

[72] Kimura H, Sakai K, Arao T, Shimoyama T, Tamura T, Nishio K. Antibody-dependent cellular cytotoxicity of cetuximab against tumor cells with wild-type or mutant epidermal growth factor receptor. Cancer Sci 2007;98:1275–80.

[73] Robert F, et al. Phase I study of anti–epidermal growth factor receptor antibody cetuximab in combination with radiation therapy in patients with advanced head and neck cancer. J Clin Oncol 2001;19:3234–43.

[74] Bonner JA, et al. Radiotherapy plus cetuximab for squamous-cell carcinoma of the head and neck. N Engl J Med 2006;354:567–78.

[75] Thienelt CD, et al. Multicenter phase I/II study of cetuximab with paclitaxel and carboplatin in untreated patients with stage IV non-small-cell lung cancer. J Clin Oncol 2005;23:8786–93.

[76] Pirker R, et al. Cetuximab plus chemotherapy in patients with advanced non-small-cell lung cancer (FLEX): an open-label randomized phase III trial. Lancet 2009;373:1525–31.

[77] Lynch TJ, et al. Cetuximab and first-line taxane/carboplatin chemotherapy in advanced non-small-cell lung cancer: results of the randomized multicenter phase III trial BMS099. J Clin Oncol 2010;28:911–7.

[78] Chung KY, et al. Cetuximab shows activity in colorectal cancer patients with tumors that do not express the epidermal growth factor receptor by immunohistochemistry. J Clin Oncol 2005;23:1803–10.

[79] Pao W, et al. EGF receptor gene mutations are common in lung cancers from "never smokers" and are associated with sensitivity of tumors to gefitinib and erlotinib. Proc Natl Acad Sci U S A 2004;101:13306–11.

[80] Lynch TJ, et al. Activating mutations in the epidermal growth factor receptor underlying responsiveness of non-small-cell lung cancer to gefitinib. N Engl J Med 2004;350:2129–39.

[81] Paez JG, et al. EGFR mutations in lung cancer: correlation with clinical response to gefitinib therapy. Science 2004;304:1497–500.

[82] Pao W, et al. Acquired resistance of lung adenocarcinomas to gefitinib or erlotinib is associated with a second mutation in the EGFR kinase domain. PLoS Med 2005;2, e73.

[83] Bertotti A, et al. The genomic landscape of response to EGFR blockade in colorectal cancer. Nature 2015; 526:263–7.

[84] Van Emburgh BO, et al. Acquired RAS or EGFR mutations and duration of response to EGFR blockade in colorectal cancer. Nat Commun 2016;7:13,665.

[85] Price T, et al. Frequency of S492R mutations in the epidermal growth factor receptor: analysis of plasma DNA from patients with metastatic colorectal cancer treated with panitumumab or cetuximab monotherapy. Cancer Biol Ther 2020;21:891–8.

[86] Moroni M, et al. Gene copy number for epidermal growth factor receptor (EGFR) and clinical response to antiEGFR treatment in colorectal cancer: a cohort study. Lancet Oncol 2005;6:279–86.

[87] Frattini M, et al. PTEN loss of expression predicts cetuximab efficacy in metastatic colorectal cancer patients. Br J Cancer 2007;97:1139–45.

[88] Personeni N, et al. Clinical usefulness of EGFR gene copy number as a predictive marker in colorectal cancer patients treated with cetuximab: a fluorescent in situ hybridization study. Clin Cancer Res 2008;14:5869–76.

[89] Cappuzzo F, et al. EGFR FISH assay predicts for response to cetuximab in chemotherapy refractory colorectal cancer patients. Ann Oncol 2008;19:717–23.

[90] Sartore-Bianchi A, et al. Epidermal growth factor receptor gene copy number and clinical outcome of metastatic colorectal cancer treated with panitumumab. J Clin Oncol 2007;25:3238–45.

[91] Hirsch FR, et al. Increased EGFR gene copy number detected by fluorescent in situ hybridization predicts outcome in non-small-cell lung cancer patients treated with cetuximab and chemotherapy. J Clin Oncol 2008;26:3351–7.

[92] Jiang Z, Li C, Li F, Wang X. EGFR gene copy number as a prognostic marker in colorectal cancer patients treated with cetuximab or panitumumab: a systematic review and meta analysis. PLoS One 2013;8, e56205.

[93] Khambata-Ford S, et al. Expression of epiregulin and amphiregulin and K-ras mutation status predict disease control in metastatic colorectal cancer patients treated with cetuximab. J Clin Oncol 2007;25:3230–7.

[94] Trahey M, McCormick F. A cytoplasmic protein stimulates normal N-ras p21 GTPase, but does not affect oncogenic mutants. Science 1987;238:542–5.

[95] Schubbert S, Shannon K, Bollag G. Hyperactive Ras in developmental disorders and cancer. Nat Rev Cancer 2007;7:295–308.

[96] Lièvre A, et al. KRAS mutation status is predictive of response to cetuximab therapy in colorectal cancer. Cancer Res 2006;66:3992–5.

[97] Di Fiore F, et al. Clinical relevance of KRAS mutation detection in metastatic colorectal cancer treated by Cetuximab plus chemotherapy. Br J Cancer 2007;96:1166–9.

[98] De Roock W, et al. KRAS wild-type state predicts survival and is associated to early radiological response in metastatic colorectal cancer treated with cetuximab. Ann Oncol 2008;19:508–15.

[99] Van Cutsem E, et al. Cetuximab and chemotherapy as initial treatment for metastatic colorectal cancer. N Engl J Med 2009;360:1408–17.

[100] Allegra CJ, et al. American Society of Clinical Oncology provisional clinical opinion: testing for KRAS gene mutations in patients with metastatic colorectal carcinoma to predict response to anti-epidermal growth factor receptor monoclonal antibody therapy. J Clin Oncol 2009;27:2091–6.

[101] Karapetis CS, et al. K-ras mutations and benefit from cetuximab in advanced colorectal cancer. N Engl J Med 2008;359:1757–65.

[102] Loupakis F, et al. PTEN expression and KRAS mutations on primary tumors and metastases in the prediction of benefit from cetuximab plus irinotecan for patients with metastatic colorectal cancer. J Clin Oncol 2009;27:2622–9.

[103] Benvenuti S, et al. Oncogenic activation of the RAS/RAF signaling pathway impairs the response of metastatic colorectal cancers to anti-epidermal growth factor receptor antibody therapies. Cancer Res 2007;67:2643–8.

[104] Marais R, Light Y, Paterson HF, Marshall CJ. Ras recruits Raf-1 to the plasma membrane for activation by tyrosine phosphorylation. EMBO J 1995;14:3136–45.

[105] Roberts PJ, Der CJ. Targeting the Raf–MEK–ERK mitogen-activated protein kinase cascade for the treatment of cancer. Oncogene 2007;26:3291–310.

[106] Niault TS, Baccarini M. Targets of Raf in tumorigenesis. Carcinogenesis 2010;31:1165–74.

[107] Di Nicolantonio F, et al. Wild-type BRAF is required for response to panitumumab or cetuximab in metastatic colorectal cancer. J Clin Oncol 2008;26:5705–12.

[108] Blanco R, et al. A gene-alteration profile of human lung cancer cell lines. Hum Mutat 2009;30:1199–206.

[109] Pratilas CA, et al. (V600E)BRAF is associated with disabled feedback inhibition of RAF–MEK signaling and elevated transcriptional output of the pathway. Proc Natl Acad Sci U S A 2009;106:4519–24.

[110] Brose MS, et al. BRAF and RAS mutations in human lung cancer and melanoma. Cancer Res 2002;62:6997–7000.

[111] Davies H, et al. Mutations of the BRAF gene in human cancer. Nature 2002;417:949–54.

[112] Weber A, Langhanki L, Sommerer F, Markwarth A, Wittekind C, Tannapfel A. Mutations of the BRAF gene in squamous cell carcinoma of the head and neck. Oncogene 2003;22:4757–9.

[113] Perrone F, et al. Molecular and cytogenetic subgroups of oropharyngeal squamous cell carcinoma. Clin Cancer Res 2006;12:6643–51.

[114] Perrotte P, et al. Anti-epidermal growth factor receptor antibody C225 inhibits angiogenesis in human transitional cell carcinoma growing orthotopically in nude mice. Clin Cancer Res 1999;5:257–65.

[115] Petit AM, et al. Neutralizing antibodies against epidermal growth factor and ErbB-2/neu receptor tyrosine kinases down-regulate vascular endothelial growth factor production by tumor cells in vitro and in vivo: angiogenic implications for signal transduction therapy of solid tumors. Am J Pathol 1997;151:1523–30.

[116] Ciardiello F, et al. Antiangiogenic and antitumor activity of anti-epidermal growth factor receptor C225 monoclonal antibody in combination with vascular endothelial growth factor antisense oligonucleotide in human GEO colon cancer cells. Clin Cancer Res 2000;6:3739–47.

[117] Ferrara N. The role of vascular endothelial growth factor in pathological angiogenesis. Breast Cancer Res Treat 1995;36:127–37.

[118] Viloria-Petit A, et al. Acquired resistance to the antitumor effect of epidermal growth factor receptor-blocking antibodies in vivo: a role for altered tumor angiogenesis. Cancer Res 2001;61:5090–101.

[119] Jung YD, et al. Effects of combination anti-vascular endothelial growth factor receptor and anti-epidermal growth factor receptor therapies on the growth of gastric cancer in a nude mouse model. Eur J Cancer 2002;38:1133–40.

[120] Ciardiello F, et al. Antitumor activity of ZD6474, a vascular endothelial growth factor receptor tyrosine kinase inhibitor, in human cancer cells with acquired resistance to antiepidermal growth factor receptor therapy. Clin Cancer Res 2004;10:784–93.

[121] Bianco R, et al. Vascular endothelial growth factor receptor-1 contributes to resistance to anti-epidermal growth factor receptor drugs in human cancer cells. Clin Cancer Res 2008;14:5069–80.

[122] Lu Y, et al. Epidermal growth factor receptor (EGFR) ubiquitination as a mechanism of acquired resistance escaping treatment by the anti-EGFR monoclonal antibody cetuximab. Cancer Res 2007;67:8240–7.

[123] Wheeler DL, et al. Mechanisms of acquired resistance to cetuximab: role of HER (ErbB) family members. Oncogene 2008;27:3944–56.

[124] Demory ML, et al. Epidermal growth factor receptor translocation to the mitochondria: regulation and effect. J Biol Chem 2009;284:36592–604.

[125] Lo HW, Cao X, Zhu H, Ali-Osman F. Constitutively activated STAT3 frequently coexpresses with epidermal growth factor receptor in high-grade gliomas and targeting STAT3 sensitizes them to Iressa and alkylators. Clin Cancer Res 2008;14:6042–54.

[126] Lo H-W, et al. Nuclear interaction of EGFR and STAT3 in the activation of the iNOS/NO pathway. Cancer Cell 2005;7:575–89.

[127] Hanada N, Lo H-W, Day C-P, Pan Y, Nakajima Y, Hung M-C. Co-regulation of B-Myb expression by E2F1 and EGF receptor. Mol Carcinog 2006;45:10–7.

[128] Kim J, et al. The phosphoinositide kinase PIKfyve mediates epidermal growth factor receptor trafficking to the nucleus. Cancer Res 2007;67:9229–37.

[129] Wheeler DL, et al. Epidermal growth factor receptor cooperates with Src family kinases in acquired resistance to cetuximab. Cancer Biol Ther 2009;8:696–703.

[130] Huynh HT, Larsson C, Narod S, Pollak M. Tumor suppressor activity of the gene encoding mammary-derived growth inhibitor. Cancer Res 1995;55:2225–31.

[131] Huynh H, Alpert L, Pollak M. Silencing of the mammary-derived growth inhibitor (MDGI) gene in breast neoplasms is associated with epigenetic changes. Cancer Res 1996;56:4865–70.

[132] Specht B, et al. Mammary derived growth inhibitor is not a distinct protein but a mix of heart-type and adipocyte-type fatty acid-binding protein. J Biol Chem 1996;271:19943–9.

[133] Nevo J, et al. Mammary-derived growth inhibitor alters traffic of EGFR and induces a novel form of cetuximab resistance. Clin Cancer Res 2009;15:6570–81.

[134] Hanahan D, Weinberg RA. The hallmarks of cancer. Cell 2000;100:57–70.

[135] Huber MA, Kraut N, Beug H. Molecular requirements for epithelial-mesenchymal transition during tumor progression. Curr Opin Cell Biol 2005;17:548–58.

[136] Thompson EW, Newgreen DF, Tarin D. Carcinoma invasion and metastasis: a role for epithelial-mesenchymal transition? Cancer Res 2005;65.

[137] Fuchs BC, et al. Epithelial-to-mesenchymal transition and integrin-linked kinase mediate sensitivity to epidermal growth factor receptor inhibition in human hepatoma cells. Cancer Res 2008;68:2391–9.

[138] Basu D, et al. Evidence for mesenchymal-like sub-populations within squamous cell carcinomas possessing chemoresistance and phenotypic plasticity. Oncogene 2010;29:4170–82.

[139] Kim SM, et al. Acquired resistance to cetuximab is mediated by increased PTEN instability and leads cross-resistance to gefitinib in HCC827 NSCLC cells. Cancer Lett 2010;296:150–9.

[140] Dunn EF, et al. Dasatinib sensitizes KRAS mutant colorectal tumors to cetuximab. Oncogene 2011;30:561–74.

[141] Hatakeyama H, et al. Regulation of heparin-binding EGF-like growth factor by miR-212 and acquired cetuximab-resistance in head and neck squamous cell carcinoma. PLoS One 2010;5, e12702.

CHAPTER

3

MM-151 overcomes acquired resistance to cetuximab and panitumumab in colorectal cancers harboring EGFR extracellular domain mutations

Zetong Ma, Changhai Lei, and Shi Hu

Department of Biophysics, College of Basic Medical Sciences, Naval Medical University (Second Military Medical University), Shanghai, China

Abstract

Cetuximab and panitumumab are used to treat RAS wild-type colorectal cancer (CRC) through the target of EGFR (epidermal growth factor receptor), but after EGFR signaling pathway blockade, approximately 20% of CRCs develop EGFR extracellular domain (ECD) mutations. Impaired antibody binding is associated with clinical recurrence. The oligoclonal antibody MM-151 developed by researchers binds to several regions of the EGFR ECD to address resistance and recurrence caused by mutations in the EGFR ECD. MM-151 blocked EGFR and inhibited cell growth in a preclinical model, including tumors with EGFR mutations. MM-151 treatment can prevent EGFR ECD mutations in CRC patients who have already developed resistance to EGFR blockade. These data provide a new strategy to treat patients who have developed resistance to anti-EGFR monoclonal antibodies due to EGFR ECD mutations.

Abbreviations

CRCs	colorectal cancers
ddPCR	droplet digital polymerase chain reaction
ECD	extracellular domain
EGF	epidermal growth factor
EGFR	epidermal growth factor receptor
ERK	extracellular signal–regulated kinase

https://doi.org/10.1016/B978-0-12-821584-5.00021-3

HaloTag-EGF	HaloTag fusion ligand
HEK–293 cell	human embryonic kidney 293 cell
mAbs	monoclonal antibodies
mCRCs	metastatic colorectal cancers
NanoLuc-EGFR	receptor fused with nanoluciferase at the N-terminal domain
PDX	patient-derived xenograft
RTK	receptor tyrosine kinase

Conflict of interest

No potential conflicts of interest were disclosed.

Introduction

The therapeutic effects of cetuximab and panitumumab are often limited by secondary resistance [1,2]. This secondary drug resistance may come from three reasons: (1) EGFR downstream effector (KRAS, NRAS, and BRAF) mutations lead to abnormal activation [3–6]; (2) alternate signaling pathway activation, such as receptor tyrosine kinase (RTK) pathways (HER2 and MET) [7–9]; (3) EGFR ECD mutations (S492R, R451C, S464 L, p. G465R, K467T, and I491 M) [10–12]. ECD-mutated colon cancer patients are often resistant to cetuximab and panitumumab at the same time [11]. However, researchers hypothesize that the progression of ECD-mutated colon cancer after anti-EGFR treatment still depends on the EGFR signaling pathway. Therefore, CRC patients with EGFR ECD mutations can still benefit from EGFR blockade after resistance to first-line treatment. Compared with cetuximab and panitumumab, the EGFR-targeted antibody cocktail induced a more comprehensive range of acceptor endocytosis and inhibition and exhibited more potent anticancer effects in mouse models [13,14]. The researchers developed a third-generation EGFR inhibitor, MM-151, a full-human IgG antibody that binds different and nonoverlapping epitopes of EGFR [15]. MM-151 has shown advantages over currently approved monoclonal antibodies in preclinical models, showing improved EGFR pathway and downstream signaling inhibition and enhanced EGFR downregulation and involvement in the innate immune response.

EGFR ECD mutations do not affect ligand–receptor binding but do affect antibody-receptor binding

Researchers have identified six EGFR ECD mutations in CRC patients resistant to cetuximab or panitumumab [10,11]. At the same time, they also identified a G465E mutation in ECD in a CRCs cell line with resistance to cetuximab (HCA-46 R5). To ensure that EGFR mutants remain competent to mediate intracellular signaling under cetuximab or panitumumab exposure, they take advantage of bioluminescent resonance energy transfer (BRET) to detect the binding ability of EGFR mutants and their ligand, epidermal growth factor (EGF). They synthesized a host of NanoLuc-EGFR, an EGFR receptor N-terminal domain fused with nanoluciferase and HaloTag-EGF as a ligand with fluorescence. These receptors are expressed in HEK-293 cells, carrying the same mutation as each ECD variant.

Meanwhile, a wild-type receptor (pNanoLuc-EGFR WT) was synthesized. They found that all synthetic receptors except EGFR R451C can bind to the ligand in a dose-dependent manner. This suggests that CRC cell growth still relies on EGFR-mediated signaling after resistance to cetuximab or panitumumab. Next, researchers tested the ability of cetuximab, panitumumab, and MM-151 to bind to various EGFR ECD mutants using the Nano-BRET assay and found that only MM-151 can bind to the mutants as well as the wild-type EGFR.

MM-151 can inhibit the growth of LIM1215 cells overexpressing EGFR ECD mutants

To determine whether cell proliferation and survival were affected by EGFR-ECD mutations, they constructed lentivirus expression vectors of each EGFR ECD mutant by site-directed mutagenesis and wild-type EGFR as a control. They used the LIM1215 cell line sensitive to anti-EGFR monoclonal antibodies [3,11] to generate cell lines that harbor each mutation. Cells expressing EGFR p. R451C are still sensitive to cetuximab, and those harboring EGFR p. K467T and p. S492R mutation are sensitive to panitumumab treatment, while cells expressing EGFR p. S464 L, p. G465R, p. G465E, and p. I491 M are resistant to cetuximab and panitumumab. All mutants were sensitive to MM-151. Biochemical analysis showed that the downstream expression levels of phosphorylated EGFR, ERK, and Akt effector proteins in drug-resistant cells were downregulated after MM-151 treatment. The results showed that MM-151 significantly inhibited the growth of all known EGFR variants compared to cetuximab and panitumumab, which only inhibited the growth of some EGFR variants.

MM-151 can inhibit the growth of cetuximab-resistant clones in spontaneous models

To simulate the EGFR ECD mutations that occur during clinical anti-EGFR antibody therapy, they used a preclinical model of spontaneous EGFR mutations following exposure to cetuximab. These spontaneously drug-resistant cells often carry multiple EGFR ECD mutations simultaneously [10,11]. They obtained several clones from those carrying EGFR ECD mutations, including LIM1215 R5 (G465R), CCK-81 R1 (S464L), and HCA-46 R5 (G465E). Meanwhile, the clone LIM1215 containing EGFR p. S492R variant was constructed using targeted homologous recombination as a control. Droplet digital polymerase chain reaction (ddPCR) showed that MM-151 still showed a remarkable inhibitory effect on cell proliferation in these cell populations with a high proportion of mutations, while biochemical analysis showed ERK signaling inhibition in these cell lines.

To further determine the effect of MM-151 on secondary resistance to anti-EGFR treatment, they built a patient-derived xenograft (PDX) CRC cell model carrying the p. G465E EGFR ECD mutation from a patient who had developed secondary cetuximab resistance, named CRC G465E-XL [16,17]. The ddPCR results showed that the occurrence frequency of the G465E mutation in CRC G465E-XL cells was high (50.9%), similar to the G465E mutation detected in PDX (53.5%).

The cetuximab-resistant cell line G465E-XL is sensitive to MM-151 treatment in a cell proliferation assay. These results are consistent with those in LIM1215 and HCA-46. Biochemical analysis showed that MM-151 significantly reduced ERK signal expression in CRC G465E-XL cell lines.

MM-151 reduced the frequency of EGFR-ECD allele mutations

To investigate the treatment effect of MM-151 on patients with EGFR ECD mutation metastatic colorectal cancer after treatment with cetuximab or panitumumab, researchers analyzed serum samples collected up to February 2015 in the phase I study of MM-151 (NCT01520389), including 11 CRC patients. The patients were selected by the availability of serum samples and stable disease or partial reactions documented in the previous anti-EGFR therapy. They detected EGFR ECD mutations using ddPCR in two patients (patients 051 and 095) whose ctDNA was isolated before the treatment with MM-151. Longitudinal analysis was performed on samples collected during the treatment of MM-151 and revealed changes in the allele frequency of EGFR ECD mutations during MM-151 treatment. Notably, after 5 weeks of treatment, there was a significant reduction in tumor volume in patient 095 and a significant reduction in the frequency of the EGFR p. G465E mutant allele. In Patient 051, EGFR p. S464 L and p. G465R mutations were reduced and stabilized, accompanied by long-term disease stabilization. Reversal of the reduced frequency of these mutated alleles predicted progress after 7 weeks.

Discussion

RTKs are critical to the pathogenesis of human tumors and have been regarded as a target for anticancer therapy. However, the application of small molecule kinase inhibitors of RTKs is often limited by drug resistance. This usually results from interference with the drug's binding site and the target protein due to the presence of secondary key mutations after the persistent application of small molecule kinase inhibitors.

For example, BCR-ABL mutations are resistant to imatinib [18–20], ALK and ROS1 mutations are resistant to crizotinib [21,22], and EGFR mutations are resistant to erlotinib and gefitinib [23,24]. Several examples have shown that the survival of drug-resistant tumor cells, despite secondary mutations, still depends on RTKs [23,25]. This provides the basis for the development of drugs that can overcome drug-resistant mutations in the first generation of tyrosine kinase inhibitor drugs. For example, dasatinib and nilotinib can operate effectively against BCR-ABL variants that are resistant to imatinib [26]; ceritinib is an ALK inhibitor that is used against ALK mutants resistant to crizotinib [27,28]; rociletinib and AZD9291 have been regarded as substitutes for treating lung cancer with acquired resistance to gefitinib or erlotinib due to EGFR p. T790 M mutations [29,30].

Although second- and third-generation small-molecule kinase inhibitors can ameliorate acquired resistance to first-generation inhibitors, little progress has been made in drug development that overcomes resistance to anti-EGFR monoclonal antibodies. A precise understanding of the resistance mechanisms is a prerequisite for further progress in drug development. The discovery that approximately 20% of CRC patients treated with anti-EGFR monoclonal antibodies had significant mutations in the EGFR ECD [10,11,13] provided an opportunity for their investigation.

They hypothesized that EGFR variants that occurred during treatment with cetuximab or panitumumab could be treated by antibodies that target different parts of the receptor. They found that the binding capacity of EGFR ECD mutants and EGF was unaffected, although it could not be interfered with by cetuximab or panitumumab. The cell model results showed that MM-151 broadly suppressed all EGFR ECD-mutated cells, demonstrating that mutated EGFR could be suppressed by recognizing other epitopes on the receptor. In addition, the therapeutic effect of MM-151 in cetuximab-resistant variants (K467T and S492R) was similar to that of panitumumab. The results indicated that a second-line EGFR blocker might overcome cetuximab resistance.

First, an in vitro experiment showed that the proliferation of CRC cells with EGFR ECD mutations induced by cetuximab can be inhibited by MM-151. Second, 11 patients who relapsed after treatment with cetuximab were treated with MM-151, and three patients benefited, 2 of whom were detected to have EGFR ECD mutations. During MM-151 treatment, EGFR ECD mutant DNA concentrations decrease or stabilize, consistent with the results assessed by radiological methods. In summary, these results suggest that in vivo treatment with MM-151 may effectively inhibit ECD variation in EGFR. The clinical trial was limited because it did not consider multiple, potentially concomitant resistance mechanisms. Patients with colorectal tumors that develop secondary resistance to EGFR blockade tend to exhibit different resistance mechanisms, including mutations in KRAS and NRAS [6,11]. Therefore, colorectal cancers with heterogeneous resistance mechanisms are not sensitive to EGFR-targeted strategies.

Further understanding of the relationship between these parallel resistance mechanisms is needed. Their findings suggest that patients harboring EGFR ECD mutations may benefit from MM-151 treatment. Therefore, MM-151 should be prospectively evaluated clinically in these patients.

References

[1] Bagchi A, et al. Molecular basis for Necitumumab inhibition of EGFR variants associated with acquired Cetuximab resistance. Mol Cancer Ther 2018;17:521–31.

[2] Misale S, Di Nicolantonio F, Sartore-Bianchi A, Siena S, Bardelli A. Resistance to anti-EGFR therapy in colorectal cancer: from heterogeneity to convergent evolution. Cancer Discov 2014;4:1269–80.

[3] Misale S, et al. Emergence of KRAS mutations and acquired resistance to anti-EGFR therapy in colorectal cancer. Nature 2012;486:532–6.

[4] Bettegowda C, et al. Detection of circulating tumour DNA in early- and late-stage human malignancies. Sci Transl Med 2014;6:224ra224.

[5] Diaz LA, et al. The molecular evolution of acquired resistance to targeted EGFR blockade in colorectal cancers. Nature 2012;486:537–40.

[6] Misale S, et al. Blockade of EGFR and MEK intercepts heterogeneous mechanisms of acquired resistance to anti-EGFR therapies in colorectal cancer. Sci Transl Med 2014;6:224ra226.

[7] Yonesaka K, et al. Activation of ERBB2 signaling causes resistance to the EGFR-directed therapeutic antibody cetuximab. Sci Transl Med 2011;3:99ra86.

[8] Bardelli A, et al. Amplification of the MET receptor drives resistance to anti-EGFR therapies in colorectal cancer. Cancer Discov 2013;3:658–73.

[9] Mohan S, et al. Changes in colorectal carcinoma genomes under anti-EGFR therapy identified by whole-genome plasma DNA sequencing. PLoS Genet 2014;10, e1004271.

[10] Montagut C, et al. Identification of a mutation in the extracellular domain of the epidermal growth factor receptor conferring cetuximab resistance in colorectal cancer. Nat Med 2012;18:221–3.

[11] Arena S, et al. Emergence of multiple EGFR extracellular mutations during Cetuximab treatment in colorectal Cancer. Clin Cancer Res 2015;21:2157–66.

[12] Morelli MP, et al. Characterizing the patterns of clonal selection in circulating tumour DNA from patients with colorectal cancer refractory to anti-EGFR treatment. Ann Oncol 2015;26:731–6.
[13] Perera RM, et al. Treatment of human tumour xenografts with monoclonal antibody 806 in combination with a prototypical epidermal growth factor receptor-specific antibody generates enhanced antitumour activity. Clin Cancer Res 2005;11:6390–9.
[14] Spiridon CI, et al. Targeting multiple Her-2 epitopes with monoclonal antibodies results in improved antigrowth activity of a human breast cancer cell line in vitro and in vivo. Clin Cancer Res 2002;8:1720–30.
[15] Kearns JD, et al. Enhanced targeting of the EGFR network with MM-151, an Oligoclonal anti-EGFR antibody therapeutic. Mol Cancer Ther 2015;14:1625–36.
[16] Bertotti A, et al. A molecularly annotated platform of patient-derived xenografts ("xenopatients") identifies HER2 as an effective therapeutic target in cetuximab-resistant colorectal cancer. Cancer Discov 2011;1:508–23.
[17] Bertotti A, et al. The genomic landscape of response to EGFR blockade in colorectal cancer. Nature 2015;526:263–7.
[18] Khorashad JS, et al. BCR-ABL1 compound mutations in tyrosine kinase inhibitor-resistant CML: frequency and clonal relationships. Blood 2013;121:489–98.
[19] Soverini S, et al. Unraveling the complexity of tyrosine kinase inhibitor-resistant populations by ultra-deep sequencing of the BCR-ABL kinase domain. Blood 2013;122:1634–48.
[20] Gibbons DL, et al. Molecular dynamics reveal BCR-ABL1 polymutants as a unique mechanism of resistance to PAN-BCR-ABL1 kinase inhibitor therapy. Proc Natl Acad Sci U S A 2014;111:3550–5.
[21] Choi YL, et al. EML4-ALK mutations in lung cancer that confer resistance to ALK inhibitors. N Engl J Med 2010;363:1734–9.
[22] Awad MM, Engelman JA, Shaw AT. Acquired resistance to crizotinib from a mutation in CD74-ROS1. N Engl J Med 2013;369:1173.
[23] Pao W, et al. Acquired resistance of lung adenocarcinomas to gefitinib or erlotinib is associated with a second mutation in the EGFR kinase domain. PLoS Med 2005;2, e73.
[24] Sequist LV, et al. Genotypic and histological evolution of lung cancers acquiring resistance to EGFR inhibitors. Sci Transl Med 2011;3:75ra26.
[25] Heidel F, et al. Clinical resistance to the kinase inhibitor PKC412 in acute myeloid leukemia by mutation of Asn-676 in the FLT3 tyrosine kinase domain. Blood 2006;107:293–300.
[26] Quintás-Cardama A, Kantarjian H, Cortes J. Imatinib and beyond- -exploring the full potential of targeted therapy for CML. Nat Rev Clin Oncol 2009;6:535–43.
[27] Shaw AT, et al. Ceritinib in ALK-rearranged non-small-cell lung cancer. N Engl J Med 2014;370:1189–97.
[28] Friboulet L, et al. The ALK inhibitor ceritinib overcomes crizotinib resistance in non-small cell lung cancer. Cancer Discov 2014;4:662–73.
[29] Jänne PA, et al. AZD9291 in EGFR inhibitor-resistant non-small-cell lung cancer. N Engl J Med 2015;372:1689–99.
[30] Sequist LV, et al. Rociletinib in EGFR-mutated non-small-cell lung cancer. N Engl J Med 2015;372:1700–9.

Further reading

Arena S, et al. MM-151 overcomes acquired resistance to cetuximab and panitumumab in colorectal cancers harboring EGFR extracellular domain mutations. Sci Transl Med 2016;8(324):324ra14.

CHAPTER

4

Sym004 anti-EGFR antibody mixture overcomes resistance to anti-EGFR antibodies in metastatic colorectal cancer

Guangyao Li[a], *Tian Li*[a], *Wenyan Fu*[b], *Changhai Lei*[a], *and Shi Hu*[a]

[a]Department of Biophysics, College of Basic Medical Sciences, Naval Medical University (Second Military Medical University), Shanghai, China [b]Department of Assisted Reproduction, Shanghai Ninth People's Hospital, Shanghai Jiao Tong University School of Medicine, Shanghai, China

Abstract

Cetuximab and panitumumab have effective treatment responses in the wild-type RAS metastatic colorectal cancer subgroup. Despite the molecular-driven option, not all patients respond to EGFR inhibitors, and thus secondary drug resistance is the critical limiting factor. By observing the in vitro and in vivo functional effects of SYM004 (an antibody combination drug synthesized by two kinds of monoclonal antibodies against ectopic epitopes of EGFR) on the cetuximab-resistant CRC model, researchers showed that the use of SYM004 in mCRC may be a new idea for acquired resistance to anti-EGFR therapy caused by MET activation.

Abbreviations

ECD	EGFR extracellular domain
EGFR	epidermal growth factor receptor
mAbs	monoclonal antibodies
mCRC	metastatic colorectal cancer
TKRs	tyrosine kinase receptors

Conflict of interest

No potential conflicts of interest were disclosed.

https://doi.org/10.1016/B978-0-12-821584-5.00026-2

EGFR and CRC

As a known therapeutic target, the vital therapeutic status of EGFR has been recognized in various tumor treatment courses. Cetuximab and panitumumab are currently two popular anti-EGFR strategies. The binding site directly binds to it on an external scale and interferes with the activation of the final signal pathway. In addition, tyrosine kinase inhibitors can be absorbed by cells, and their phosphorylation inhibition is exerted by activating the intracellular tyrosine kinase domain of EGFR. Although the clinical efficacy of cetuximab and panitumumab is exciting, acquired drug resistance significantly limits the further effects of these drugs. The various mechanisms leading to acquired resistance have been described: mutation of KRAS, NRAS, and BRAF is the most common event [1–3]. It is speculated that these mutations either did not begin to appear in the cloned subpopulations of the tumor before the drug was administered or that these mutations continued to progress as the treatment progressed. In the absence of RAS modification or its direct and final effects, there are other mechanisms for targeting EGFR. The study of HER2 and MET found that the excitation of the MAPK chain comes from the gene distortion of this kind of TKR, and its path of action has nothing to do with EGFR signaling [4–7].

Colorectal cancer (CRC) ranks third on a worldwide scale, and it also ranks fourth among cancer deaths [8]. The common EGFR activation and signaling pathways in cancer are caused by increased related expression and the enhancement of the EGFR gene [1–3,9]. Therefore, EGFR is a hot anticancer treatment target with great potential [10,11].

Two monoclonal antibodies, cetuximab and panitumumab, have been included in the clinical diagnosis of RCR WT mCRC [12,13]. The expected growth, reproduction, survival, metabolism, and differentiation of cells are inseparable from transmembrane tyrosine receptors, including EGFR or HER2. Although the clinical efficacy of cetuximab and panitumumab has been confirmed, acquired drug resistance significantly limits these drugs' further in-depth clinical efficacy [14,15]. SYM004 is an antibody combination drug synthesized by two kinds of monoclonal antibodies (human & mouse) against ectopic epitopes of EGFR [16].

SYM004 compared with cetuximab

Cetuximab is a chimeric monoclonal IgG antibody that binds to the third domain on EGFR. A dose that inhibits the development of cancer cells by 50% (IC_{50}) is initially given. The drug dose was increased to 1/5/10 μg/mL every 2 months for approximately 6 months.

The mature cetuximab-resistant cancer cell lines were cultured for a long time, and the cell lines were observed to proliferate at the maximum dose of cetuximab. Experiments show that the IC_{50} in parental cells is only one-tenth of cetuximab-resistant cancer cell lines.

Cell proliferation and induction of apoptosis

The cancer cells were cultured in a 24-well plate and exposed to cetuximab with a concentration gradient for 96h. Note the IC_{50} value after drawing the dose–response curve. The growth inhibition by SYM004 and cetuximab observed in CRC cells sensitive to cetuximab was the same. In contrast, in cetuximab-resistant CRC cells, SYM004 has a better antiproliferative effect due to its IC_{50} value, which is approximately 10 times lower than that of cetuximab. In addition, Annexin V-FITC has a weakening effect on drugs to inhibit cell proliferation and induce apoptosis. The LIM1215, CACO2, GEO, SW48 and LIM1215-CR, CACO2-CR, GEO-CR, SW48-CR cell lines were cultured in six-well plates, the latter of which are resistant to cetuximab. SYM004 was exposed to various concentration gradients for 72h and finally dyed. After counting and evaluating the number of apoptotic cells, it was found that in cancer cell lines sensitive to cetuximab, both cetuximab and SYM004 exerted an excellent effect on inducing apoptosis and death of cancer cells regardless of whether they were used alone or in combination. For cancer cell lines resistant to cetuximab, exposure to SYM004 alone significantly increased apoptotic cell death.

EGFR-dependent intracellular signaling/MET activation and ERBB2 amplification

The cancer cell lines were cultivated and exposed to a particular concentration gradient of SYM004 and cetuximab for 24h. After exposure to cetuximab alone, decreased values of MAPK and phosphorylated AKT protein were observed in cell lines sensitive to cetuximab but not in cells resistant to cetuximab. In this phenomenon, in contrast, in all cancer cell lines exposed to SYM004, its inhibitory effect on cell proliferation did not conflict with the low phosphorylation levels of MAPK and Akt and even had a synergistic effect. The above experimental results indicate that the MAPK and PIK3CA/AKT pathways may be the hidden mechanism by which SYM004 overcomes drug resistance.

MET is the key to their survival and proliferation [6,7]. EGFR and MET exist in combination in resistant cell lines, but no interaction is observed in nonresistant cell lines. The resistant cancer cell lines were exposed or not to SYM004 and cetuximab and treated with TGFα at the same time. Treatment with TGFα in nonresistant cancer cell lines can promote the occurrence of MET-EGFR heterogeneity. After exposure to cetuximab, this effect is enhanced in all cancer cell lines and has nothing to do with whether it is drug-resistant. However, it was not observed in cells exposed to SYM004, indicating that SYM004 and TGFα have an antagonistic effect and can further interfere with the failure of the cross-reaction of MET-EGFR heterogeneity. MET activation and enhancement may provide new ideas for overcoming cell resistance.

Tumor xenograft models of human colorectal cancer

SYM004 showed more potent antitumor activity and long-term effects in these CRC models and significantly reduced tumor size compared with mice treated with cetuximab.

After ten weeks of treatment, only one of the ten mice in the SYM004 treatment group had unmistakable evidence of progress. In contrast, five out of ten mice had significant clinical treatment progress in the cetuximab treatment group. However, SYM004 had a more prominent prolongation effect at the 30th week, and the xenograft model was still affected by therapeutic drugs. The xenograft model mice in the cetuximab-exposed group were not affected by the apparent effect of the drug, and disease progression was normally observed. However, after adding SYM004 drug treatment, the model again showed a response to the treatment. In fact, at the 30th week, four-sevenths of the mice continued to be positively affected by SYM004 treatment, which is the basis for the long-term efficacy of the drug. Experiments show that the resistance model may have a positive treatment response to SYM004.

The noninterfering binding sites of the two antibodies in the outer structure of EGFR affect each other, resulting in effective EGFR uptake and inactivation, thereby inhibiting the cell proliferation of cetuximab.

Therefore, even in ECD mutations of EGFR, SYM004 can be effective [17]. As mentioned earlier, the two antibodies target noninterference sites on the outer structure of EGFR and have a synergistic effect, so their effect of inhibiting cell growth and proliferation is also significantly better than that of a single antibody [18]. Compared with single antibodies (such as cetuximab), SYM004 inactivation of EGFR can significantly inhibit receptor activity and significantly impact its external structure expression and heterogeneity. In addition, SYM004 proved that specific cancer cells are resistant to cetuximab [19,20].

Sanchez Martin FG et al. showed that in a CRC model resistant to cetuximab, the efficacy of SYM004 was evaluated in vivo and in vitro, and it was found that SYM004 is a strategy for the treatment of mutant CRC tumors on ECF-EGFR [17]. In addition, this indicates that the high synergy of EGF and TGFα bonds improves the growth inhibitory effect of SYM004 on CRC cells, and it is known that there are still potential determinants for the increase in EGF and TGFα bonds [20].

These results indicate that SYM004 has a more prominent antitumor power than cetuximab, which is clearly expressed in cells, indicating EGFR ligand dependence.

HER2

Another mechanism leading to EGFR resistance is HER2 amplification [9, 10]. Similarly, different concentrations of SYM004 and cetuximab were used to treat cancer cell lines containing the HER2 gene, and SYM004 did not show an inhibitory effect superior to cetuximab. Experimental results show that SYM004 is powerless against cetuximab resistance to HER2 gene expression. During the clinical treatment of cetuximab resistance, the amplification of HER2 in less than one-fifth of the cancer cells (14%) before treatment increased significantly in the samples after treatment (71%) [4]. These results indicate that different kinds of drug resistance are related to the expression of the HER2 gene and can be significantly enhanced by it. The disappearance of SYM004's ability to overcome drug resistance can be caused by acquired drug resistance caused by HER2 and does not depend on the destruction of the direct pathway of EGFR signaling. Therefore, the choice of clinical application of SYM004 remains to be discussed in terms of therapeutic drugs for the entire line of drug-resistant cancer cells.

However, MEK and PI3KCA treatment has better antitumor activity. These results were confirmed in tumor xenografts obtained from three HER2-amplified patients, indicating that combination therapy in the treatment of HER2-enhanced CRC may not be a successful attempt.

Final thoughts and future directions

The latest clinical trials indicate that combining two noncompetitive antibodies against EGFR may benefit many types of cancer patients. There are many ways for EGFR to enter cells, and the application of monoclonal antibodies can lead to further uptake of EGFR. The rapid degradation of EGFR under the action of Sym004 is inhibited by it, thereby exerting the cell transplantation effect of the drug. There is evidence that Sym004 can provide survival or stability benefits [21]. Therefore, the discovery of anti-EGFR therapeutic drugs for cancer can be further researched and tested on the rapid degradation and regeneration of EGFR, highlighting the superiority and importance of developing new anti-EGFR targets.

References

[1] Huang Z, Brdlik C, Jin P, Shepard HM. A pan-HER approach for cancer therapy: background, current status and future development. Expert Opin Biol Ther 2009;9:97–110.

[2] Wong AJ, et al. Structural alterations of the epidermal growth factor receptor gene in human gliomas. Proc Natl Acad Sci U S A 1992;89:2965–9.

[3] Damstrup L, et al. Amphiregulin acts as an autocrine growth factor in two human polarizing colon cancer lines that exhibit domain selective EGF receptor mitogenesis. Br J Cancer 1999;80:1012–9.

[4] Yonesaka K, et al. Activation of ERBB2 signaling causes resistance to the EGFR-directed therapeutic antibody cetuximab. Sci Transl Med 2011;3:99ra86.

[5] Bertotti A, et al. A molecularly annotated platform of patient-derived xenografts ("xenopatients") identifies HER2 as an effective therapeutic target in cetuximab-resistant colorectal cancer. Cancer Discov 2011;1:508–23.

[6] Troiani T, et al. Increased TGF-α as a mechanism of acquired resistance to the anti-EGFR inhibitor cetuximab through EGFR-MET interaction and activation of MET signaling in colon cancer cells. Clin Cancer Res 2013;19:6751–65.

[7] Bardelli A, et al. Amplification of the MET receptor drives resistance to anti-EGFR therapies in colorectal cancer. Cancer Discov 2013;3:658–73.

[8] Siegel RL, Miller KD, Fuchs HE, Jemal A. Cancer statistics, 2021. CA Cancer J Clin 2021;71:7–33.

[9] Peghini PL, et al. Overexpression of epidermal growth factor and hepatocyte growth factor receptors in a proportion of gastrinomas correlates with aggressive growth and lower curability. Clin Cancer Res 2002;8:2273–85.

[10] Ciardiello F, Tortora G. EGFR antagonists in cancer treatment. N Engl J Med 2008;358:1160–74.

[11] Troiani T, et al. Therapeutic value of EGFR inhibition in CRC and NSCLC: 15 years of clinical evidence. ESMO Open 2016;1, e000088.

[12] Van Cutsem E, et al. Cetuximab and chemotherapy as initial treatment for metastatic colorectal cancer. N Engl J Med 2009;360:1408–17.

[13] Douillard JY, et al. Randomized, phase III trial of panitumumab with infusional fluorouracil, leucovorin, and oxaliplatin (FOLFOX4) versus FOLFOX4 alone as first-line treatment in patients with previously untreated metastatic colorectal cancer: the PRIME study. J Clin Oncol 2010;28:4697–705.

[14] Troiani T, et al. Molecular aspects of resistance to biological and non-biological drugs and strategies to overcome resistance in colorectal cancer. Curr Med Chem 2014;21:1639–53.

[15] Sforza V, et al. Mechanisms of resistance to anti-epidermal growth factor receptor inhibitors in metastatic colorectal cancer. World J Gastroenterol 2016;22:6345–61.

[16] Pedersen MW, et al. Sym004: a novel synergistic anti-epidermal growth factor receptor antibody mixture with superior anticancer efficacy. Cancer Res 2010;70:588–97.

[17] Sánchez-Martín FJ, et al. The first-in-class anti-EGFR antibody mixture Sym004 overcomes Cetuximab resistance mediated by EGFR extracellular domain mutations in colorectal Cancer. Clin Cancer Res 2016;22:3260–7.

[18] Koefoed K, et al. Rational identification of an optimal antibody mixture for targeting the epidermal growth factor receptor. mAbs 2011;3:584–95.

[19] Iida M, et al. Sym004, a novel EGFR antibody mixture, can overcome acquired resistance to cetuximab. Neoplasia (New York, NY) 2013;15:1196–206.
[20] Dienstmann R, et al. Safety and activity of the first-in-class Sym004 anti-EGFR antibody mixture in patients with refractory colorectal Cancer. Cancer Discov 2015;5:598–609.
[21] Troiani T, et al. Maintenance treatment with Cetuximab and BAY86-9766 increases antitumor efficacy of irinotecan plus Cetuximab in human colorectal Cancer xenograft models. Clinical Cancer Res 2015;21:4153–64.

CHAPTER

5

Broad RTK-targeted therapy overcomes molecular heterogeneity-driven resistance to cetuximab via vectored immunoprophylaxis in colorectal cancer

Guangyao Li, Changhai Lei, and Shi Hu

Department of Biophysics, College of Basic Medical Sciences, Naval Medical University (Second Military Medical University), Shanghai, China

Abstract

Cetuximab is a chimeric monoclonal antibody widely used in the clinical guidelines of metastatic colorectal cancer. Its key mechanism lies in its targeting effect on human epidermal growth factor receptor (EGFR). However, with the progress of treatment, intensification of acquired drug resistance greatly limits its efficacy. The mechanism may be due to the activation of special interactions between EGFR family receptors and RTK. To verify our innovative treatment strategy based on the speculation of the potential mechanism of cetuximab resistance acquisition, researchers identified RTK signals by establishing relevant CRC cell lines and enabling RNAi screening methods. First, IGF-1R, Axl, and met were identified, and a variety of targeted vector immunoprophylaxis (VIP) markers were characterized. Next, in vivo experiments of parental cells and xenografts based on different VIP methods were carried out. After drug resensitization, RTK signal changes were mainly detected. The blocking of the above three potential influencing factors is the key to the wide effect of VIP therapy. There is a cross influence on this pathway in the experimental model. Compared with conventional tumor inhibition, VIP innovative procedural therapy has a more exciting effect on the reduction of cancer recurrence rate and the delay of drug resistance. In conclusion, the VIP ladder treatment strategy and its combination based on the new mechanism of cetuximab resistance (RTK signal network) may be an innovative and powerful supplement to the current treatment scheme and have the opportunity to solve the persistent problem of drug resistance.

https://doi.org/10.1016/B978-0-12-821584-5.00029-8

Abbreviations

AAV adeno-associated virus
CRC colorectal cancer
EGFR epidermal growth factor receptor
mAb monoclonal antibody
RTKs receptor tyrosine kinases
VIPs vectored immunoprophylaxis

Conflict of interest

No potential conflicts of interest were disclosed.

Study on the related functional signaling network pathway based on the drug resistance of targeted monoclonal antibodies in CRC

Colorectal cancer ranks third in the ranking of common cancers in the West and maintains an annual growth rate of 200,000 cases [1]. The routine treatment of CRC, including drugs, surgery, and other early means, has proven that its late effect is not satisfactory. Therefore, successively developed targeted drugs, such as cetuximab, have made new progress in improving the survival rate of patients. Unfortunately, the inevitable emergence of drug resistance greatly limits the benefits of this treatment [2].

Studies have shown that some exposure therapies for cancer inevitably have targeted pressure, which may aggravate mutation randomness and value-added selectivity to achieve evolutionary purposes, showing prominent molecular heterogeneity [3–7]. In reports of targeted drug therapy, such as cetuximab, the negative heterogeneity of CRC genes is largely distributed in the structural genes downstream of EGFR activation. In addition, KRAS mutation analysis is a widely recognized detection method based on the combination of chemotherapy and monoclonal antibody treatment of wild-type CRC.

When EGFR Ras or its downstream pathway is not affected, other potential mechanisms of drug resistance should be considered from a genetic perspective. Although the RTK signaling pathway has been proven to be critical in the negative effects of targeted drug application [8–11], there are few joint studies on the stimulation of drug resistance by monoclonal antibodies based on related mechanisms. Therefore, it is necessary to study the mechanism of secondary drug resistance to guide the combination of clinical monoclonal antibodies. Considering the complex molecular heterogeneity encountered in drug resistance research and the many classifications of RTK in different subgroups and stages, it is inefficient to evaluate RTK inhibition with targeted drug combinations with high cost and workload.

Phased therapeutic strategy of antibody transformation in vivo based on VIP background

Alternatively, gene transfer using vectors is an effective method, and the advantage is that it can provide continuously secreted antibodies to the circulation. Vector immunoprophylaxis (VIP) is a mature practical strategy. It can increase the secretion of antibody in muscle tissue through a specific AAV vector [12]. A single injection into mice can achieve a high concentration and continuous expression of antibody. Therefore, researchers consider the cross design of the vector and targeted antibody in the scheme as an efficient tool for research experiments. Through the conventional strategy of screening functional genes, we selected the part highly compatible with human RTK from the siRNA library and transfected it into the targeted drug resistance model to further analyze its effect. For the verified RTK, the corresponding VIP is designed to achieve a variety of targeted application schemes. After further in vivo experimental research, the molecular mechanism of its bypass network was characterized. In addition, the phased VIP treatment scheme shows great potential, which not only has therapeutic and protective effects in vivo but also improves the survival of mouse models of drug-resistant tumors.

In the clinical progression of CRC, human epidermal growth factor receptor (EGFR) and related signaling networks are at the core [13,14]. Cetuximab is the core of CRC-targeted drug combinations [15] and has significant effects on metastatic colorectal cancer (mCRC) [16], nonsmall cell lung cancer (NSCLC) [17], and head and neck squamous cell carcinoma (HNSCC) [18]. Studies have shown that among the molecular genetic mutations based on targeted monoclonal antibody resistance, some mutations of KRAS and NRAS account for more than half, while met, HER2, and KRAS are amplified in approximately one-fifth of patients with resistance efficacy. BRAF mutations are closely associated with poor prognosis exist in less than 10% of cases [2,9,19,20].

Therefore, the combination of chemotherapy and cetuximab has certain clinical benefits, which is reflected in the expansion of treatment scope and the improvement of survival rate, but it is still limited to patients without mutation, which is only a small part of the population [21]. Although the cases suitable for this combination scheme can be determined through the study of the EGFR pathway at the gene level, the time limit of clinical effect is the key that cannot be ignored. Usually, it will not last longer than 1 year at most, and secondary drug resistance will appear. Research on the mechanism of acquiring drug resistance has progressed smoothly from different angles, such as the S4 92r [22], T790 M mutation in the application of kinase inhibitors [23], pathway bypass mutations such as KRAS or BRAF [5], met [24], or erbB2 [10] parallel pathways, and so on. It is possible that the underlying mechanism of drug resistance is similar to that of the secondary mechanism.

In vivo experiments were conducted to characterize the relationship between pathway crosstalk of key gene effectors and drug resistance of targeted monoclonal antibodies

Lm1215 [22], SW48 [13], and dif cells [10] were found to be resistant to cetuximab and sensitive to cetuximab in clinical trials. After the drug resistance model was established by routine means, further in vivo and in vitro experiments were carried out on their phenotypes. The data showed that the level of EGFR phosphorylation decreased significantly under the action of drugs. Based on the previously proposed key factor RTK affecting drug resistance, the molecular level of the signaling pathway was evaluated. Functional RNAi was used to screen the effective crosstalk path in a human shRNA library [23,24]. Next, by means of gene knockout, the changes in the effectors of the Axl, met, and IGF1R genes in relevant cell lines were verified, and the sensitivity effect of the drug resistance model was observed. Its role suggests that new changes may occur in the drug resistance of targeted monoclonal antibodies in the process of treatment.

Through research, it was confirmed that the occurrence and development of targeted monoclonal antibody resistance were closely related to the receptor crosstalk between EGFR and met and Axl RTK, and the therapeutic effect of the increased combination of Axl-targeted drugs and met [8] on tumors resistant to cetuximab was tested. In previous studies, these two have been proven to be the key development points of EGFR inhibitors. In particular, it has promoted the progress of antidrug resistance research in TKIs [25–29] and other clinical drugs [30,31], including cetuximab [32]. In addition, the results showed that the human antibody RTK based on the VIP strategy inhibited the development of the population against cetuximab, and the phased treatment of VIP showed a prominent tumor persistent inhibitory response in CRC. Therefore, large-scale selective blocking of these related RTKs is an effective treatment for cancer resistant to targeted monoclonal antibodies. The key to its development lies in the screening of the correct combination, and it is not advisable to block RTK without selectivity. The acquisition of monoclonal antibodies with high activity and high affinity is a key step, and a wide genetic background is an important condition. Relevant teams have been engaged in the production of therapeutic antibodies for a long time. Gene transfer is a conventional technical means that is applied in different clinical stages. The production of AAV antibodies has made good progress in tumor treatment [33,34]. In this experimental animal model, antibodies were continuously expressed in vivo for up to 10 months, including different specific monoclonal antibodies against Axl, IGF1R, and EGFR. Unexpectedly, although the resensitivity effect was observed in the drug resistance model in vitro and successfully appeared in the means of blocking several RTKs with siRNA, the VIP strategy did not successfully achieve the synergy of the expected inhibitory effect in the in vivo experiment. This suggests that the specific inhibition of Axl by monoclonal antibodies may have a positive impact on the progression of cetuximab resistance. However, the possibility that met activation has a deeper impact in acquiring drug resistance cannot be ruled out.

Cross design and effect presentation of VIP therapeutic drugs and carriers

The phased strategy based on VIP adopts four therapeutic monoclonal antibodies, VIP Dalo, vip-58c46, VIP cetu, and VIP ficl, which are blocked by HGF to eliminate the affinity effect of conventional met technology. Through the cross design of the AAV vector, the corresponding antibody drugs were tested in the CHO cell line in vitro. The results showed the expected binding specificity and the high concentration of antibody, which confirmed the feasibility of the transfection method. In addition, in the relevant experiments of animal models, a sustained high level of human IgG was observed in the serum of nude mice injected with the genome, and its peak value was stable for more than 40 weeks.

To evaluate the effect difference of different VIP treatment methods, relevant experiments were carried out in the drug-resistant cell model. The data showed that VIP ficl/VIP Dalo showed an excellent inhibitory effect compared with VIP cetu, which almost doubled. The treatment combination of VIP cetu/VIP Dalo had a significant inhibitory effect on xenografts, while the combination with the highest effect was VIP cetu/VIP ficl, which achieved an inhibition rate of approximately 90% in the two drug resistance models and showed a unique long-term effect. Interestingly, vip-58c46, which can delay or resist drug resistance, did not show the expected tumor inhibition effect, and its low inhibition rate showed that the effect of RTK changed dynamically in the development of cetuximab drug resistance.

Final summary and assumption

In the face of the tense situation that targeted monoclonal antibodies are blocked by drug resistance and promote the process of clinical treatment, our research provides a direct and effective way to combine therapeutic antibodies with preventive immune antitumor effects in vivo. The relevant data of this study also strongly support this hypothesis, that is, based on the blocking idea of multiple functional signaling pathways to solve a drug resistance mechanism or have better clinical effects. In the context of VIP, interfering or blocking the relevant signal networks of Axl, met, and EGFR is a promising anticancer means to delay or solve the occurrence and development of cetuximab resistance.

References

[1] Siegel R, DeSantis C, Jemal A. Colorectal cancer statistics. CA Cancer J Clin 2014;64(2014):104–17.
[2] Misale S, Di Nicolantonio F, Sartore-Bianchi A, Siena S, Bardelli A. Resistance to anti-EGFR therapy in colorectal cancer: from heterogeneity to convergent evolution. Cancer Discov 2014;4:1269–80.
[3] McGranahan N, Swanton C. Biological and therapeutic impact of intratumor heterogeneity in cancer evolution. Cancer Cell 2015;27:15–26.
[4] Kandoth C, McLellan MD, Vandin F, Ye K, Niu B, Lu C, et al. Mutational landscape and significance across 12 major cancer types. Nature 2013;502:333–9.
[5] Bettegowda C, Sausen M, Leary RJ, Kinde I, Wang Y, Agrawal N, et al. Detection of circulating tumor DNA in early-and late-stage human malignancies. Sci Transl Med 2014;6:224ra224.
[6] Ahronian LG, Sennott EM, Van Allen EM, Wagle N, Kwak EL, Faris JE, et al. Clinical acquired resistance to RAF inhibitor combinations in BRAF-mutant colorectal cancer through MAPK pathway alterations. Cancer Discov 2015;5:358–67.
[7] Russo M, Siravegna G, Blaszkowsky LS, Corti G, Crisafulli G, Ahronian LG, et al. Tumor heterogeneity and lesion-specific response to targeted therapy in colorectal cancer. Cancer Discov 2015;6:147–53.
[8] Hu S, Fu W, Xu W, Yang Y, Cruz M, Berezov SD, et al. Four-in-one antibodies have superior cancer inhibitory activity against EGFR, HER2, HER3, and VEGF through disruption of HER/MET crosstalk. Cancer Res 2015;75:159–70.
[9] Bertotti A, Migliardi G, Galimi F, Sassi F, Torti D, Isella C, et al. A molecularly annotated platform of patient-derived xenografts ("xenopatients") identifies HER2 as an effective therapeutic target in cetuximab-resistant colorectal cancer. Cancer Discov 2011;1:508–23.
[10] Yonesaka K, Zejnullahu K, Okamoto I, Satoh T, Cappuzzo F, Souglakos J, et al. Activation of ERBB2 signaling causes resistance to the EGFR-directed therapeutic antibody cetuximab. Sci Transl Med 2011;3:99ra86.
[11] Bardelli A, Corso S, Bertotti A, Hobor S, Valtorta E, Siravegna G, et al. Amplification of the MET receptor drives resistance to anti-EGFR therapies in colorectal cancer. Cancer Discov 2013;3:658–73.
[12] Balazs AB, Chen J, Hong CM, Rao DS, Yang L, Baltimore D. Antibody-based protection against HIV infection by vectored immunoprophylaxis. Nature 2012;481:81–4.
[13] Hynes NE, MacDonald G. ErbB receptors and signaling pathways in cancer. Curr Opin Cell Biol 2009;21:177–84.
[14] Yarden Y, Sliwkowski MX. Untangling the ErbB signalling network. Nat Rev Mol Cell Biol 2001;2:127–37.
[15] Brand TM, Iida M, Wheeler DL. Molecular mechanisms of resistance to the EGFR monoclonal antibody cetuximab. Cancer Biol Ther 2011;11:777–92.

[16] Jonker DJ, O'Callaghan CJ, Karapetis CS, Zalcberg JR, Tu D, Au H-J, et al. Cetuximab for the treatment of colorectal cancer. NEJM 2007;357:2040–8.

[17] Pirker R, Pereira JR, Szczesna A, Von Pawel J, Krzakowski M, Ramlau R, et al. Cetuximab plus chemotherapy in patients with advanced non-small-cell lung cancer (FLEX): an open-label randomised phase III trial. Lancet 2009;373:1525–31.

[18] Vermorken JB, Mesia R, Rivera F, Remenar E, Kawecki A, Rottey S, et al. Platinum-based chemotherapy plus cetuximab in head and neck cancer. NEJM 2008;359:1116–27.

[19] Vaughn CP, ZoBell SD, Furtado LV, Baker CL, Samowitz WS. Frequency of KRAS, BRAF, and NRAS mutations in colorectal cancer. Genes Chromosomes Cancer 2011;50:307–12.

[20] Valtorta E, Misale S, Sartore-Bianchi A, Nagtegaal ID, Paraf F, Lauricella C, et al. KRAS gene amplification in colorectal cancer and impact on response to EGFR-targeted therapy. Int J Cancer 2013;133:1259–65.

[21] Saltz LB, Meropol NJ, Loehrer PJ, Needle MN, Kopit J, Mayer RJ. Phase II trial of cetuximab in patients with refractory colorectal cancer that expresses the epidermal growth factor receptor. J Clin Oncol 2004;22:1201–8.

[22] Diaz Jr LA, Williams RT, Wu J, Kinde I, Hecht JR, Berlin J, et al. The molecular evolution of acquired resistance to targeted EGFR blockade in colorectal cancers. Nature 2012;486:537–40.

[23] Pao W, Miller VA, Politi KA, Ricly G, Somwar R, Zakowski MF, et al. Acquired resistance of lung adenocarcinomas to gefitinib or erlotinib is associated with a second mutation in the EGFR kinase domain. PLoS Med 2005;2, e73.

[24] Mohan S, Heitzer E, Ulz P, Lafer I, Lax S, Auer M, et al. Changes in colorectal carcinoma genomes under anti-EGFR therapy identified by whole-genome plasma DNA sequencing. PLoS Genet 2014;10, e1004271.

[25] Byers LA, Diao L, Wang J, Saintigny P, Girard L, Peyton M, et al. An epithelial– mesenchymal transition gene signature predicts resistance to EGFR and PI3K inhibitors and identifies AXL as a therapeutic target for overcoming EGFR inhibitor resistance. Clin Cancer Res 2013;19:279–90.

[26] Giles KM, Kalinowski FC, Candy PA, Epis MR, Zhang PM, Redfern AD, et al. Axl mediates acquired resistance of head and neck cancer cells to the epidermal growth factor receptor inhibitor erlotinib. Mol Cancer Ther 2013;12:2541–58.

[27] Meyer AS, Miller MA, Gertler FB, Lauffenburger DA. The receptor AXL diversifies EGFR signaling and limits the response to EGFR-targeted inhibitors in triple-negative breast cancer cells. Sci Signal 2013;6:ra66.

[28] Zhang Z, Lee JC, Lin L, Olivas V, Au V, LaFramboise T, et al. Activation of the AXL kinase causes resistance to EGFR-targeted therapy in lung cancer. Nat Genet 2012;44:852–60.

[29] Rho JK, Choi YJ, Kim SY, Kim TW, Choi EK, Yoon S-J, et al. MET and AXL inhibitor NPS-1034 exerts efficacy against lung cancer cells resistant to EGFR kinase inhibitors because of MET or AXL activation. Cancer Res 2014;74:253–62.

[30] Linger RM, Cohen RA, Cummings CT, Sather S, Migdall-Wilson J, Middleton DH, et al. Mer or Axl receptor tyrosine kinase inhibition promotes apoptosis, blocks growth and enhances chemosensitivity of human non-small cell lung cancer. Oncogene 2013;32:3420–31.

[31] Dunne PD, McArt DG, Blayney JK, Kalimutho M, Greer S, Wang T, et al. AXL is a key regulator of inherent and chemotherapy-induced invasion and predicts a poor clinical outcome in early-stage colon cancer. Clin Cancer Res 2014;20:164–75.

[32] Brand TM, Iida M, Stein AP, Corrigan KL, Braverman CM, Luthar N, et al. AXL mediates resistance to cetuximab therapy. Cancer Res 2014;74:5152–64.

[33] Ho D, Wykoff-Clary S, Gross C, Schneider D, Jin F, Kretschmer P, et al. Growth inhibition of an established A431 xenograft tumor by a full-length anti-EGFR antibody following gene delivery by AAV. Cancer Gene Ther 2009;16:184–94.

[34] Watanabe M, Boyer JL, Crystal RG, AAVrh. 10-mediated genetic delivery of bevacizumab to the pleura to provide local anti-VEGF to suppress growth of metastatic lung tumors. Gene Ther 2010;17:1042–51.

CHAPTER

6

Pan-HER, an antibody mixture, simultaneously targeting EGFR, HER2, and HER3 effectively overcomes resistance to anti-EGFR antibodies

Tian Li, Changhai Lei, and Shi Hu

Department of Biophysics, College of Basic Medical Sciences, Naval Medical University (Second Military Medical University), Shanghai, China

Abstract

Many studies have shown that members of the human epidermal growth factor receptor (HER) family are not only closely related to the onset, progression, or resistance of cancer cells but are also highly plastic and can play an important role in the compensatory transduction of some signaling pathways. HER family members interfere with treatment and affect the mid- and long-term efficacy of treatment regimens. This chapter introduces Pan-HER, a therapeutic drug developed by researchers targeting members of the HER family. Pan-HER is not a novel drug targeting a member of the HER family but a cocktail of six antibodies capable of targeting EGFR, HER2, and HER3. By targeting three targets simultaneously, Pan-HER can effectively prevent compensatory tumor-promoting mechanisms in vivo. Preclinical results demonstrate that Pan-HER is effective in a variety of cancer cell lines and xenograft models, including cell lines that have acquired resistance to a single therapeutic antibody. Pan-HER can overcome acquired resistance due to increased ligand production, and it also has a certain curative effect in the presence of HER family ligands. Compared with single receptor targeting, therapy targeting multiple HER family targets simultaneously has greater advantages. Pan-HER represents a new class of cancer treatment strategies that may be able to address both primary and acquired drug resistance due to tumor heterogeneity and plasticity of HER family members. Therefore, this treatment strategy has the potential value to carry out large-scale trials and clinical promotion.

Abbreviations

ADCC antibody-dependent cell cytotoxicity
CDC complement-dependent cytotoxicity
FcRn neonatal Fc receptor

Novel Sensitizing Agents for Therapeutic Anti-EGFR Antibodies
https://doi.org/10.1016/B978-0-12-821584-5.00022-5

HER human epidermal growth factor receptor
PDX patient-derived xenograft
TKIs tyrosine kinase inhibitors

Conflict of interest

No potential conflicts of interest were disclosed.

The human epidermal growth factor receptor (HER) family is distributed simultaneously on both the cell membrane and the nucleus membrane and is responsible for the transmission of information such as growth factors. It also plays a significant role in normal cell growth, metabolism, proliferation, survival, and differentiation. However, many events, such as the onset, deterioration, or resistance of cancer cells, are closely related to disorders caused by mutations, overexpression, or gene amplification of the human epidermal growth factor receptor [1]. According to current research results, it is suggested that the mechanism of this process is that the combination of ligands in the EGF growth factor family induces homolysis or isomerization, leading to the cross phosphorylation of the disaggregation partner, which ultimately triggers intracellular signal transduction [2,3].

Existing reports show that the HER family contains four kinds of receptors: EGFR, HER2, HER3, and HER4. After clinical data verification, EGFR and HER2 were found to play the most important roles in head and neck squamous cell carcinoma, colorectal cancer, breast cancer, gastric cancer, and lung cancer. Increasing experimental and clinical data indicate that HER3 is also an important target in related cancers [2,4–6]. However, there is not enough evidence to show what role HER4 plays in the pathogenesis of cancer. Small-molecule inhibitors of receptor tyrosine kinases (TKIs) against single or multiple HER members have been approved to be beneficial for cancer treatment in clinical practice, and some therapeutic monoclonal antibodies have also been developed for many different cancer treatments. It is worth noting that the targets of currently approved antibody therapeutics are all single receptors, and the therapeutic effect can only be guaranteed when the tumor highly expresses the specific receptor. There are some highly effective antibody drugs, including cetuximab, panitumumab, trastuzumab, and pertuzumab. The first two antibody drugs are against EGFR and the latter two antibody drugs are against HER2. It is generally believed that pertuzumab prevents HER2 from interacting with other HER family members to inhibit dimerization, but it is still regarded as a single-target drug [7]. To provide better therapeutic effects, bispecific antibodies and combination antibody drugs that have the ability to target two members of the HER family at the same time are still being developed clinically [8,9].

A large number of recent clinical and preclinical data show that HER members have high plasticity, and it is easy to have compensation signals between them in the process of cancer treatment so that the tumor can produce acquired drug resistance and antitherapeutic intervention [10,11]. Specific examples include HER3 and, consequently, the PI(3)K/Akt signaling pathway evade inhibition by TKIs of the HER family in HER2-driven breast cancer, which is related to the high expression of HER3 [12]. Drug resistance occurs in the treatment of HER2-overexpressing tumors with trastuzumab targeting the HER2 receptor in breast cancer due to the increased expression of EGFR and EGFR ligands or overexpression of HER3 in cell lines [13,14]. HER3 upregulation leads to resistance to lapatinib, and the resistance of breast cancer to TKI, which targets EGFR/HER2, and the resistance of colorectal cancer to cetuximab are both mediated by the upregulation of HER2 or ligand neuromodulating proteins of HER3 [15,16]. Targeting multiple HER family receptors at the same time usually reverses cancer resistance to earlier treatment drugs and is usually more effective than drugs that target a single receptor alone [15–17]. Therefore, inhibiting two or more HER family members together may be crucial to limit acquired resistance and even achieve the effect of completely suppressing cancer.

Moreover, tumor cells may exhibit unexpected heterogeneity and then respond differently to drug treatments. This heterogeneity between tumor cells includes not only clinically important features (such as the probability of cancer cell metastasis and drug resistance) but also the level of expression of tumor-related markers and potential therapeutic targets (such as HER members) [18,19]. Many studies have reported the heterogeneous expression pattern of EGFR in glioblastoma and the heterogeneous expression pattern of HER2 in breast carcinoma, stomach cancer, and salivary duct carcinoma [18–24]. In addition, according to biomarker studies of primary breast cancer, the average coefficient of variation of the protein is as high as 31%, and the relevant results also reveal the extensive heterogeneous expression characteristic of all HER members [25]. Therefore, even if multiple biopsy specimens are obtained in the clinic, it is still uncertain whether the single targeted therapy selected after biomarker analysis is the best choice.

Based on related findings, researchers have proposed a new treatment model: therapeutic drugs target EGFR, HER2, and HER3 at the same time and exert a downregulating effect, which can provide effective tumor growth inhibitory effects in a variety of cancers and overcome the plasticity or coordination due to receptors [26]. Acquired drug resistance caused by increased body expression and corresponding preclinical results has been achieved. In the earliest research, monoclonal antibody combinations or recombinant antibody mixtures were used to treat tumors to achieve better therapeutic effects [27–31]. Specifically, in multiple studies, combinations of two drugs against nonoverlapping regions on EGFR or HER2 have a synergistic effect. It is better than a single antibody in eliminating the target and suppressing tumor growth. In the presence of EGFR ligands, this receptor elimination also leads to better activity than a single antibody (such as cetuximab). The latest research results show that Pan-HER (a hybrid antibody that is composed of EGFR, HER2, and HER3 works synergistically) has further verified this concept [26].

Members of the HER family have the ability to provide compensatory signals to each other in human cancers, and all HER family members have sufficient plasticity [12,15–17]. These factors lead to acquired resistance to tumor cell-targeted therapies, which have been generally recognized by clinical and scientific researchers. Based on this discovery, researchers have also developed targeted measures, and several treatment methods have been well established. These methods target different receptors and/or survival pathways by combining treatments at the same time. There are a number of studies that combine targeted drugs targeting two of the three targets of EGFR, HER2, and HER3, and produce positive effects. For example, in pancreatic cancer, the combined use of cetuximab/trastuzumab has a dual targeting effect on EGFR/HER2 compared with the treatment effect of trastuzumab/erlotinib or lapatinib alone [28]. However, the combination of two antibodies or bispecific antibody treatments still cannot capture the full possibility of HER member plasticity because only two specific receptors or pathways are targeted at the same time.

The existence of small molecule TKIs can usually inhibit multiple kinases at the same time, but because this member of the HER family has very limited kinase activity associated with it, they cannot directly target HER3.

To study and solve these challenges and limitations, researchers have upgraded the number of antibody drugs that can be targeted at the same time and developed three kinds of synergistic antibodies that target EGFR, HER2, and HER3 into an effective mixture of antibodies targeting the HER family, Pan-HER [32]. As previously said, the combination of two antibodies that bind to nonoverlapping regions on the targets by the antibody mixture can effectively inhibit the function of receptors and degradation of receptors [33]. In most of the expected cases, the combination of two antibodies can better inhibit the growth of cancer cells and the binding of receptors and ligands. Using a pair of antibodies to target two receptors can also prevent the emergence of mutations in the target's extracellular domain or weaken its effect and improve the situation that tumor cells are resistant to monoclonal antibodies for this reason and cannot achieve the desired therapeutic effect [34–36].

In vitro experimental data show that Pan-HER can effectively inhibit a variety of cancer cell lines that are from different tissue origins and have different genetic backgrounds. Some of these cancer cell lines are drug-resistant cell lines, including those with acquired resistance to cetuximab, trastuzumab, or pertuzumab. Because of the presence of HER family ligands, the effect of the antibody against the single receptor is weakened, while Pan-HER is still highly effective. Because of the presence of EGF, the combined targeting of EGFR and HER2 is comparable to Pan-HER targeting all three receptors, while in the presence of neuregulin, the dual-receptor targeting mixture is not as efficient as Pan-HER. The sustained antiproliferative effect of Pan-HER may be attributed to the effectiveness of all three receptors while downregulating them, preventing them from binding to ligands and being unable to form signal transduction competent heterodimers. Although HER2 has not been found to have a high-affinity ligand, it can still form dimers in a permanently open conformation [37,38]. HER2 is the preferred heterodimer partner of all HER family receptors and a mediator of lateral signal transduction [39–41]. The HER2-HER3 dimer was previously thought to be the most effective signal transduction module. Perhaps because of the limited effect of HER3 targeting in vitro, it is not obvious from the newer experimental data that the enhancement effect of targeting both HER2 and HER3 in the absence of ligands is not obvious. More than expected, the removal of HER2 has greatly enhanced the overall efficacy of the Pan-HER mixture in vitro and is likely to be associated with the presence of HER3 ligand neuromodulation proteins. Therefore, Pan-HER can effectively inhibit cell proliferation in cancer cell lines induced by ligands due to exposure to EGF and neuromodulating proteins, indicating that acquired tolerance due to increased ligand production may be enough to be overcome through clinical treatment.

Pan-HER targets EGFR at the same time, and HER2 and HER3 not only lead to receptor downregulation but also prevent compensatory receptor upregulation, which is another resistance to single-receptor targeted therapy observed in preclinical models and clinics [12–17] and known mechanisms of drugs. In the study described in this chapter, compensatory receptor upregulation was observed in multiple cell lines in vitro, and the cetuximab-resistant cell line used here also demonstrated other receptors, including HER3. A consensus has now been reached that the upregulation of EGFR and/or HER3 causes breast cancer cells to develop resistance after trastuzumab treatment [14]. In the current study, the effects of trastuzumab treatment were also observed compensatory response. The unpredictability of compensatory responses makes it difficult to determine the most effective target combination. However, Pan-HER effectively prevented the compensatory response of all cell lines examined and appeared to be able to control the plasticity of the HER family.

There is no correlation between RAS status and Pan-HER activity because Pan-HER is active on cell lines with RAS mutations and xenografts. The related findings have certain value but still did not exceed expectations, because in preclinical studies, RAS status is not a key predictor of tumor resistance to anti-EGFR, anti-HER2, and anti-HER3 antibody drugs. Whether Pan-HER will benefit the tumors of patients with RAS gene mutations needs to be confirmed in clinical trials.

The inhibitory effect of Pan-HER on cancer cells is mediated through apoptosis and cell cycle arrest. In Calu-3 cells, a considerable degree of cell cycle arrest was observed, which was closely related to the significant decrease in Ki67 expression in Pan-HER-treated Calu-3 tumor xenografts. It seems that the dependence of the HER family receptor in each cell line determines the outcome of the treatment. Some cell lines are more prone to cell death, while others are more prone to cell cycle arrest. Obviously, in the model established on the basis of Calu-3 cells, whether in in vitro or in vivo experiments, pan-HER can kill tumor cells to the greatest extent and inhibit the proliferation of the remaining cells. These results are consistent with the observations of previous studies, indicating that the simultaneous targeting of antibodies to two HER family receptors can significantly increase the level of tumor cell apoptosis and cycle arrest [31,42]. Although the immunohistochemical analysis results did not show improved apoptosis in Pan-HER-treated BxPC3 or Calu-3 tumors, it still cannot be ruled out that apoptosis in the early stage of treatment affects the observation of the treatment effect.

Pan-HER's excellent ability can simultaneously downregulate EGFR, HER2, and HER3, and inhibit the growth of cancer cells in vitro, and this expected ability successfully inhibits tumor growth in vivo. Pan-HER strongly inhibits tumor growth in various tumor xenograft models that depend on EGFR, HER2, and HER3 to varying degrees (including PDX models of ovarian cancer, colorectal cancer, lung cancer, and pancreatic cancer), thus further indicating that it has a wide range of cancer treatment activities in vivo. According to previous observations, BxPC3 and Calu-3 xenograft studies have shown that compared with single-receptor targeting, binding to the HER family target has obvious advantages [27–30]. It is worth noting that these three specific targets have all been shown to help Pan-HER perform better in vivo. In the BxPC3 model, compared with single receptor targeting, the mixtures made of three dual-targeting antibodies showed higher efficacy, while the EGFR + HER3 targeting mixture and Pan-HER had the same efficacy. In the model established by Calu-3 cells, all dual-receptor targeting mixtures have a good synergistic effect, but the triple-targeted Pan-HER drug has the best therapeutic effect and is completely better than all other therapies in this model method. The BxPC3 study also provides an example of compensatory upregulation in vivo due to the targeting effect of HER2 leading to a significant increase in HER2 levels. Pan-HER effectively prevented this upregulation by simultaneously lowering all three targets. The in vivo results indicate that pan-HER can effectively deal with family-dependent human cancers in her family with extensive plasticity, as these xenograft models show.

Preclinical and clinical data indicate that the development of chimeric, humanized, and fully humanized antibodies has made "second-generation" therapeutic antibodies a great success, such as rituximab, trastuzumab, cetuzumab, coximab, and bevacizumab. Further modification may be able to better improve the efficacy of current antibody therapeutics. According to the clinical results analysis and mechanism research of current therapeutic antibodies, there is some general consensus among researchers that Fc-mediated functions (such as antibody-dependent cellular cytotoxicity, ADCC; complement-dependent cytotoxicity, CDC; neonatal Fc receptor [FcRn]-mediated storage) have the possibility to further improve the antitumor effect of various therapies using antibodies [43]. At the same time, multiple antibodies that bind to cell targets may also lead to enhanced effector functions, presumably by increasing the density of antibodies on the cell surface. Pan-HER induced ADCC levels in vitro similar to cetuximab or trastuzumab, while CDC levels were significantly increased. According to the aforementioned report, the nonengineered antiher3 antibody induced by the IgG1 subtype was found to induce relatively limited ADCC, which may be due to the low expression of HER3 in the tumor cell line of the initial state used in the experiment [44].

In summary, studies have shown that targeting EGFR with an antibody mixture, HER2 and HER3, can simultaneously induce the downregulation of these three receptors and effectively prevent the upregulation of compensatory receptors, which is an anti-HER family-targeted therapy, the well-known mechanism of sex. This mechanism of action translates into a wide range of effective activities in a variety of cancer models, and these expressions have different expression and dependence on these receptors in vitro and in vivo. In the presence of HER family ligands, the broader inhibitory activity is even more pronounced, which may contribute to the ligand-dependent drug resistance that is known to occur in some patients treated with HER-targeted therapies. In addition, due to the heterogeneity and plasticity of the expression of the HER family, the simultaneous inhibitory effect of Pan-HER on EGFR, HER2, and HER3 will be better than the existing targeted therapies, thus performing both in the treatment of primary and acquired drug resistance. As a result, Pan-HER may offer a more comprehensive and valuable strategy for treating a variety of HER family-dependent epithelial tumors. The characteristics of Pan-HER drugs have been evaluated in primates to determine whether there are potential adverse reactions to joint HER-targeted drug therapy.

References

[1] Bublil EM, Yarden Y. The EGF receptor family: spearheading a merger of signaling and therapeutics. Curr Opin Cell Biol 2007;19:124–34.

[2] Baselga J, Swain SM. Novel anticancer targets: revisiting ERBB2 and discovering ERBB3. Nat Rev Cancer 2009;9:463–75.

[3] Hynes NE, Lane HA. ERBB receptors and cancer: the complexity of targeted inhibitors. Nat Rev Cancer 2005;5:341–54.

[4] Kol A, et al. HER3, serious partner in crime therapeutic approaches and potential biomarkers for effect of HER3-targeting. Pharmacol Ther 2014;143:1–11.

[5] Seshacharyulu P, Ponnusamy MP, Haridas D, Jain M, Ganti AK, Batra SK. Targeting the EGFR signaling pathway in cancer therapy. Expert Opin Ther Targets 2012;16:15–31.

[6] Tebbutt N, Pedersen MW, Johns TG. Targeting the ERBB family in cancer: couples therapy. Nat Rev Cancer 2013;13:663–73.

[7] Franklin MC, Carey KD, Vajdos FF, Leahy DJ, de Vos AM, Sliwkowski MX. Insights into ErbB signaling from the structure of the ErbB2-pertuzumab complex. Cancer Cell 2004;5:317–28.

[8] Demarest SJ, Hariharan K, Dong J. Emerging antibody combinations in oncology. MAbs 2011;3:338–51.

[9] May C, Sapra P, Gerber HP. Advances in bispecific biotherapeutics for the treatment of cancer. Biochem Pharmacol 2012;84:1105–12.
[10] Vlacich G, Coffey RJ. Resistance to EGFR-targeted therapy: a family affair. Cancer Cell 2011;20:423–5.
[11] Brand TM, Iida M, Wheeler DL. Molecular mechanisms of resistance to the EGFR monoclonal antibody cetuximab. Cancer Biol Ther 2011;11:777–92.
[12] Sergina NV, et al. Escape from HER-family tyrosine kinase inhibitor therapy by the kinase-inactive HER3. Nature 2007;445:437–41.
[13] Ritter CA, et al. Human breast cancer cells selected for resistance to trastuzumab in vivo overexpress epidermal growth factor receptor and ErbB ligands and remain dependent on the ErbB receptor network. Clin Cancer Res 2007;13:4909–19.
[14] Narayan M, Wilken JA, Harris LN, Baron AT, Kimbler KD, Maihle NJ. Trastuzumab-induced HER reprogramming in "resistant" breast carcinoma cells. Cancer Res 2009;69:2191–4.
[15] Garrett JT, et al. Transcriptional and posttranslational up-regulation of HER3 (ErbB3) compensates for inhibition of the HER2 tyrosine kinase. Proc Natl Acad Sci U S A 2011;108:5021–6.
[16] Yonesaka K, et al. Activation of ERBB2 signaling causes resistance to the EGFR-directed therapeutic antibody cetuximab. Sci Transl Med 2011;3.
[17] Garrett JT, Sutton CR, Kuba MG, Cook RS, Arteaga CL. Dual blockade of HER2 in HER2-overexpressing tumour cells does not completely eliminate HER3 function. Clin Cancer Res 2013;19:610–9.
[18] Marusyk A, Almendro V, Polyak K. Intra-tumour heterogeneity: a looking glass for cancer? Nat Rev Cancer 2012;12:323–34.
[19] Almendro V, Marusyk A, Polyak K. Cellular heterogeneity and molecular evolution in cancer. Annu Rev Pathol 2013;8(1):277–302.
[20] Snuderl M, et al. Mosaic amplification of multiple receptor tyrosine kinase genes in glioblastoma. Cancer Cell 2011;20:810–7.
[21] Szerlip NJ, et al. Intratumoural heterogeneity of receptor tyrosine kinases EGFR and PDGFRA amplification in glioblastoma defines subpopulations with distinct growth factor response. Proc Natl Acad Sci U S A 2012;109:3041–6.
[22] Davila E, Amazon K. The clinical importance of the heterogeneity of HER2 neu. Case Rep Oncol 2010;3:268–71.
[23] Allison KH, Dintzis SM, Schmidt RA. Frequency of HER2 heterogeneity by fluorescence in situ hybridization according to CAP expert panel recommendations time for a new look at how to report heterogeneity. Am J Clin Pathol 2011;136:864–71.
[24] Yang JY, et al. Intratumoural heterogeneity determines discordant results of diagnostic tests for human epidermal growth factor receptor (HER) 2 in gastric cancer specimens. Cell Biochem Biophys 2012;62:221–8.
[25] Malinowsky K, et al. Common protein biomarkers assessed by reverse phase protein arrays show considerable intratumoural heterogeneity in breast cancer tissues. PLoS One 2012;7.
[26] Jacobsen HJ, et al. Pan-HER, an antibody mixture simultaneously targeting EGFR, HER2, and HER3, effectively overcomes tumour heterogeneity and plasticity. Clin Cancer Res 2015;21:4110–22.
[27] Larbouret C, et al. In vivo therapeutic synergism of anti-epidermal growth factor receptor and anti-HER2 monoclonal antibodies against pancreatic carcinomas. Clin Cancer Res 2007;13:3356–62.
[28] Larbouret C, et al. In pancreatic carcinoma, dual EGFR/HER2 targeting with cetuximab/trastuzumab is more effective than treatment with trastuzumab/erlotinib or lapatinib alone: implication of receptors' down-regulation and dimers' disruption. Neoplasia 2012;14:121–30.
[29] Schaefer G, et al. A two-in-one antibody against HER3 and EGFR has superior inhibitory activity compared with monospecific antibodies. Cancer Cell 2011;20:472–86.
[30] Maron R, Schechter B, Mancini M, Mahlknecht G, Yarden Y, Sela M. Inhibition of pancreatic carcinoma by homo- and heterocombinations of antibodies against EGF-receptor and its kin HER2/ErbB-2. Proc Natl Acad Sci U S A 2013;110:15389–94.
[31] McDonagh CF, et al. Antitumour activity of a novel bispecific antibody that targets the ErbB2/ErbB3 oncogenic unit and inhibits heregulin-induced activation of ErbB3. Mol Cancer Ther 2012;11:582–93.
[32] Wang M, Hu Y, Yu T, Ma X, Wei X, Wei Y. Pan-HER-targeted approach for cancer therapy: mechanisms, recent advances and clinical prospect. Cancer Lett 2018;439:113–30.
[33] Pedersen MW, et al. Sym004: a novel synergistic anti-epidermal growth factor receptor antibody mixture with superior anticancer efficacy. Cancer Res 2010;70:588–97.

[34] Montagut C, et al. Identification of a mutation in the extracellular domain of the epidermal growth factor receptor conferring cetuximab resistance in colorectal cancer. Nat Med 2012;18:221–3.
[35] Bose R, et al. Activating HER2 mutations in HER2 gene amplification negative breast cancer. Cancer Discov 2013;3:224–37.
[36] Jaiswal BS, et al. Oncogenic ERBB3 mutations in human cancers. Cancer Cell 2013;23:603–17.
[37] Garrett TPJ, et al. The crystal structure of a truncated ErbB2 ectodomain reveals an active conformation, poised to interact with other ErbB receptors. Mol Cell 2003;11:495–505.
[38] Cho HS, et al. Structure of the extracellular region of HER2 alone and in complex with the herceptin fab. Nature 2003;421:756–60.
[39] GrausPorta D, Beerli RR, Daly JM, Hynes NE. ErbB-2, the preferred heterodimerization partner of all ErbB receptors, is a mediator of lateral signaling. EMBO J 1997;16:1647–55.
[40] PinkasKramarski R, et al. Diversification of Neu differentiation factor and epidermal growth factor signaling by combinatorial receptor interactions. EMBO J 1996;15:2452–67.
[41] Tzahar E, et al. A hierarchical network of interreceptor interactions determines signal transduction by neu differentiation factor/neuregulin and epidermal growth factor. Mol Cell Biol 1996;16:5276–87.
[42] Jiang N, et al. Combination of anti-HER3 antibody MM-121/SAR256212 and cetuximab inhibits tumour growth in preclinical models of head and neck squamous cell carcinoma. Mol Cancer Ther 2014;13:1826–36.
[43] Kaneko E, Niwa R. Optimizing therapeutic antibody function progress with fc domain engineering. BioDrugs 2011;25:1–11.
[44] Mirschberger C, et al. RG7116, a therapeutic antibody that binds the inactive HER3 receptor and is optimized for immune effector activation. Cancer Res 2013;73:5183–94.

CHAPTER

7

Four-in-one antibodies have superior cancer inhibitory activity against EGFR, HER2, HER3, and VEGF through disruption of HER/MET crosstalk

Zetong Ma[a], *Tian Li*[a], *Wenyan Fu*[b], *Changhai Lei*[a], *and Shi Hu*[a]

[a]Department of Biophysics, College of Basic Medical Sciences, Naval Medical University (Second Military Medical University), Shanghai, China [b]Department of Assisted Reproduction, Shanghai Ninth People's Hospital, Shanghai Jiao Tong University School of Medicine, Shanghai, China

Abstract

In metastatic cancer, activation of the HER receptor and crosstalk with other RTKs, especially HER-MET crosstalk, are important reasons for tumor resistance to anti-HER receptor therapy. In this chapter, FL518 (four-in-one antibody) and DVD-Ig CRTB6 (tetravalent specific antibody) were developed, which can recognize EGFR, HER2, HER3, and VEGF while simultaneously inhibiting HER-MET crosstalk. Compared with bispecific antibodies, four-in-one antibodies can inhibit the growth of drug-resistant tumor cells with MET overactivation more effectively. In conclusion, this study provides a new idea to use a single antibody to inhibit HER activation and limit the occurrence of drug resistance.

Abbreviations

EGFR epidermal growth factor receptor
HER human epidermal growth factor receptor
KiH knobs-into-holes
MET mesenchymal epithelial transition
RTK receptor tyrosine kinase
SPR surface plasmon resonance

https://doi.org/10.1016/B978-0-12-821584-5.00012-2

Conflict of interest

No potential conflicts of interest were disclosed.

Introduction

Abnormal activation of the human epidermal growth factor receptor (HER) family is one of the important reasons for the occurrence and development of tumors [1]. Tyrosine kinase inhibitors erlotinib and gefitinib and monoclonal antibodies are important strategies for treating tumors associated with HER family activation [2,3]. In addition, the upregulation of HER3, another member of the HER family, is considered a poor prognostic factor [4–6] and has also become a new drug target. However, patients receiving anti-HER-targeted therapy can only benefit for a period of time [7], mainly due to the occurrence of shift signaling of tumor growth, such as HER3 substitution activation caused by inhibition of AKT pathway [8–11], HER1 isodimer signal transduction to HERI/HER3 heterodimer leads to reactivation of the PI3K/AKT pathway. Another mechanism is to induce HER3 phosphorylation and PI3K activation through MET amplification [11,12].

Combined inhibition strategies for the key signaling network of the HER family, such as using different monoclonal antibody combinations or developing bispecific antibodies that simultaneously bind two different antigens, are novel strategies for antitumor therapy. Two-in-one antibodies use a single antibody binding site to form a specific binding to the surface of two high-affinity target antigens [13,14] and have pharmacokinetics and stability similar to those of traditional monoclonal antibodies. In 2009, the first specific antibody drug, catumaxomab (Removab), was approved for the treatment of malignant ascites [15]. Studies have also shown that bispecific antibodies targeting EGFR or HER2 are also potential therapeutic strategies [16,17]. In this chapter, current antibody engineering methods were used to create IgG-based tetra-specificity antibodies.

Design and characterization of four-in-one antibodies

The researchers first constructed a four-in-one antibody using two two-in-one antibodies, MEHD7945A [13] for EGFR and HER3 and bH1-44 [14] for VEGF and HER2. Two heavy chains are heterodimerized by the "knobs-into-holes" (KiH) method during the production of four-in-one antibody (FL518) [18,19], and the distinction between two light/heavy chain interactions is achieved by the CrossMab method [20]. The resulting four-in-one antibody can simultaneously bind EGFR, HER3, HER2, and VEGF at certain concentrations. However, since one molecule of a tetrad antibody can only target two simultaneously, the researchers then used the dual-variable domain antibody method [21], KiH method, and CrossMab method to prepare a tetra antibody called DVD-Ig CRTB6. Tetravalent can bind to four targets via a single antibody molecule. The researchers used the variable region in the Fab segment of the EGFR-targeted cetuximab antibody and HER3 antibody RG7116 to block EGF- and HRG-dependent homologous or heterologous dimerization signals. Variable regions of trastuzumab affect epidermal growth factor receptor ligand/HER2, HER3 or/HER2 signaling [22], and the anti-VEGF antibody bevacizumab.

Variable regions of tetravalent antibodies are derived from four classic antibodies, cetuximab, trastuzumab, and bevacizumab HER3 antibody RG7116 [22,23]. The inhibition of dimers and communication between MEMBERS of the HER family by the tetravalence antibody DVD-Ig CRTB6 may be more pronounced than that by FL518. The binding ability of FL518 to the extracellular domain of the HER3 or HER2 membrane decreases with the increase in EGFR-ECD or VEGF, suggesting that the neutralization of FL518 to these targets is competitive. Similar results were observed for MEHD7945A and bH1-44 but not for DVD-Ig CRTB6. Competitive binding analysis showed that the epitopes of the FL518 binding target overlapped with MEHD7945A and bH1-44. Surface plasmon resonance (SPR) results showed that the affinity of the FL518 binding target was not significantly different from that of MEHD7945A and bH1-44. Meanwhile, the pharmacokinetic properties of FL518 and DVD-Ig CRTB6 IgG molecules are similar to those of traditional IgG molecules, showing favorable stability.

Four-in-one antibodies can effectively inhibit EGFR, HER2, HER3, and VEGFR2

To evaluate the inhibitory effect of four-in-one antibody on downstream phosphorylation of EGFR and ERK1/2, Caco-2 cells with high EGFR expression (38) were pretreated with parental antibody, FL518, or DVD-Ig CRTB6 and then stimulated by EGF stimulation. The phosphorylation of EGFR and ERK1/2 was inhibited by FL518 and DVD-Ig CRTB6. The effect was similar to that of MEHD7945A and cetuximab. No significant inhibition was observed for bH1-44, trastuzumab, or bevacizumab. Second, pretreatment of McF-7 cells with parental antibodies, FL518 or DVD-Ig CRTB6, and subsequent treatment of McF-7 cells with HRG showed that both FL518 and DVD-Ig CRTB6 significantly inhibited HER3 phosphorylation, similar to that of the parental antibodies, and reduced AKT and ERK1/2 phosphorylation, albeit less effectively. In BT474 cells, FL518 or DVD-Ig CRTB6 effectively inhibited HER2 autophosphorylation and VEGFR2 phosphorylation stimulated by VEGF and reduced AKT phosphorylation. MEHD7945A or bH1-44 does not inhibit either HER2 or VEGFR2 phosphorylation. The ability of these antibodies to inhibit AKT phosphorylation is inferior to that of four-in-one antibodies.

The four-in-one antibody has excellent antitumor activity both in vitro and in vivo

Four bispecific antibodies, including cetuximab/trastuzumab (CT9), cetuximab/bevacizumab (CB12), RG7116/trastuzumab (RT6), and RG7116/bevacizumab (RB4), were generated using the KiH and CrossMab methods. EGF-, HRG-, and VEGF-stimulated cell proliferation tests were compared with the four-in-one antibody. The nonsmall cell lung cancer cell line H1666 [11] was used as the target cell. CB12 and bH1-44 show little cell growth inhibition effect, but the inhibition by RB6, CT9, and RB4 in cell proliferation is dose-dependent. MEHD7945A also showed a significant inhibitory effect on cell proliferation. However, the four-in-one antibody inhibited cell growth the most, and the concentration of effective inhibition of cell proliferation was lower than MEHD7945A combined with bH1-44. The inhibitory effect of DVD-IG CRTB6 on cell growth was better than that of FL518. Similar results were observed in another cell line, H1993, with high expression of EGFR, HER2, and HER3 [24].

Next, similar proliferation tests were performed on MDA-MB-231 (high EGFR expression) [23], BT474 (high HER2 expression) and McF-7 (low HER2 expression) breast cancer cells, and BxPC3 cells (pancreatic cancer cell line with HER3 activation) [13]. In MDA-MB-231 cells, DVD-Ig CRTB6 significantly inhibited cell proliferation, MEHD7945 and bH1-44 showed considerable inhibitory effects, and the bispecific antibody did not significantly inhibit cell proliferation. RT6, RB4, and MEHD7945A inhibited tumor cell (BT474 cells, MCF-7 cells, and BxPC cells) proliferation more effectively than anti-EGFR/HER2 antibody and anti-VEGF antibody, while FL518 was more effective than bispecific antibody or MEHD7945A combined with bH1-44. DVD-Ig CRTB6 had the strongest inhibitory effect on all the above cell lines. Taken together, these data suggest an increased in vitro antiproliferative activity of tetrad antibodies.

The BxPC3 cell-burdened nude mouse model was treated with four-in-one antibody or bispecific antibody once a week. Compared with the control group, CT9, CB12, RB4, and BH1-44 were less likely to restrict tumor growth. RT6, RB4, and MEHD7945A have certain inhibitory ability, but the effect is not as good as FL518. In this experiment, the inhibition of tumor growth by the four-in-one antibody was more than 90%, which was even more effective than the parent antibody combination (MEHD7945A + bHI-44). Tumor growth was almost completely suppressed in the DVD-Ig CRTB6-treated group compared to the control group, and tumors established in the tumor-bearing mice were eventually eradicated. In the MDA-MB-231 and MCF-7-cell xenograft female BALB/C nude mouse models, FL518 and DVD-Ig CRTB6 showed stronger tumor inhibition activity than the bispecific antibody, and DVD-Ig CRTB6 showed the best effect.

Only four-in-one antibodies interfere with HER/MET crosstalk

These results suggest that simultaneous targeting of multiple cell growth-dependent signaling pathways can better inhibit tumor growth and improve patient healing. The researchers also found that four-in-one antibodies at the same concentration inhibited AKT and ERKI/2 signaling pathways more significantly than any other antibody. At the same time, DVD-IG CRTB6 monotherapy provided stronger inhibition than other antibody combinations, suggesting that the inhibition by the four-in-one antibody may have other mechanisms of action beyond the inhibition of phosphorylation of these membrane receptors. ERK1/2 or STAT3 signaling in MKN45 cells [25] and HCC827 cells [26] is continuously activated by EGF and HRG, even though MEHD7945A or bH1-44 inhibits EGFR, HER2, and HER3 phosphorylation. Phosphorylation of MET, an oncogene often co-expressed with the EGF family, was inhibited only in combination with high doses of MEHD7945A and bH1-44. Four-in-one antibodies against F1518 and CRTB6 effectively inhibited MET activation at low concentrations and successfully inhibited ERK1/2 and STAT3. Therefore, in MKN45 cells, only four-in-one antibodies obstruct MET/EGFR or MET/HER3 heterodimer formation and block MET/HER crosstalk.

MET/HER signaling is critical for resistance to HER-targeted therapy

Using a previously described method, researchers established cetuximab-resistant HCC827 cells (ReHCC827) [26]. The xenografts of HCC1954 (breast cancer cell line) were treated with increased concentrations of trastuzumab and pertuzumab to produce double drug-resistant ReHCC1954 cells. HER3 and ERK1/2 phosphorylation was enhanced, and the overexpression of MET or increased heterodimer of EGFR/MET and HER3/MET was observed in drug-resistant cells. At the same time, the expression levels of EGF-like ligands such as EGF, HRG, BTC, and HGF are higher in drug-resistant cells.

Next, they used siRNA knockout to investigate the effect of MET on drug-resistant cell lines. In ReHCC827 cells, MET knockout effectively inhibited EGF-induced MET/EGFR signaling and HRG-mediated HER3 signaling and significantly inhibited cell growth. In addition, MET knockout also re-sensitizes cells to cetuximab or to both trastuzumab and pertuzumab, further suggesting that MET/EGFR and MET/HER3 crosstalk and signal transduction may be mechanisms of acquired antibody resistance.

Four-in-one antibodies can overcome resistance to HER therapy

Cetuximab-resistant ReHCC827 cells were dependent on EGFR/MET rather than EGFR signaling, and similarly, ReHCC1954 cells were dependent on HER3/MET rather than HER2/HER3 signaling. Because of the tetrad's ability to inhibit crosstalk between MET and EGFR or HER3, the researchers tested whether the four-in-one antibodies could overcome acquired resistance. The results show that only four-in-one antibodies can effectively inhibit MET/EGFR, MET/HER3 heterodimerization, and EGF HRG-dependent signal transduction and ERK1/2, STAT3, and AKT phosphorylation in drug-resistant cells. Four-in-one antibody also significantly inhibited the growth of both parent and drug-resistant cell tumors in all tumor-bearing mice.

Discussion

In this chapter, the researchers synthesized FL518, a four-in-one antibody produced by MEHD7945A and bH1-44, and CRTB6, a four-in-one DVD-IgG-like antibody with four different variable regions. Four-in-one antibodies not only block the HER family and VEGF signals but also impede HER/MET crosstalk. Similar results were achieved in MEHD7945A and bH1-44 combinations of high-dose applications. In HER-targeted therapy, the effectiveness of adding bevacizumab remains controversial. In addition, the cardiac effect of bevacizumab may also limit its application. Combined therapy with other targets, such as CTLA4 and CD3, can enhance the immune response and is one of the multitarget combination therapy schemes.

Activation or amplification of MET as well as activation of downstream signals is considered to be one of the key causes of resistance to EGFR family inhibitors in a variety of tumors, such as NSCLC [11,26,27], gastric cancer [28], and breast cancer [29,30]. The crosstalk between MET and HER2 may be the mechanism of cancer progression [30]. Met inhibitors can inhibit phosphorylation of the downstream molecules, and HER3/Met crosstalk and signal transduction were observed directly in HER2 blockade treatment, which has played a key role in drug resistance, especially for bead sheet resistance against duly added para bead single double cancer cells for the treatment of drug-resistant strains. It is now in clinical assessment and has promoted clinical results [2]. To effectively block HER/MET crosstalk, it is necessary to target EGFR, HER2, and HER3 simultaneously. Although blocking the combination of EGFR, HER2, and HER3 antibodies can also inhibit this crosstalk, it requires an extremely high dose. Four-in-one antibodies can target EGFR, HER2, and HER3, and inhibit HER/MET crosstalk at low dose levels, suggesting that four-in-one antibodies have obvious advantages in the treatment of egF-family-driven tumors.

References

[1] Hynes NE, MacDonald G. ErbB receptors and signaling pathways in cancer. Curr Opin Cell Biol 2009;21:177–84.
[2] Baselga J, et al. Pertuzumab plus trastuzumab plus docetaxel for metastatic breast cancer. N Engl J Med 2012;366:109–19.
[3] Sliwkowski MX, Mellman I. Antibody therapeutics in cancer. Science 2013;341:1192–8.
[4] Beji A, Horst D, Engel J, Kirchner T, Ullrich A. Toward the prognostic significance and therapeutic potential of HER3 receptor tyrosine kinase in human colon cancer. Clin Cancer Res 2012;18:956–68.
[5] Lee-Hoeflich ST, et al. A central role for HER3 in HER2-amplified breast cancer: implications for targeted therapy. Cancer Res 2008;68:5878–87.
[6] Funayama T, Nakanishi T, Takahashi K, Taniguchi S, Takigawa M, Matsumura T. Overexpression of c-erbB-3 in various stages of human squamous cell carcinomas. Oncology 1998;55:161–7.
[7] Jain A, et al. HER kinase axis receptor dimer partner switching occurs in response to EGFR tyrosine kinase inhibition despite failure to block cellular proliferation. Cancer Res 2010;70:1989–99.
[8] Wheeler DL, et al. Mechanisms of acquired resistance to cetuximab: role of HER (ErbB) family members. Oncogene 2008;27:3944–56.
[9] Nahta R, Yu D, Hung MC, Hortobagyi GN, Esteva FJ. Mechanisms of disease: understanding resistance to HER2-targeted therapy in human breast cancer. Nat Clin Pract Oncol 2006;3:269–80.
[10] Kruser TJ, Wheeler DL. Mechanisms of resistance to HER family targeting antibodies. Exp Cell Res 2010;316:1083–100.
[11] Engelman JA, et al. MET amplification leads to gefitinib resistance in lung cancer by activating ERBB3 signaling. Science 2007;316:1039–43.
[12] Linggi B, Carpenter G. ErbB receptors: new insights on mechanisms and biology. Trends Cell Biol 2006;16:649–56.
[13] Schaefer G, et al. A two-in-one antibody against HER3 and EGFR has superior inhibitory activity compared with monospecific antibodies. Cancer Cell 2011;20:472–86.
[14] Bostrom J, et al. Variants of the antibody herceptin that interact with HER2 and VEGF at the antigen binding site. Science 2009;323:1610–4.
[15] Heiss MM, et al. The trifunctional antibody catumaxomab for the treatment of malignant ascites due to epithelial cancer: results of a prospective randomized phase II/III trial. Int J Cancer 2010;127:2209–21.
[16] Li B, et al. Bispecific antibody to ErbB2 overcomes trastuzumab resistance through comprehensive blockade of ErbB2 heterodimerization. Cancer Res 2013;73:6471–83.
[17] Wang S, et al. Effective suppression of breast tumor growth by an anti-EGFR/ErbB2 bispecific antibody. Cancer Lett 2012;325:214–9.
[18] Ridgway JB, Presta LG, Carter P. 'Knobs-into-holes' engineering of antibody CH3 domains for heavy chain heterodimerization. Protein Eng 1996;9:617–21.
[19] Atwell S, Ridgway JB, Wells JA, Carter P. Stable heterodimers from remodeling the domain interface of a homodimer using a phage display library. J Mol Biol 1997;270:26–35.
[20] Schaefer W, et al. Immunoglobulin domain crossover as a generic approach for the production of bispecific IgG antibodies. Proc Natl Acad Sci U S A 2011;108:11187–92.
[21] Wu C, et al. Simultaneous targeting of multiple disease mediators by a dual-variable-domain immunoglobulin. Nat Biotechnol 2007;25:1290–7.
[22] Junttila TT, et al. Ligand-independent HER2/HER3/PI3K complex is disrupted by trastuzumab and is effectively inhibited by the PI3K inhibitor GDC-0941. Cancer Cell 2009;15:429–40.
[23] Mirschberger C, et al. RG7116, a therapeutic antibody that binds the inactive HER3 receptor and is optimized for immune effector activation. Cancer Res 2013;73:5183–94.

[24] Tang Z, et al. Dual MET-EGFR combinatorial inhibition against T790M-EGFR-mediated erlotinib-resistant lung cancer. Br J Cancer 2008;99:911–22.
[25] Khoury H, et al. HGF converts ErbB2/Neu epithelial morphogenesis to cell invasion. Mol Biol Cell 2005;16: 550–61.
[26] Shattuck DL, Miller JK, Carraway KL, Sweeney C. Met receptor contributes to trastuzumab resistance of Her2-overexpressing breast cancer cells. Cancer Res 2008;68:1471–7.
[27] Rho JK, et al. MET and AXL inhibitor NPS-1034 exerts efficacy against lung cancer cells resistant to EGFR kinase inhibitors because of MET or AXL activation. Cancer Res 2014;74:253–62.
[28] Bachleitner-Hofmann T, et al. HER kinase activation confers resistance to MET tyrosine kinase inhibition in MET oncogene-addicted gastric cancer cells. Mol Cancer Ther 2008;7:3499–508.
[29] Nath D, Williamson NJ, Jarvis R, Murphy G. Shedding of c-Met is regulated by crosstalk between a G-protein coupled receptor and the EGF receptor and is mediated by a TIMP-3 sensitive metalloproteinase. J Cell Sci 2001;114:1213–20.
[30] Lindemann K, et al. Differential expression of c-Met, its ligand HGF/SF and HER2/neu in DCIS and adjacent normal breast tissue. Histopathology 2007;51:54–62.

CHAPTER

8

Overcoming acquired resistance to cetuximab by combining EGFR- and HER3-neutralizing monoclonal antibodies

Kewen Qian, Changhai Lei, and Shi Hu

Department of Biophysics, College of Basic Medical Sciences, Naval Medical University (Second Military Medical University), Shanghai, China

Abstract

The HER3 signaling pathway plays an important role in acquired resistance to EGFR inhibitors. In the process of exploring the resistance mechanism of EGFR inhibitors, the researchers selected sensitive tumor cell lines that did not contain EGFR mutations and KRAS mutations to establish a long-term EGFR inhibitor resistance cell model and found that the expression of EGFR and HER3 was abnormally increased, and siRNA knockout was used. HER3 can restore the sensitivity of cells to cetuximab. Further studies on these resistant strains found that EGFR-HER3 dimerization and EGFR-dependent HER3 activation were significantly increased. The above evidence suggests that one of the mechanisms of EGFR inhibitor resistance is to upregulate HER3-related signals as a supplementary bypass after EGFR signal inhibition. Based on the heterodimerization of EGFR-HER3, we were inspired to target both HER3 and EGFR to eliminate resistance to EGFR inhibitors. In this regard, there have been the following research results: on the one hand, the use of monoclonal antibodies such as MEHD7945A that target both HER3 and EGFR is more effective than the combined use of EGFR-targeting monoclonal antibodies cetuximab and HER3-targeted monoclonal antibodies; on the other hand, studies have found that the use of HER3, the targeted monoclonal antibody mAb33, can enhance the antitumor activity of HER2- and EGFR-targeted monoclonal antibodies and reduce their drug resistance.

Abbreviations

ADCC antibody-dependent cell-mediated cytotoxicity.
EGFR epidermal growth factor receptor.
EGFRi EGFR kinase inhibitors.
HER3 human epidermal growth factor receptor 3

https://doi.org/10.1016/B978-0-12-821584-5.00015-8

HNSCC human head and neck squamous cell carcinoma.
NSCLC human nonsmall cell lung carcinoma.
PDX patient-derived xenograft
SCCHN squamous cell carcinoma of head and neck.
TKI small molecule kinase inhibitor.

Conflict of interest

No potential conflicts of interest were disclosed.

EGFR inhibitor tolerance is related to both EGFR and HER3, which MEHD7945A completely targets

To explore the resistance mechanism of EGFR inhibitors, the researchers used two EGFR inhibitors, monoclonal antibody cetuximab and small molecule kinase inhibitor (TKI) erlotinib, and constructed a series of tumor-derived EGFR-inhibitor resistant clones that do not contain EGFR and KRAS mutations [1,2]. They found that both EGFR and HER3 signaling pathways showed significant activation, and the use of siRNA depleted HER3 could restore sensitivity to cetuximab [3,4]. The results of multiple studies have shown that the activation of HER3 triggered by EGFR activation may serve as a bypass to support insensitivity to EGFR inhibition. MEHD7945A is a polyclonal antibody that can double-target EGFR and HER3 by both inhibiting ligand-dependent EGFR and HER3 downstream signals [5,6]. Compared with the corresponding monoclonal antibody, MEHD7945A shows antitumor activity in vitro and in vivo against a variety of tumor cell lines. In addition, MEHD7945A can effectively promote antibody-dependent cell-mediated cytotoxicity (ADCC), and its skin toxicity is less than that of cetuximab. The researchers also tested the radiosensitivity of MEHD7945A to EGFR inhibitor-resistant cell lines.

MEHD7945A inhibits the growth of two tumor cell lines that are resistant to cetuximab

The researchers constructed two drug-resistant cell lines, H226-CetR and SCC6-CetR, derived from human nonsmall cell lung carcinoma (NSCLC) and human head and neck squamous cell carcinoma (HNSCC). In vitro experiments obtained the following results. First, although both cetuximab and MEHD7945A can inhibit parental H226 and SCC6 cells, only MEHD7945A has an inhibitory effect on H226-CetR and SCC6-CetR cells. Second, MEHD7945A alone has a stronger inhibitory effect on the two drug-resistant cell lines than the combined use of EGFR-targeted monoclonal antibody and HER3-targeted monoclonal antibody, as it inhibits EGFR-MAPK and HER3-PI3K-AKT signaling pathways more effectively. Among them, HER3-PI3K-AKT is abnormally activated in EGFR inhibitor-resistant strains, which is one of the important mechanisms for drug resistance.

MEHD7945A inhibits the growth of human tumor xenografts that are resistant to EGFR inhibitors

Next, the researchers transplanted SCC6 cells resistant to cetuximab or erlotinib into mice, administered the same dose of cetuximab, erlotinib, or MEHD7945A after the tumor mass reached 150–200 mm [3], and found that MEHD7945A can effectively inhibit cetuximab or erlotinib resistance and the proliferation of tumors. However, the use of cetuximab or erlotinib alone has no obvious effectiveness.

MEHD7945A inhibits radiation-induced survival and overcomes cross-resistance to radiation

In addition, based on the discovery that EGFR inhibitor-resistant strains have cross-resistance to radiation, the researchers then explored whether MEHD7945A can enhance the radiosensitivity of drug-resistant strains and improve the lethality of radiation to drug-resistant strains. The administration of MEHD7945A induced cell cycle arrest after exposure to radiation, and simultaneously inhibited DNA repair in the cell cycle and enhanced the killing effect of radiation by inducing apoptosis of these unrepaired cells.

mAb33: A neutralizing anti-HER3 antibody that prevents resistance to an EGF kinase inhibitor

The researchers prepared a neutralizing antibody mAb33 that can effectively block HER3 signaling. When the antibody is combined with the protein kinase inhibitor osimertinib and the EGFR-targeting mAb cetuximab and HER2-targeting mAb trastuzumab, it can completely block EGFRi resistance in xenograft tumors derived from medicinal patients.

Combining osimertinib with mAb33 overcomes resistance by upregulating HER3 and inhibiting the release of the soluble extracellular domain

The researchers constructed two patient-derived xenograft (PDX) models derived from patients with nonsmall cell lung cancer. Both tumor grafts carry two EGFR mutations. The models were divided into four groups: vehicle control, osimertinib, a combination of cetuximab and trastuzumab and an additional group with a mixture of the two antibodies and osimertinib [7]. When osimertinib is used alone or in combination, tumor growth can be affected. However, the overall growth trend is still present. When protease inhibitors are used in combination with the two antibodies, the inhibitory effect on the proliferation of tumor masses is the strongest.

In fact, when the protease inhibitor osimertinib was used alone, HER3 expression levels were upregulated. Therefore, the researchers explored the mechanism by which two mAbs combined with mAb33 inhibit TKI-induced HER3 upregulation. The findings may be related to the promotion of the endocytosis and intracellular degradation of HER3 by mAb33. The results of RNA analysis showed that osimertinib's regulation of HER3 signaling was mainly at the transcriptional level, while mAb33's regulation of HER3 signaling was at the posttranscriptional level. The secreted form of HER3 (sHER3) is an important indicator of cancer progression [8]. Osimertinib alone can activate the metalloprotease ADAM10, thereby increasing the level of sHER3 in the blood [9].

Combination of mAb33, cetuximab, and osimertinib inhibits growth of human tumor xenografts by simultaneously inhibiting other RTKs

To further explore the molecular biological mechanism, the researchers rebuilt a mouse tumor transplantation model and found that mAb33 can not only downregulate HER3 but also cause a decrease in signals such as AXL and MET. In two-thirds of mice, EGFR, HER2, HER3, MET, and AXL are all inhibited. Therefore, the reason why mAb33 can retain good efficacy for a period of time after treatment may be related to the simultaneous inhibition of multiple drug resistance-related pathways.

The drug combination comprising mAb33 delays relapse of a model derived from a patient's metastatic lesion

Since then, in response to the problem of tumor heterogeneity and easy recurrence, the researchers also established a PDX model derived from patients with primary tumors and patients with metastatic tumors and divided the mice into five groups according to different dosing regimens. After 21 days of administration, the treatment was stopped, and the mouse body weight and tumor volume were continuously observed until the 160th day. It was found that although osimertinib alone can inhibit tumor growth in a short period of time, the tumor recurs within 2–3 weeks of treatment, and the combination of osimertinib + cetuximab + mAb33 or osimertinib + cetuximab + trastuzumab + mAb33 can effectively prolong tumor growth. The recurrence time improves the overall survival rate of mice.

In another malignant PDX model, the number of groups was increased to eight, and the treatment time was extended to 66 days, in which the osimertinib alone administration group was administered for 120 days. Similar to the above tumor transplantation models, tumors can quickly respond to osimertinib but relapse later and during the treatment period. However, when osimertinib + cetuximab + trastuzumab or osimertinib + cetuximab + mAb33 is used during treatment, continuous administration can effectively inhibit tumor growth in real time, but it relapsed immediately or was delayed after the treatment was stopped.

In addition, using the same regimen to retreat recurrent tumors after a period of time can still achieve good inhibitory effectiveness, indicating that the tumor remains sensitive to the drug combination regimen during the withdrawal period.

Discussion

EGFR and HER3 are both key factors that regulate trastuzumab resistance. MEHD7945A demonstrates the capacity to overcome acquired resistance to EGFR inhibitors and radiation by dually targeting EGFR and HER3. Several clinical trials are also in development with MEHD7945A to further investigate its safety. A phase I study indicated that MEHD7945A is well tolerated as a single agent with evidence of tumor pharmacodynamic modulation and antitumor activity in squamous cell carcinoma of the head and neck (SCCHN) [10]. The use of anti-HER3 neutralizing antibody mAb33 can block the drug-induced upregulation of HER3 and the release of soluble HER3. In addition, when the antibody is used in combination with EGFR inhibitors, it can downregulate other drug resistance-related receptors, such as AXL and MET, thereby blocking AKT signaling and promoting cell apoptosis [6]. Although the combination of antibodies and EGFR inhibitors can delay the drug resistance of NSCLC patients with tumor transplantation mice and improve their survival rate, it still cannot solve the tumor recurrence after drug withdrawal. Therefore, a large number of safety evaluations and preclinical trials are needed.

References

[1] Huang S, Armstrong EA, Benavente S, Chinnaiyan P, Harari PM. Dual-agent molecular targeting of the epidermal growth factor receptor (EGFR): combining anti-EGFR antibody with tyrosine kinase inhibitor. Cancer Res 2004;64:5355–62.

[2] Benavente S, et al. Establishment and characterization of a model of acquired resistance to epidermal growth factor receptor targeting agents in human cancer cells. Clin Cancer Res 2009;15:1585–92.

[3] Wheeler DL, et al. Mechanisms of acquired resistance to cetuximab: role of HER (ErbB) family members. Oncogene 2008;27:3944–56.

[4] Sergina NV, et al. Escape from HER-family tyrosine kinase inhibitor therapy by the kinase-inactive HER3. Nature 2007;445:437–41.

[5] Engelman JA, et al. MET amplification leads to gefitinib resistance in lung cancer by activating ERBB3 signaling. Science 2007;316:1039–43.

[6] Huang S, et al. Dual targeting of EGFR and HER3 with MEHD7945A overcomes acquired resistance to EGFR inhibitors and radiation. Cancer Res 2013;73:824–33.

[7] Romaniello D, et al. Targeting HER3, a catalytically defective receptor tyrosine kinase, prevents resistance of lung cancer to a third-generation EGFR kinase inhibitor. Cancer 2020;12.

[8] Yonesaka K, et al. Circulating heregulin level is associated with the efficacy of patritumab combined with erlotinib in patients with non-small cell lung cancer. Lung Cancer 2017;105:1–6.

[9] Liu PCC, et al. Identification of ADAM10 as a major source of HER2 ectodomain sheddase activity in HER2 overexpressing breast cancer cells. Cancer Biol Ther 2006;5:657–64.

[10] Juric D, et al. Safety and pharmacokinetics/pharmacodynamics of the first-in-class dual action HER3/EGFR antibody MEHD7945A in locally advanced or metastatic epithelial tumors. Clin Cancer Res 2015;21:2462–70.

CHAPTER

9

Dual targeting of EGFR and HER3 with MEHD7945A overcomes acquired resistance to EGFR inhibitors and radiation

Guangyao Li and Shi Hu

Department of Biophysics, College of Basic Medical Sciences, Naval Medical University (Second Military Medical University), Shanghai, China

Abstract

EGFR is a popular target for cancer treatment today and provides a new treatment direction for refractory tumors. However, the other role of inhibitory drugs targeting EGFR targets in clinical treatment has been dramatically restricted in drug development due to acquired drug resistance. MEHD7945A, as an antibody against EGFR and HER3 simultaneously, can effectively inhibit the downstream signals conducted by the above target ligands. The effect is even better than the combination of antibodies against two targets. Tumor cells all show strong antitumor activity. In addition, MEHD7945A also has a critical regulatory role in the radiation response of the acquired radiation cross-resistance of drug-resistant tumor cells. Therefore, MEHD7945A may be a potentially effective strategy to solve the resistance of EGFR inhibitors.

Abbreviations

DSB double-strand breaks
EGFR epidermal growth factor receptor
mAb monoclonal antibody
TKI tyrosine kinase inhibitor

Conflict of interest

No potential conflicts of interest were disclosed.

https://doi.org/10.1016/B978-0-12-821584-5.00006-7

HER3 is an essential regulatory target for acquired drug resistance

The regulation of ErbB/HER receptor family members is inseparable from the process of tumor occurrence, development, and prognosis. Among the many strategies to block EGFR, small-molecule drugs, such as TKIs and mAbs, are highly promising choices. However, the clinical application of EGFR inhibitors has not achieved long-term expected results, even for people who are initially sensitive to the drug [1–3].

Therefore, the underlying occurrence, development, and outcome of EGFR inhibitor resistance need to be understood, and complementary strategies need to be implemented. Cetuximab and erlotinib are different types of common EGFR inhibitors. Their resistant clones are derived from sensitive tumor cell lines that do not have EGFR and KRAS mutations for a long time and have been systematically screened and identified. Later, p-EGFR and p-HER3 increased significantly. After siRNA depletion of HER3, the sensitivity to the EGFR inhibitor cetuximab was restored [4].

Further analysis showed that EGFR-HER3 dimerization increased, followed by HER3-dependent EGFR activation. Some studies have shown that HER3 may activate part of the acquired resistance to EGFR inhibitors, thereby effectively circumventing the exertion of EGFR inhibition [5–7]. Since HER3 is a specific heterodimer structure, the experimental results provide a theoretical basis for how to adopt further EGFR/HER3 combined targeted therapy on tumors with poor clinical effects of EGFR inhibitors due to acquired drug resistance.

Increasing evidence shows that crosstalk between the HER family affects the clinical efficacy of HER as a target [8–10]. The blockade of one HER family member that can compensate for the blockade of another HER receptor has been the consensus of long-term breast cancer research that trastuzumab can inhibit the signal transduction of HER2, but the activation of dimerization among HER family members is not affected by it [7]. A large amount of evidence shows that HER3 and EGFR are indispensable for developing acquired anti-HER2 targeted drugs [11]. Related research proves the critical status of HER3 in acquired resistance [4,5,12]. Many studies have shown that HER3 is in a critical position in the signal transduction chain of the HER family, and targeted strategies for acquired resistance to EGFR inhibitors should be considered from the inhibition of the HER family signal transduction chain. Although MEHD7945A is an antibody, it has two targets, HER3 and EGFR, to achieve a combined therapeutic effect [13].

Interestingly, tumors and signaling pathways, such as PI3K/AKT or MAPK, can be effectively inhibited by MEHD7945A, which is better than using it alone. Moreover, it has a more substantial dual role than EGFR. Due to the gap between the monospecific antibody and MEHD7945A in terms of affinity and its dual targeting ability, MEHD7945A is superior to the combination of the two antibodies. Numerous experimental results have also proven that EGFR dimers still exist in cells [14,15]. The activation of receptor activity, including dimer transformation, plays an important role [16]. It is known that the heterodimer of EGFR produces the deepest downstream signal transduction. Therefore, the antiproliferation effect of MEHD7945A through the EGFR/HER3 cluster may be more effective in combination with a single antibody. In addition, it has recently been determined that the combination of noncompetitive anti-EGFR Abs can synergistically reduce the level of surface receptors in all kinds of mouse models, resulting in tumor growth inhibition and life extension [17,18]. Given that MEHD7945A has the potential to promote the interaction between EGFR and HER3, how to enhance the internalization and degradation of EGFR and HER3 is the key next step in the study of the way MEHD7945A works.

The mechanism of MEHD7945A in overcoming acquired drug resistance through tumor xenografts and cells

After continuously treating NSCLC and HNSCC tumors with cetuximab or erlotinib, acquired drug-resistant clones against the two inhibitors were formed. First, comparing the in vitro antiproliferative effects of the drug in SCC6 cells and cetuximab and MEHD7945A are comparable in inhibiting the proliferation of H226 parental generation. However, it is worth noting that MEHD7945A still has an inhibitory effect on H226-CetR, while cetuximab is powerless. From the perspective of cell cycle progression, the induction of G0/G1 phase blockade is crucial for MEHD7945A to inhibit proliferation, and the percentage of S phase cells is also reduced. These results strongly support the antiproliferative effect of MEHD7945A. Further immunoblotting analysis showed that MEHD7945A significantly inhibited MAPK that was not inhibited by cetuximab. In previous studies, it is known that one of the potential mechanisms by which EGFR inhibitors acquire resistance is the PI3K-AKT signaling pathway [19]. The experimental results indicate that the ability of MEHD7945A to overcome acquired resistance may be achieved by inhibiting HER3-PI3K-AKT signaling.

Then, we compared the differences between single antibodies and MEHD7945A (corresponding to HER3) in anti-cell proliferation, including pairwise combinations of single antibodies [13].

Keeping the antibody concentration at the same level, MEHD7945A has a more powerful antitumor cell proliferation effect, and its advantages also exist in the signal inhibition of HER3-PI3K-AKT and EGFR-MAPK. The above findings illustrate that the study of this dual-target antibody method has better clinical potential than the multidrug combination therapy of monospecific EGFR and HER3 antibodies.

To expand the results of the in vitro study, the mice were inoculated with tumor cells resistant to cetuximab or erlotinib after their thymus was removed and maintained long-term exposure to the same level of cetuximab or MEHD7945A for treatment. Researchers compared the untreated control group and the tumor growth under the action of MEHD7945A, finding that the latter's effect was more prominent and even induced the regression of highly drug-resistant tumors. In contrast, cetuximab has no growth inhibitory effect on tumors previously treated with MEHD7945A. Similarly, the initial attack of MEHD7945A still had a slower response to erlotinib-resistant tumors that did not respond to erlotinib treatment and caused those tumors that were ineffective to erlotinib treatment to recede very quickly.

The above studies prove that MEHD7945A shows great potential for overcoming acquired resistance in vivo and in vitro.

Study on the mechanism of MEHD7945A in radiation-damaged cells

It has been found that cells resistant to EGFR inhibitors are cross-resistant to radiation after long-term treatment with EGFR inhibitors [20]. To determine whether MEHD7945A can overcome this cross-resistance, the experimental comparison of the radiation response of MEHD7945A or cetuximab to cetuximab-resistant clones and the analysis of clone survival showed that pretreatment with cetuximab compared with radiation alone (NT) did not change the response spectrum. In contrast, in the cetuximab-resistant H226 and SCC6 clones, the survival rate of radiation-exposed cells decreased significantly after treatment with MEHD7945A. In addition, the DNA damage after 3 Gy radiation is characterized by inspection; it was shown that compared with the untreated control, ΓH2AX lesions increased significantly in cells exposed to MEHD7945A but not observed in cells exposed to cetuximab. Interestingly, the DNA damage caused by radiation was also significantly enhanced after MEHD7945A treatment. This finding is similar to the more substantial killing effect in SCC6-CetR cells treated with MEHD7945A. To further expand the results of these in vitro studies, after establishing a drug-resistant xenograft model with cetuximab, cetuximab and MEHD7945A were used to treat cells to observe the changes in reflex response. The results unsurprisingly show that the therapeutic combination of cetuximab plus radiotherapy has no apparent therapeutic benefit in antidrug-resistant tumors compared with the corresponding two separate applications. However, after radiotherapy and MEHD7945A, tumor growth was significantly stagnated, accompanied by many growth delays. MEHD7945A can enhance the radiation response of anti-cetuximab tumor xenografts.

To determine the change mechanism of the radiation response after MEHD7945A treatment, western blot analysis was performed on the relevant functional proteins in the process of repairing DNA damage, focusing on their functional expression and activity changes.

The survival signals generated by cells after radiation exposure can be effectively suppressed, including p-Rb levels, a functional defect of cetuximab. Forty-eight hours after irradiation, a significant increase in lysed phosphorylated p53, PARP and Caspase 3 caused by apoptosis activation appeared significantly in cells exposed to MEHD7945A.

Next, proteins with the critical ability to repair broken double strands are closely watched, and whether MEHD7945A can inhibit repairability will be checked. The formation of the DSB complex is inseparable from Rad50, Nbs1, and Mre11, which are also involved in cell cycle detection [21]. It is known that inhibition of these molecules can make tumor cells sensitive to radiation. After the control cells were irradiated, most of the repair proteins were increased in the nucleus. The level of repair protein tested in all cells after irradiation decreased more significantly after MEHD7945A treatment than after cetuximab treatment. In summary, these results indicate that the mechanism of MEHD7945A's inhibitory effect on the DNA damage repair mechanism may increase cell apoptosis, death, cycle arrest, and radiation response.

Resistance to EGFR drugs may be related to sensitivity to radiation and drugs. Current research shows that MEHD7945A overcomes radiation-induced cross-resistance by hindering survival signals from cell apoptosis initiated by DNA damage and radiation-activated transduction. The DNA repair mechanism of most unrepaired cells exposed to MEHD7945A for 48 h after blocked radiation and then entered the apoptotic state. In addition, p53 levels were also identified as significantly increased in cells treated with MEHD7945A. Recent studies have shown that both radiation-induced and EGFR inhibitor drug resistance are subject to the critical regulatory role of p53 [19]. H226 cells respond strongly to drugs and radiation, but their sensitivity is significantly reduced after knocking out p53.

In contrast, after p53 function is reactivated, resistance does not exist [19]. Therefore, p53 regulation may also be one of the potential mechanisms by which acquired resistance loses its effect in the face of MEHD7945A. The development of MEHD7945A's potential to solve the problem of resistance and combination chemotherapy is of great significance.

Final thoughts and future directions

Overall, MEHD7945A, an antibody that targets not only EGFR but also HER3, has solid potential to solve acquired resistance. The research direction of MEHD7945A is not only that, but the clinical application in combination with other resistant drugs also has broad prospects in small molecule targeted therapy. With the gradual clarity of the interaction mechanism of HER family signals, drugs such as MEHD7945A may be of great significance in improving the level of clinical treatment effects of anti-EGFR. MEHD7945A is developing clinical trials, and it will gradually be applied to the in-depth aspects of these issues.

References

[1] Riely GJ, Kris MG, Zhao B, Akhurst T, Milton DT, Moore E, et al. Prospective assessment of discontinuation and reinitiation of erlotinib or gefitinib in patients with acquired resistance to erlotinib or gefitinib followed by the addition of everolimus. Clin Cancer Res 2007;13:5150–5.

[2] Jackman D, Pao W, Riely GJ, Engelman JA, Kris MG, Janne PA, et al. Clinical definition of acquired resistance to epidermal growth factor receptor tyrosine kinase inhibitors in non-small-cell lung cancer. J Clin Oncol 2010;28:357–60.

[3] Chaft JE, Oxnard GR, Sima CS, Kris MG, Miller VA, Riely GJ. Disease flare after tyrosine kinase inhibitor discontinuation in patients with EGFR-mutant lung cancer and acquired resistance to erlotinib or gefitinib: implications for clinical trial design. Clin Cancer Res 2011;17:6298–303.

[4] Wheeler DL, Huang S, Kruser TJ, Nechrebecki MM, Armstrong EA, Benavente S, et al. Mechanisms of acquired resistance to cetuximab: role of her (ERBB) family members. Oncogene 2008;27:3944–56.

[5] Sergina NV, Rausch M, Wang D, Blair J, Hann B, Shokat KM, et al. Escape from her-family tyrosine kinase inhibitor therapy by the kinase-inactive her3. Nature 2007;445:437–41.

[6] Engelman JA, Zejnullahu K, Mitsudomi T, Song Y, Hyland C, Park JO, et al. Met amplification leads to gefitinib resistance in lung cancer by activating erbb3 signaling. Science 2007;316:1039–43.

[7] Jain A, Penuel E, Mink S, Schmidt J, Hodge A, Favero K, et al. Her kinase axis receptor dimer partner switching occurs in response to EGFR tyrosine kinase inhibition despite failure to block cellular proliferation. Cancer Res 2010;70:1989–99.

[8] Shepard HM, Brdlik CM, Schreiber H. Signal integration: a framework for understanding the efficacy of therapeutics targeting the human EGFR family. J Clin Invest 2008;118:3574–81.

[9] Huang Z, Brdlik C, Jin P, Shepard HM. A pan-her approach for cancer therapy: background, current status and future development. Expert Opin Biol Ther 2009;9:97–110.

[10] Vlacich G, Coffey RJ. Resistance to EGFR-targeted therapy: a family affair. Cancer Cell 2011;20:423–5.

[11] Narayan M, Wilken JA, Harris LN, Baron AT, Kimbler KD, Maihle NJ. Trastuzumab-induced her reprogramming in "resistant" breast carcinoma cells. Cancer Res 2009;69:2191–4.

[12] Campbell MR, Amin D, Moasser MM. Her3 comes of age: new insights into its functions and role in signaling, tumor biology, and cancer therapy. Clin Cancer Res 2010;16:1373–83.

[13] Schaefer G, Haber L, Crocker LM, Shia S, Shao L, Dowbenko D, et al. A two-in-one antibody against her3 and EGFR has superior inhibitory activity compared with monospecific antibodies. Cancer Cell 2011;20:472–86.

[14] Saffarian S, Li Y, Elson EL, Pike LJ. Oligomerization of the egf receptor investigated by live cell fluorescence intensity distribution analysis. Biophys J 2007;93:1021–31.
[15] Chung I, Akita R, Vandlen R, Toomre D, Schlessinger J, Mellman I. Spatial control of EGF receptor activation by reversible dimerization on living cells. Nature 2010;464:783–7.
[16] Gan HK, Walker F, Burgess AW, Rigopoulos A, Scott AM, Johns TG. The epidermal growth factor receptor (EGFR) tyrosine kinase inhibitor ag1478 increases the formation of inactive untethered EGFR dimers: implication for combination therapy with monoclonal antibody 806. J Biol Chem 2007;282:2840–50.
[17] Friedman LM, Rinon A, Schechter B, Lyass L, Lavi S, Bacus SS, et al. Synergistic down-regulation of receptor tyrosine kinases by combinations of mAbs: implications for cancer immunotherapy. Proc Natl Acad Sci U S A 2005;102:1915–20.
[18] Pedersen MW, Jacobsen HJ, Koefoed K, Hey A, Pyke C, Haurum JS, et al. Sym004: a novel synergistic anti-epidermal growth factor receptor antibody mixture with superior anticancer efficacy. Cancer Res 2010;70:588–97.
[19] Huang S, Benavente S, Armstrong EA, Li C, Wheeler DL, Harari PM. P53 modulates acquired resistance to EGFR inhibitors and radiation. Cancer Res 2011;71:7071–9.
[20] Benavente S, Huang S, Armstrong EA, Chi A, Hsu K-T, Wheeler DL, et al. Establishment and characterization of a model of acquired resistance to epidermal growth factor receptor targeting agents in human cancer cells. Clin Cancer Res 2009;15:1585–92.
[21] Bolderson E, Richard DJ, Zhou B-BS, Khanna KK. Recent advances in cancer therapy targeting proteins involved in DNA double-strand break repair. Clin Cancer Res 2009;15:6314–20.

CHAPTER

10

Targeting ERBB2 overcomes resistance to the anti-EGFR therapeutic antibody cetuximab

Chuqi Wang[a,b], *Tian Li*[a], *Wenyan Fu*[c], *and Shi Hu*[a]

[a]Department of Biophysics, College of Basic Medical Sciences, Naval Medical University (Second Military Medical University), Shanghai, China [b]Department of Pharmacy, National University of Singapore, Singapore [c]Department of Assisted Reproduction, Shanghai Ninth People's Hospital, Shanghai Jiao Tong University School of Medicine, Shanghai, China

Abstract

Cetuximab, an EGFR-directed antibody, has shown therapeutic promise in colorectal, head and neck, and nonsmall cell lung cancers. However, only a small fraction of patients who undergo cetuximab treatment will enjoy durable benefits, and even those who respond well initially will eventually develop acquired resistance. In addition to KRAS, NRAS, and BRAF mutations, whose roles in anti-EGFR resistance have been defined previously, ERBB2 signaling activation is another possible reason for resistance to cetuximab. The underlying mechanisms for ERBB2-mediated cetuximab resistance include ERBB2 amplification, heregulin upregulation, and increased phosphorylation of the HER2 C-terminal fragment (611-CTF). Simultaneous inhibition of ERBB2 and EGFR has proven effective in overcoming cetuximab resistance in a series of preclinical models and patients. These findings collectively highlight ERBB2 as a valuable target to restore anti-EGFR sensitivity and suggest that ERBB2 inhibition represents a rational therapeutic strategy against cetuximab-resistant cancers.

Abbreviations

BRAF	B-Raf proto-oncogene
CR	cetuximab resistance
CRC	colorectal cancer
CTF	carboxy-terminal fragment
DLTs	dose-limiting toxicities
EGFR	epidermal growth factor receptor
FISH	fluorescence in situ hybridization
MTDs	maximum tolerated doses
NSCLC	nonsmall cell lung carcinoma
TKIs	tyrosine kinase inhibitors

Novel Sensitizing Agents for Therapeutic Anti-EGFR Antibodies
https://doi.org/10.1016/B978-0-12-821584-5.00031-6

Conflict of interest

No potential conflicts of interest were disclosed.

Introduction

In unselected patients with EGFR-positive tumors, the response rates of cetuximab as a monotherapy are approximately 10% [1,2], which only accounts for a small subset of all patients. The response rates rise to 40%–60% in different types of tumors when KRAS is taken into consideration [3–6]. Other studies identified NRAS, BRAF, and PIK3CA as emerging biomarkers of response to cetuximab therapy. Although success has been made in finding predictive factors of cetuximab resistance, their reliability is still not high enough, highlighting a necessity to search for alternative biomarkers and underlying mechanisms in anti-EGFR antibody resistance.

ERBB receptor tyrosine kinases play a crucial role in the tumorigenesis of many tumor types. This family consists of four members, EGFR (HER1, ERBB1), ERBB2 (Her2, Neu), ERBB3 (Her3), and ERBB4 (Her4). EGFR and HER4 are two classical family members that can homo- or heterodimerize with other ERBBs upon binding with ligands and subsequently activate downstream signaling, including the RAS-ERK and PI3K-AKT pathways. In contrast, ERBB2 cannot bind any known ligand, and ERBB3 is generally considered kinase deficient [7,8]. Despite not binding to any ligand, ERBB2 could readily interact with other ligand-bound ERBB receptors or form homodimerization in a ligand-independent manner when overexpressed [8]. The most potent mitogenic signals are created by ERBB2-ERBB3 heterodimers [8,9]. Compensatory signaling through one ERBB family member can result in resistance to another receptor's inhibitors. For example, elevated levels of EGFR are associated with decreased benefit from trastuzumab, an anti-ERBB2 antibody [10,11], and heterodimerization of ERBB3 with other ERBB receptors may impair the sensitivity to cetuximab [12]. Co-targeting, alternatively activated tyrosine kinase receptors, has demonstrated advantages over monotherapy in various studies, indicating that dual inhibition of different ERBB members might become a plausible strategy. This chapter summarizes the underlying mechanisms of ERBB2-related cetuximab resistance, including ERBB2 amplification, heregulin upregulation, and activation of a C-terminal fragment of HER2 (611-CTF). Researchers also discussed using ERBB2 inhibitors, either antibodies or tyrosine kinase inhibitors (TKIs), as sensitizers to cetuximab treatment.

ERBB2 amplification mediates cetuximab resistance through Erk1/2 signaling

Several groups have reported the association between ERBB2 amplification and cetuximab resistance, mainly contributed by Yonesaka et al. and Bertotti et al. To study ERBB2 amplification in cetuximab-resistant cells, Yonesaka et al. first generated the cetuximab-resistant (CR) cell line HCC827 CR by exposing cetuximab-sensitive HCC827 cells to increasing drug concentrations [13]. Using genome-wide copy analyses, they found copy number gain in an area on chromosome [14], which contains the ERBB2 oncogene. They confirmed ERBB2 amplification in HCC827 CR with fluorescence in situ hybridization (FISH). Western blot also suggested elevated levels of both total and phosphorylated ERBB2 in HCC827 CR cells compared with parental cells. In addition to the nonsmall cell lung cancer (NSCLC) cell line HCC827, colorectal cancer (CRC) GEO cells showed similar results: GEO CR cells expressed higher levels of ERBB2, and cetuximab failed to downregulate pERK 1/2 in these cells.

To further confirm the causal relationship between ERBB2 and cetuximab resistance, they depleted ERBB2 in HCC827 CR cells and found that ERBB2 knockdown effectively restored cetuximab sensitivity and the inhibition of ERK1/2 phosphorylation. Combining an ERRB2 antibody, trastuzumab, or ERBB2 inhibitor lapatinib with cetuximab also restored the sensitivity of CR cells to cetuximab. HCC827 cells transfected with wild-type ERBB2, but not kinase-dead (K753) ERBB2, exhibited resistance to cetuximab, indicating that ERBB2 kinase activity is essential for mediating cetuximab resistance. These results together indicate that ERBB2 amplification leads to resistance to cetuximab.

As ERBB2 may influence cetuximab activity via several different pathways, the researchers then investigated whether activation of ERK 1/2 signaling is the leading cause of cetuximab resistance. They introduced the BRAF V600E mutation, which could activate Erk1/2 signaling, into two cetuximab-sensitive cell lines, HCC827 and HN11, and found that both cell lines developed resistance to cetuximab. MEK inhibitors, which intercept Erk1/2 signaling, are equally effective for cetuximab-sensitive and cetuximab-resistant GEO cells. Furthermore, co-precipitation showed that GRB2 was recruited to ERBB2 in HCC827 CR cells, suggesting that ERK1/2 signaling is activated in cetuximab-resistant cells. These findings reveal that activation of ERK1/2 signaling is the principal mechanism of cetuximab resistance caused by ERBB2 amplification.

In addition to the findings above by Yonesaka et al., Bertotti et al. discovered ERBB2 amplification in EGFR-resistant tumors by using xenograft cohorts from 85 patient-derived metastatic colorectal cancer samples (xenopatients) [1]. Genome-wide expression profiling indicated ERBB2 amplification in a subset of cetuximab-resistant, KRAS/NRAS/BRAF/PIK3CA wild-type cases [1]. According to their results, 3 of 137 metastatic colorectal cancers (2.2%) showed ERBB2 overexpression. This percentage is consistent with other studies in which ERBB2 overexpression prevalence ranged between 2% and 3.5% among genetically unselected colorectal cancers [15–17]. All three outliers fell into the KRAS/NRAS/BRAF/PIK3CA wild-type subset, and all showed apparent resistance to cetuximab. ERBB2 amplification increased from 2.2% in genetically unselected cancers to 36% in cetuximab-resistant KRAS/NRAS/BRAF/PIK3CA wild-type cancers. Moreover, the researchers found ERBB2-amplified cases accounted for 17.6% (3/17) in all 17 cetuximab- or panitumumab-resistant KRAS wild-type metastatic colorectal cancers, whereas the percentage was only 0% (0/14) in 14 anti-EGFR-sensitive KRAS wild-type tumors. The results collectively suggest that ERBB2 amplification is a predictive biomarker of resistance to anti-EGFR antibodies in KRAS wild-type metastatic colorectal cancers, although this research did not investigate the mechanisms, and the number of ERBB2-overexpressing cases is limited.

Heregulin activates ERBB2 signaling and mediates resistance to cetuximab without ERBB2 amplification

One hallmark of cancer cells is that they may produce growth factor ligands to activate cognate receptors on themselves, thus sustaining proliferative signaling [14]. Such autocrine proliferative stimulation has been observed to mediate cetuximab resistance by several groups [18]. For example, metastatic CRCs could increase TGF-alpha secretion to bypass the inhibition of cetuximab [19].

After revealing the role of ERBB2 amplification in resistance to cetuximab, Yonesaka et al. uncovered upregulated heregulin as yet another mechanism to lead to resistance [13]. Heregulin is a soluble secreted growth factor that promotes preferential dimerization of ERBB3 with ERBB2 and consequently phosphorylates both ERBB proteins [20]. The researchers found that heregulin concentrations were approximately 2.5 times higher in A431 CR cells than in parental cells. Although there was no evidence of ERBB2 amplification, the pERBB2 and pERBB3 levels were upregulated. The addition of heregulin induced dose-dependent pERBB2 and pERBB3 expression in A431 cells. Immunoprecipitation results suggested more ERBB2/ERBB3 dimers in A431 CR cells than in A431 parental cells. Gain- and loss-of-function experiments further confirmed the hypothesis: depletion of heregulin in A431 CR cells restores cetuximab sensitivity, while exogenous heregulin caused dose-dependent increases in cetuximab resistance in a panel of cetuximab-sensitive cells; not surprisingly, in the absence of heregulin, cetuximab strongly reduced pERK 1/2, whereas when heregulin was added, cetuximab's effect on ERK 1/2 phosphorylation was negligible. Collectively, these results demonstrated that heregulin is a mediator of cetuximab resistance by activating ERBB2 signaling.

611-CTF, a C-terminal fragment of HER2, is phosphorylated in cetuximab-resistant cells

The HER2 carboxy-terminal fragments (611-CTF) have been demonstrated to be potent oncogenes capable of promoting tumor progression and metastasis [21]. However, lacking the extracellular domain, 611-CTF is a hyperactive form of ERBB2 and can establish homodimers by disulfide bonds in the absence of ligands [22]. Quesnelle et al. induced cetuximab resistance in the bladder cancer cell line T24 and found that these resistant cells had higher levels of phosphorylated 611-CTF than parental T24 cells. To investigate the role of phosphorylated 611-CTF in conferring resistance to cetuximab, the researchers knocked down full-length HER2 or 611-CTF in cetuximab-resistant clones. Cetuximab decreased the invasion of HER2 knockdown cells robustly, indicating restored cetuximab sensitivity. Trastuzumab, an antibody exclusively targeting full-length HER2 but not 611-CTF, weakened the invasion of cetuximab-resistant T24PR3 cells by 14.5% when added alone and 43.8% when in combination with cetuximab. Irreversible EGFR and the ERBB2 inhibitor afatinib reduced the invasion of T24PR3 cells by 38.1% and 62.1% when used alone or in combination with cetuximab, respectively.

Moreover, in cetuximab-resistant xenograft models, cetuximab plus afatinib induced more pronounced tumor regression than cetuximab or afatinib. These results suggest that co-targeting HER2 and 611-CTF yields high efficacy in vitro and in vivo. However, since there are currently no specific inhibitors of 611-CTF, this study did not determine to what extent inhibiting 611-CTF plays a role in contrast to inhibiting full-length HER2 [23].

Targeting ERBB2 signaling overcomes resistance to cetuximab: Preclinical studies

Two therapeutic antibodies against ERBB2, trastuzumab and pertuzumab, and two FDA-approved tyrosine kinase inhibitors, lapatinib and afatinib, were combined with cetuximab to combat cetuximab resistance in previous preclinical and clinical studies. In the paper by Yonesaka et al., pertuzumab was used in combination with cetuximab in A431 and A431 CR xenograft models [13]. While both cetuximab monotherapy and cetuximab + pertuzumab dramatically hampered the growth of A431 in vivo, only the combinatorial therapy could lead to regression of A431CR xenografts, consistent with their in vitro findings.

To assess to what extent ERBB2 blockade could restore sensitivity to anti-EGFR therapies, Bertotti et al. selected pertuzumab and lapatinib in combination with cetuximab and evaluated the efficacy of this dual inhibition in xenografts from two representative cases (M077 and M091) [1]. Three weeks after treatment, pertuzumab alone showed no inhibition of MO77 (tumor volume variation compared with volume at baseline: +196%); pertuzumab plus cetuximab only produced a mild growth delay (+123%); lapatinib alone showed a more substantial antitumor effect and induced disease stabilization (+10%); lapatinib plus cetuximab led to significant tumor regression (−41%); surprisingly, the regression was more remarkable in the lapatinib plus pertuzumab group (−75%) than in the lapatinib plus cetuximab group. After 6 weeks of treatment, tumor regression became massive in the pertuzumab plus lapatinib group (−93%) and cetuximab plus lapatinib (−79%), whereas the efficacy of lapatinib alone tended to decline (+37%). In the other case, M091, the response to the same treatment modalities was similar. After 3 weeks, pertuzumab alone was completely ineffective (+237%); pertuzumab plus lapatinib resulted in marked tumor regression (−65% after 3 weeks and −64% after 6 weeks); the combination of cetuximab and lapatinib produced a considerable effect, but less significant than that caused by pertuzumab plus lapatinib (−53% after 3 weeks and −45% after 6 weeks), which was in agreement with case M077. In contrast to case M077, pertuzumab and cetuximab produced disease stabilization (−11% after 3 weeks and +15% after 6 weeks) but not lapatinib alone (+163% after 3 weeks).

Quesnelle et al. engrafted cetuximab-sensitive resistant tumors on each flank of the same mouse. They treated the mice with vehicle control, cetuximab alone, afatinib alone, or cetuximab plus afatinib [23]. After 3 weeks of treatment, the cetuximab plus afatinib group yielded the most pronounced response among all treatment regimens in cetuximab-resistant tumors (tumor volumes compared with vehicle control-treated tumors: −76.5%), in contrast to afatinib, which produced a moderate reduction, and cetuximab, which was completely ineffective. In cetuximab-sensitive tumors, the effect of cetuximab plus afatinib was similar to that of cetuximab alone (−89.7%), probably due to the already strong antitumor effects of cetuximab on these tumors.

Dual inhibition of ERBB2 and EGFR in cetuximab-resistant patients: Clinical trials

In one phase I clinical trial, patients were treated with standard weekly cetuximab and progressively increased doses of lapatinib for 3 weeks [24]. Of the 18 patients evaluated, 3 partial responses, 9 stable diseases, and 6 disease progressions were observed. The overall response rate was 17%, and the clinical benefit rate was 67%. The clinical benefit rate in patients who received prior anti-EGFR therapy was 70%. Common side effects included rash and diarrhea. No dose-limiting toxicity was observed at the highest dose level, and no patient experienced grade 4 toxicity. These results indicated that the cetuximab and lapatinib combination was tolerable and showed pronounced clinical activity.

In another phase I study, pertuzumab plus cetuximab was tested in cetuximab-resistant KRAS wild-type metastatic CRC [25]. Among the seven patients evaluable for response, one had a partial response that lasted for 6 months, and the objective response rate was 14%. Two had stable disease (29%), and four had progressive disease (57%). Although pertuzumab and cetuximab showed potential antitumor effects, this combinatorial treatment caused DLTs in 6 of the 13 patients (46%) and was therefore terminated early. Grade 3 or higher DLTs included dermatitis with desquamation and acneiform rash ($n=6$), mucositis or stomatitis ($n=5$), and diarrhea ($n=2$).

Conclusion

This chapter first introduces the currently discovered mechanisms of ERBB2-mediated cetuximab resistance. Then, researchers reviewed the progress of co-targeting ERBB2 and EGFR to combat anti-EGFR resistance. Both preclinical and clinical studies confirmed the potential antitumor activity of combining ERBB2 and EGFR inhibitors. However, a proportion of patients did not respond to such regimens, indicating that other unknown mechanisms might play a principal role in cetuximab resistance. To further increase the percentage of responders, we need to understand how cetuximab resistance forms and identify novel biomarkers or validate existing biomarkers to refine patient stratification.

Although potent activity has been seen to restore cetuximab sensitivity both in vitro and in animal models, the therapeutic effects of ERBB2 inhibitors in combination with cetuximab are still hard to predict in patients, partially owing to the complex mechanisms of ERBB inhibitors, especially therapeutic antibodies in human bodies [8]. In addition, whether such combinatorial treatments are tolerable by patients remains unclear due to the serious side effects that occurred in a previous clinical trial [25]. In conclusion, more evidence is needed to validate the effects of dual inhibition of ERBB2 and EGFR in cetuximab-resistant tumors, and the risk-benefit ratio should be further evaluated.

References

[1] Bertotti A, et al. A molecularly annotated platform of patient-derived xenografts ("xenopatients") identifies HER2 as an effective therapeutic target in cetuximab-resistant colorectal cancer. Cancer Discov 2011;1:508–23.

[2] Bardelli A, Siena S. Molecular mechanisms of resistance to cetuximab and panitumumab in colorectal cancer. J Clin Oncol 2010;28:1254–61.

[3] De Roock W, et al. KRAS wild-type state predicts survival and is associated to early radiological response in metastatic colorectal cancer treated with cetuximab. Ann Oncol 2008;19:508–15.

[4] Lièvre A, et al. KRAS mutation status is predictive of response to cetuximab therapy in colorectal cancer. Cancer Res 2006;66:3992–5.

[5] Hsu H-C, et al. Mutations of KRAS/NRAS/BRAF predict cetuximab resistance in metastatic colorectal cancer patients. Oncotarget 2016;7:22257–70.
[6] Linardou H, et al. Assessment of somatic k-RAS mutations as a mechanism associated with resistance to EGFR-targeted agents: a systematic review and meta-analysis of studies in advanced non-small-cell lung cancer and metastatic colorectal cancer. Lancet Oncol 2008;9:962–72.
[7] Casalini P, Iorio MV, Galmozzi E, Ménard S. Role of HER receptors family in development and differentiation. J Cell Physiol 2004;200:343–50.
[8] Tebbutt N, Pedersen MW, Johns TG. Targeting the ERBB family in cancer: couples therapy. Nat Rev Cancer 2013;13:663–73.
[9] Zhang Q, Park E, Kani K, Landgraf R. Functional isolation of activated and unilaterally phosphorylated heterodimers of ERBB2 and ERBB3 as scaffolds in ligand-dependent signaling. Proc Natl Acad Sci U S A 2012;109:13237–42.
[10] Dua R, Zhang J, Nhonthachit P, Penuel E, Petropoulos C, Parry G. EGFR over-expression and activation in high HER2, ER negative breast cancer cell line induces trastuzumab resistance. Breast Cancer Res Treat 2010;122:685–97.
[11] Cheng H, et al. EGFR expression is associated with decreased benefit from trastuzumab in the NCCTG N9831 (Alliance) trial. Br J Cancer 2014;111:1065–71.
[12] Wang D, et al. HER3 targeting sensitizes HNSCC to cetuximab by reducing HER3 activity and HER2/HER3 dimerization: evidence from cell line and patient-derived xenograft models. Clin Cancer Res 2017;23:677–86.
[13] Yonesaka K, et al. Activation of ERBB2 signaling causes resistance to the EGFR-directed therapeutic antibody cetuximab. Sci Transl Med 2011;3, 99ra86.
[14] Hanahan D, Weinberg RA. Hallmarks of cancer: the next generation. Cell 2011;144:646–74.
[15] Ooi A, et al. Protein overexpression and gene amplification of HER-2 and EGFR in colorectal cancers: an immunohistochemical and fluorescent in situ hybridization study. Mod Pathol 2004;17:895–904.
[16] Al-Kuraya K, et al. HER2, TOP2A, CCND1, EGFR and C-MYC oncogene amplification in colorectal cancer. J Clin Pathol 2007;60:768–72.
[17] Marx AH, et al. Heterogenous high-level HER-2 amplification in a small subset of colorectal cancers. Hum Pathol 2010;41:1577–85.
[18] Parseghian CM, Napolitano S, Loree JM, Kopetz S. Mechanisms of innate and acquired resistance to anti-EGFR therapy: a review of current knowledge with a focus on rechallenge therapies. Clin Cancer Res 2019;25:6899–908.
[19] Troiani T, et al. Increased TGF-α as a mechanism of acquired resistance to the anti-EGFR inhibitor cetuximab through EGFR-MET interaction and activation of MET signaling in colon cancer cells. Clin Cancer Res 2013;19:6751–65.
[20] Baselga J, Swain SM. Novel anticancer targets: revisiting ERBB2 and discovering ERBB3. Nat Rev Cancer 2009;9:463–75.
[21] Pedersen K, et al. A naturally occurring HER2 carboxy-terminal fragment promotes mammary tumor growth and metastasis. Mol Cell Biol 2009;29:3319–31.
[22] García-Castillo J, et al. HER2 carboxyl-terminal fragments regulate cell migration and cortactin phosphorylation. J Biol Chem 2009;284:25302–13.
[23] Quesnelle KM, Grandis JR. Dual kinase inhibition of EGFR and HER2 overcomes resistance to cetuximab in a novel in vivo model of acquired cetuximab resistance. Clin Cancer Res 2011;17:5935–44.
[24] Deeken JF, et al. A phase 1 study of cetuximab and lapatinib in patients with advanced solid tumor malignancies. Cancer 2015;121:1645–53.
[25] Rubinson DA, et al. Multi-drug inhibition of the HER pathway in metastatic colorectal cancer: results of a phase I study of pertuzumab plus cetuximab in cetuximab-refractory patients. Invest New Drugs 2014;32:113–22.

CHAPTER

11

CT16: A dual antibody targeting both EGFR and Notch suppresses EGFR blockage and radiation resistance by decreasing tumor-initiating cell frequency

Kewen Qian, Changhai Lei, and Shi Hu

Department of Biophysics, College of Basic Medical Sciences, Naval Medical University (Second Military Medical University), Shanghai, China

Abstract

EGFR-targeted treatment and radiation have certain curative effects on cancer, but the accompanying resistance is an intractable problem. Existing studies have shown that abnormal Notch signaling and the expansion of cancer stem cells may be closely related to this problem. Therefore, in this chapter, researchers conducted a dual-targeting antibody, CT16, targeting both EGFR and Notch2/3 receptors, in addition to blocking the activation of these two receptors while inhibiting the expansion of stem-like tumor cells. The results of in vitro and in vivo experiments showed that CT16 effectively inhibited the resistance of nonsmall cell lung cancer (NSCLC) cell lines and patient-derived xenograft (PDX) tumor models to EGFR inhibitors and radiation and enhanced the sensitivity to radiation. The enhancement effect is better than that of EGFR inhibitors or tarextumab combined with radiation. Significantly, CT16 was able to effectively reduce the proliferation of the stem-like cell subpopulation. This study aimed to characterize a dual antibody targeting EGFR and Notch2/3 signaling and recommend its potential clinical value in combination with radiotherapy for the treatment of malignant tumors.

Abbreviations

CSC cancer stem cell
EGFR epidermal growth factor receptor
EMT epithelial to mesenchymal transition
FACS fluorescence-activated cell sorting

https://doi.org/10.1016/B978-0-12-821584-5.00004-3

NSCLC nonsmall cell lung cancer
OR overall survival
PDX patient-derived xenograft
PFS progression-free survival
SPR surface plasmon resonance
TIC tumor-initiating cell

Conflict of interest

No potential conflicts of interest were disclosed.

Introduction

Cancer stem cells (CSCs) have been proven to have stem cell-like characteristics, with strong self-renewal and differentiation ability due to their initial state of differentiation [1]. However, CSCs are relatively resistant to chemotherapy and radiotherapy; they are induced to release cytokines and initiate DNA repair, promoting self-renewal [2,3].

Blockade of EGFR enhances the sensitivity of tumor cells to radiation, as activation of EGFR signaling after radiation limits apoptosis and promotes repair of double-stranded DNA breaks, thereby promoting growth of the tumor cells [4]. However, clinical trials have shown that the combination of EGFR blockers with chemotherapy and radiation therapy does not reveal an ideal curative effect and can even further induce resistance in patients to these EGFR inhibitors [5–7]. Interestingly, these resistant cells had stronger "stem cell" properties and resistance to other antitumor treatments than their parental cells, possibly because of the expansion of CSCs [8].

The crosstalk of Notch and EGFR signaling can assist cancer cells in developing antidrug resistance and promote epithelial to mesenchymal transition (EMT) and other bioactivities that are beneficial to tumor cell survival and accelerate disease progression [9]. Notably, EGFR can directly regulate Notch signaling. In human lung cancer, EGFR and Notch signaling can physically react through kinase reactions, and EGFR inhibition can promote the expression of Notch signaling [10]. These results collectively suggest a therapeutic strategy for NSCLC: dual targeting of EGFR and Notch receptors and regulation of CSCs.

Combined use of EGFR inhibitors and irradiation can promote EMT and stem cell-like properties in NSCLC cells

Researchers first clustered 30 NSCLC cell lines into four subpopulations: $CD133^+ ALDH^+$, $CD133^+ ALDH^-$, $CD133^- ALDH^+$, and $CD133^- ALDH^-$. By evaluating the tumorigenicity of the four subsets of several cancer cell lines, researchers found that only the $CD133^- ALDH^-$ subpopulation is nontumorigenic. Based on the IC_{50} values for erlotinib and cetuximab, researchers selected several EGFR inhibitor-responsive cell lines and found that the number of $ALDH^+$ cells was relatively increased. By fluorescence-activated cell sorting (FACS), researchers found that EGFR inhibitors can only inhibit the proliferation of the $CD133^- ALDH^-$ subpopulation but have no effect on other cell populations with stem cell markers and promote the heterogeneity of several other cell populations. For instance, the $CD133^+ ALDH^-$ subpopulation showed decreased EGFR expression and phosphorylation levels. The above results suggest that the growth rate of the $ALDH^+$ cell population is faster and may originate from the transition of bulk tumor cells to stem-like tumor cells.

These EGFR inhibitor-responsive cell lines were then treated with radiation. There was no significant change in the number of the other two single-positive cell populations, except for the $CD133^+ALDH^+$ subpopulation, which was increased. According to the following experiments, CSCs are indeed the most tolerant to radiation. Furthermore, the $CD133^-ALDH^+$ subpopulation was the most sensitive to irradiation.

Due to the association between EMT and CSCs, researchers tested the expression of epithelial- and mesenchymal-related genes in each subpopulation. Mesenchymal and EMT markers were upregulated in the $CD133^+$ cell population, and epithelial-related genes were upregulated in the $ALDH^+$ cell population, while bulk tumor cells showed upregulation of mesenchymal markers and suppressed expression of epithelial markers.

Enrichment of the $ALDH^+$ cell population depends on Notch2/3 receptors

Researchers next tested the role of Notch signaling in stem cell-like expansion induced by EGFR inhibitors or XRT. Notch receptors and their downstream signaling were enhanced upon EGFR or XRT treatment, and blockade of Notch2 and Notch3 signaling using siRNA effectively suppressed the number of $ALDH^+$ cell populations.

Dual blockade of EGFR and Notch2/3 inhibits resistance to EGFR inhibitors in vitro

Researchers used the EGFR-targeted antibody cetuximab and the Notch2/3-targeted antibody tarextumab to construct a dual-targeting antibody CT16 that targets both EGFR and Notch2/3. First, a heterodimeric heavy chain was generated between cetuximab and tarextumab by the "knobs into hole" method, and then an anti-EGFR/Notch CrossMab was constructed by using the CrossMab method to avoid light chain mismatch. The affinity of CT16 for EGFR and Notch2/3 was found to be similar to that of the parental antibodies as determined by surface plasmon resonance (SPR).

Researchers first performed preliminary characterization of the biological activity of CT16. The results of an enzyme-linked immunosorbent assay showed that CT16 could bind to EGFR and human Notch receptors at the same time, inhibit the binding of EGF to EGFR and TGF-α-induced EGFR phosphorylation and proliferation of EGFR-NR6 cells, and block the binding of the ligands DLL4 and JAG1 to Notch2 and Notch3. Interestingly, the competitive binding assay showed that the binding site of CT16 was the same as that of cetuximab and tarextumab. In the A431 tumor xenograft mouse model, CT16 exerted a similar antitumor effect as cetuximab. The pharmacokinetic parameters of CT16 were similar to those of cetuximab and tarextumab. In terms of safety testing, a 13-week repeated dose toxicity experiment was conducted, with no toxic effects and unreached MTD.

Next, researchers tested the antitumor biological activity. Both CT16 and the combination of cetuximab and tarextumab effectively inhibited the proliferation of HCC827 and H1648 cells, and the CT16 treatment simultaneously mediated the inhibition of $CD133^+ALDH^-$ and $ALDH^+$ cell populations.

Furthermore, the antistem cell effect of CT16 was tested. Here, researchers used a cross-linker to construct a simple "double antibody" called cross-linked cetuximab/tarextumab. After other cell populations were induced using $CD133^{+}ALDH^{-}$-derived cultures, the use of CT16 or cross-linked cetuximab/tarextumab both effectively reduced the $CD133^{+}ALDH^{-}$ cell population and increased the $CD133^{-}ALDH^{+}$ cell population. As researchers thought, due to the bidirectional binding of crossMab to both targets, although the surface EGFR of the $CD133^{+}ALDH^{-}$ cell population is downregulated, the anti-EGFR arm of CT16 can bind to EGFR on the surface of other cell populations and then recruit these EGFR high-level cells, thus acting synergistically to kill $CD133^{+}ALDH^{-}$ cells. Experiments found that cross-linked cetuximab/tarextumab did not have a good antistem cell effect and could not even exert their respective neutralizing effects. However, as CT16 neutralized EGFR and Notch signals at the same time, it played a significant role in the inhibition of $CD133^{+}ALDH^{-}$ cells.

Dual blockade of EGFR and Notch2/3 inhibits resistance to EGFR inhibitors in vivo

Next, the antitumor activity of CT16, an EGFR inhibitor, and tarextumab was compared in animal models. Two cancer cell lines, HCC827 and H1648, were used to construct PDX models. It was found that when cetuximab or erlotinib was used alone for HCC827 PDX mice, they could achieve a certain tumor inhibition within 30 and 50 days, respectively, but they were all accompanied by tumor recurrence. CT16 delayed this recurrence by 35–75 days, significantly improving the therapeutic efficacy.

However, when CT16 was administered to several cetuximab- or erlotinib-resistant cancer cell lines, it was found to be less effective in inhibiting EMT, suggesting that the killing effect of CT16 may be related to cell responsiveness to EGFR inhibitors. A mouse tumor model with relapse after secondary transplantation was constructed, and it was found that cancer cells that were resistant to one EGFR inhibitor were also less sensitive to other EGFR inhibitors, and CT16 did not solve such resistance very well.

CT16 enhances the radiosensitivity of cancer cells and reduces the $ALDH^{+}$ cell populations

Since EGFR inhibitor-resistant cells simultaneously exhibited resistance to radiation and $ALDH^{+}$ CSCs were associated with radioresistance, researchers next tested the effect of CT16 on the radioresponse. It was found that CT16 can block XRT-induced EGFR/Notch activation and EMT. Pretreatment of cells with CT16 before XRT effectively reduced cell viability compared with pretreatment with cetuximab.

The effects on DNA damage and repair were observed under XRT treatment. After EGFR inhibitor treatment, the DNA repair marker phosphorylated histone variant γH2AX after DNA double-stranded breaks were significantly increased, and CT16 even caused more severe DNA damage. When XRT was combined with an EGFR inhibitor, $ALDH^{+}$ cell markers were increased, and tarextumab effectively suppressed this change. The DNA repair markers of $CD133^{+}$ cells increased after XRT, indicating that $CD133^{+}$ cells are an important cell subpopulation to initiate DNA repair and restore cancer cell growth after DNA damage. Compared to the combination of cetuximab and tarextumab, CT16 not only inhibited stem cell-like expansion but also reduced the $CD133^{+}ALDH^{-}$ cell population.

CT16 can reduce the proliferation of tumor cells and shows good antitumor activity

Next, researchers transplanted patient-derived xenografts into NOD-SCID mice to further evaluate the antitumor activity of CT16. When administered prophylactically 2 days before modeling, tarextumab was found to be less effective against tumors, but when combined with cetuximab, it was able to achieve better efficacy than either cetuximab or erlotinib alone. It was also found that each PDX model had different inhibitory effects, indicating that Notch signaling was expressed differently in different cancer cell lines. For each EGFR blockade-responsive PDX model, CT16 can have a good growth inhibitory effect, while for the EGFR blockade-nonresponsive PDX models, there were still high expression levels of the mesenchymal markers VIM and CDH2, which did not exert antitumor effects. Tarextumab-responsive PDX models showed higher expression of Notch3, indicating that both EMT markers and Notch3 can be used as markers for the potential therapeutic efficacy of CT16 in NSCLC.

Histopathological analysis of N3 among a series of PDX models found that CT16 treatment inhibited cell proliferation better than the EGFR inhibitors cetuximab, erlotinib, or tarextumab alone. Compared with tarextumab, the tumor cell density decreased by 22%, and cell proliferation decreased by 30%. Immunohistochemical results showed that the EGFR inhibitor caused the upregulation of VIM and downregulation of CDH1, while CT16 reduced the expression level of VIM.

Dual blockade of EGFR and Notch2/3 reduces the number of CSCs and delays tumor recurrence after radiotherapy

Next, researchers constructed a cancer recurrence model by secondary transplantation of human-derived cancer cells without further treatment. It was found that tarextumab reduced the number of tumor-initiating cells (TICs) by 1.4- to 2.4-fold, while no significant change in the cetuximab or erlotinib group was observed, with tumorigenic enhancement. However, CT16 reduced tumorigenicity by 2.6- to 10.6-fold.

Next, to test the effect of CT16 on possible recurrence after radiotherapy, after 4 weeks of high-dose irradiation of $10\,Gy \times 3$ to the tumor model, it was found that the number of CSCs in the surviving tumor tissue was approximately twice as high as it was before radiation treatment. Compared to XRT treatment alone, the combined use of EGFR inhibitors resulted in a doubling of CSC marker expression. Compared with the XRT control group, CT16 reduced the number and tumorigenicity of CSCs by 5–10-fold. The above results comprehensively show that the dual blockade of EGFR and Notch by CT16 can effectively inhibit the growth of tumor cells and reduce the population of stem-like tumor cells.

Discussion

In a phase III clinical study, gefitinib and erlotinib were included in concurrent chemoradiotherapy or maintenance therapy after chemoradiotherapy, but they only showed progression-free survival (PFS) and overall survival (OR) prolongation in patients with EGFR overexpression or mutation, while drugs targeting EGFR inevitably face the intractable problem of drug resistance [11,12]. Existing studies suggest that CSCs may be related to resistance to EGFR blockade as well as radiotherapy. Correspondingly, our experiments also show that either EGFR inhibition or irradiation leads to the expansion of CSCs in several PDX models. Researchers also found that erlotinib induced less conversion of EGFR-mutant cell populations to $ALDH^{+}$ cells than wild-type cells, which corrects the extent that erlotinib is only effective in EGFR-overexpressing patients.

As CT16 has high affinity and specificity for EGFR and Notch receptors, this antibody not only shows good antitumor properties but also effectively inhibits EGFR blockade and irradiation-induced CSC expansion, which also corresponds to the key roles of the two signaling pathways in tumor cells and CSCs, respectively [13,14]. However, although CT16 was able to delay resistance to EGFR blockade, it did not exert curative effects on cell lines that had already developed cetuximab or erlotinib resistance, which may be related to the expansion of CSCs and the complex suppression and plasticity of tumor cells.

It is worth noting that researchers found that the differences in the expression of EMT markers were reflected not only in differences in cell lines but also in different cell subpopulations, and combined treatment with EGFR blockade and irradiation resulted in subpopulations of cells with high expression of mesenchymal markers. The upregulation of DNA repair-related proteins is particularly evident in the $CD133^{+}$ cell population. Combined with tarextumab, it can inhibit the initiation of EMT and DNA repair to a certain extent, while CT16 can almost completely prevent this change.

Here are some existing problems of our research. First, there were limitations of the in vivo experiments. Our animal models may not be a good representation of the human NSCLC status, and the number of animals that provide final data is relatively limited. Second, the mechanism of the action of antibodies on CSCs remains unclear. A possible mechanism is that EGFR inhibitors or irradiation can induce the release of intracellular antigens or ligands from stressed or apoptotic cells, which in turn activates Notch signaling and is critical for the proliferation and self-renewal of neighboring cells or CSCs. This may be the reason why only CT16 but not the combination of cetuximab and tarextumab effectively inhibited the $CD133^{+}ALDH^{-}$ cell population.

Overall, our study suggests the potential therapeutic value of targeting CSCs in combination with molecularly targeted therapy and radiotherapy for malignancies. Researchers also suggest the potential value of inhibiting Notch signaling for overcoming resistance to EGFR inhibitors and radiation. Therefore, CT16 is promising as an adjuvant therapy for targeted therapy and radiotherapy.

References

[1] Visvader JE, Lindeman GJ. Cancer stem cells: current status and evolving complexities. Cell Stem Cell 2012;10:717–28.

[2] Maugeri-Saccà M, Bartucci M, De Maria R. DNA damage repair pathways in cancer stem cells. Mol Cancer Ther 2012;11:1627–36.
[3] Li X, et al. Intrinsic resistance of tumorigenic breast cancer cells to chemotherapy. J Natl Cancer Inst 2008;100:672–9.
[4] Nyati MK, Morgan MA, Feng FY, Lawrence TS. Integration of EGFR inhibitors with radiochemotherapy. Nat Rev Cancer 2006;6:876–85.
[5] Montagut C, et al. Identification of a mutation in the extracellular domain of the epidermal growth factor receptor conferring cetuximab resistance in colorectal cancer. Nat Med 2012;18:221–3.
[6] Pao W, et al. Acquired resistance of lung adenocarcinomas to gefitinib or erlotinib is associated with a second mutation in the EGFR kinase domain. PLoS Med 2005;2, e73.
[7] Ang KK, et al. Randomized phase III trial of concurrent accelerated radiation plus cisplatin with or without cetuximab for stage III to IV head and neck carcinoma: RTOG 0522. J Clin Oncol Off J Am Soc Clin Oncol 2014;32:2940–50.
[8] Shien K, et al. Acquired resistance to EGFR inhibitors is associated with a manifestation of stem cell-like properties in cancer cells. Cancer Res 2013;73:3051–61.
[9] Baker AT, Zlobin A, Osipo C. Notch-EGFR/HER2 bidirectional crosstalk in breast cancer. Front Oncol 2014;4:360.
[10] Arasada RR, Amann JM, Rahman MA, Huppert SS, Carbone DP. EGFR blockade enriches for lung cancer stem-like cells through Notch3-dependent signaling. Cancer Res 2014;74:5572–84.
[11] Herbst RS, et al. TRIBUTE: a phase III trial of erlotinib hydrochloride (OSI-774) combined with carboplatin and paclitaxel chemotherapy in advanced non-small-cell lung cancer. J Clin Oncol Off J Am Soc Clin Oncol 2005;23:5892–9.
[12] Ready N, et al. Chemoradiotherapy and gefitinib in stage III non-small cell lung cancer with epidermal growth factor receptor and KRAS mutation analysis: cancer and leukemia group B (CALEB) 30106, a CALGB-stratified phase II trial. J Thorac Oncol 2010;5:1382–90.
[13] Arteaga CL, Engelman JA. ERBB receptors: from oncogene discovery to basic science to mechanism-based cancer therapeutics. Cancer Cell 2014;25:282–303.
[14] Takebe N, Nguyen D, Yang SX. Targeting notch signaling pathway in cancer: clinical development advances and challenges. Pharmacol Ther 2014;141:140–9.

CHAPTER

12

BET inhibition overcomes receptor tyrosine kinase-mediated cetuximab resistance in HNSCC

Zetong Ma[a], *Tian Li*[a], *Wenyan Fu*[b], *and Shi Hu*[a]

[a]Department of Biophysics, College of Basic Medical Sciences, Naval Medical University (Second Military Medical University), Shanghai, China [b]Department of Assisted Reproduction, Shanghai Ninth People's Hospital, Shanghai Jiao Tong University School of Medicine, Shanghai, China

Abstract

Surrogate receptor tyrosine kinase (RTK) activation, such as HER3, MET, and AXL, is one of the major causes of cetuximab resistance in head and neck squamous cell carcinoma (HNSCC). Researchers have identified a protein that coregulates the expression of multiple RTKs and its potential to overcome cetuximab resistance as a molecular target. Studies have shown that bromodomain-containing protein-4 (BRD4) can regulate the expression of RTKs in breast cancer models under the background of PI3K and HER2 inhibition of drug resistance. The researchers hypothesized that BDR4 inhibition could overcome cetuximab resistance by inhibiting the expression of complementary RTK. RTK and BRD4 expression in HNSCC cell lines resistant to cetuximab showed that several RTKs, such as MET and AXL, were upregulated in the cetuximab-resistant clones, and the percentage of cells expressing BRD4 was increased. However, targeted inhibition of BRD4 resulted in the loss of cell viability, which was associated with decreased expression of alternative RTK. The combination of cetuximab and the bromodomain inhibitor JQ1 treatment downregulates RTK and overcomes cetuximab resistance in the HNSCC PDX model. These results suggest that cetuximab combined with bromine domain inhibition may provide a novel therapeutic strategy for HNSCC treatment.

Abbreviations

BET	bromodomain and extra terminal
BRD4	bromodomain-containing protein-4
CTL	control
CTX	cetuximab
CTXR	cetuximab-resistant
EGFR	epidermal growth factor receptor
EMT	epithelial to mesenchymal transition
HNSCC	head and neck squamous cell carcinoma

https://doi.org/10.1016/B978-0-12-821584-5.00011-0

HPV	human papilloma virus
NSCLC	nonsmall cell lung cancer
PDX	patient-derived xenograft
PFS	progression-free survival
PIN1	prolyl isomerase 1
RTK	receptor tyrosine kinases
SFKs	Src family kinases

Conflict of interest

No potential conflicts of interest were disclosed.

Introduction

Secondary resistance to cetuximab in head and neck squamous cell carcinoma (HNSCC) is often caused by surrogate receptor tyrosine kinase (RTK) upregulation and activation [1–3]. In addition, in lung or colorectal cancer, RTK is also one of the reasons for poor response to anti-EGFR in patients [4,5]. The activation of RTK strongly stimulates the dedifferentiation and rapid proliferation of tumor cells [5], and the combination of multiple RTK can improve patients' resistance to cetuximab [6]. In this chapter [7], the researchers explored the coregulation mechanism of RTK expression and sought a broad-spectrum drug target. Previously, it was believed that the expression of RTK depends on bromo-containing domain protein 4 (BRD4) [8]. At the same time, another group of researchers found that RTK upregulation also leads to breast cancer resistance to PI3K inhibition. BRD4 inhibition can reduce RTK expression and improve the efficacy of PI3K blockade [9]. According to the above, this study proposed the hypothesis that blocking BRD4 could help to overcome cetuximab resistance in HNSCC by inhibiting substituted RTKs.

BRD4 contains two bromine domain regions and is a member of the bromodomain and extra terminal (BET) protein family, which binds to acetylated lysine or some proteins on chromatin through their bromine domain and promotes the recruitment and activation of CDK9 (part of the p-TeFb complex) [10]. Then, gene transcription can be initiated due to RNA polymerase phosphorylation [11,12]. BRD4 has been regarded as an attractive therapeutic target in recent years because of its ability to influence chromatin status. Meanwhile, some bromodomain inhibitors have shown good antitumor efficacy in preclinical models [13–15]. Previously, researchers demonstrated the presence of BRD4-driven RTK activation in a cetuximab-resistant HNSCC preclinical model. In addition, RTK expression, such as AXL and ROR2 expression, was increased in HNSCC patients treated with cetuximab. Inhibition of BRD4 blocks the upregulation of RTK, overcomes cetuximab resistance in xenotransplantation (PDX) models, and delays the regeneration of patient-derived HNSCC cells treated with cetuximab. Therefore, it is necessary to evaluate the clinical efficacy of bromodomain inhibitors in HNSCC patients.

The HNSCC model of acquired cetuximab resistance relies on surrogate RTK activity

Overactivation of multiple RTKs in HNSCC is associated with cetuximab resistance [1–4,16]. To uncover the shared mechanism of RTK upregulation in cetuximab-resistant (CTXR) HNSCC models, the researchers established three new acquired HNSCC cell lines resistant to cetuximab (CTXR), PECAPJ49, FADU, and CAL33. A phospho-RTK array was used to determine whether changes in RTK expression were responsible for cetuximab resistance. The results of different RTK activations were observed in CTXR cell lines, suggesting that the HNSCC cell line model was heterogeneous. The PECAPJ49 CTXR clones showed higher levels of phosphorylated Axl and Ryk, while the FADU CTXR clones showed higher levels of Axl, Met, and Ryk. The CAL33 CTXR clone showed overactivation of HER3, MET, and ROR2. These findings are consistent with previous reports, suggesting that secondary resistance to cetuximab relates to the upregulation of multiple RTKs [1–3,16].

BDR4 has been found to play an irreplaceable role in drug resistance in other malignancies [8,9]. Based on previous reports, this study evaluated BRD4 expression levels and function in CTXR HNSCC models to verify the hypothesis that the activation of BRD4 drives cetuximab-induced RTK upregulation. There are three subtypes of BRD4, one long subtype and two short subtypes [17]. Using an antibody capable of detecting all three subtypes of BRD4, compared with the parent counterparts, increased expression of one or both short subtypes of BRD4 was observed in PECAPJ49 and FADU CTXR clones. The specificity of BRD4 antibodies was tested by treating parents and CTXR clones with MZ1, a BRD4 inhibitor [18]. There were differences in BRD4 expression levels between PECAPJ49 and FADU CTXR clones of parental cells.

Next, to determine how BRD4 mediates RTK expression, BDR4 was silenced with siRNA genes. When treated with cetuximab, PECAPJ49 parents and CTXR clones were silenced by siRNA oligonucleotides or transfected with interference controls. After 96 h of transfection with BRD4 siRNA, the survival status of cells was assessed to determine whether BRD4 deletion could resensitize the cetuximab resistance model. Both parents and CTXR clones showed increased sensitivity to BRD4 deletion, and BRD4 siRNA significantly resensitized the cetuximab-resistant clones to cetuximab.

It was also observed that BRD4 siRNA downregulated the expression of AXL in CTXR # 4 to a greater extent. BRD4 is considered to be a strong activator of MYC expression. Thus, as a control, western blotting of p21 (CDKN1A), which MYC can inhibit, revealed [13] that p21 was upregulated when BRD4 was knocked down. The study then assessed the ability of BRD4 to bind to the RTK promoter. Since AXL expression is associated with poor prognosis in HNSCC patients and is upregulated in both PECAPJ49 and FADU CTXR clones, researchers chose to detect the ability of BRD4 to bind to AXL promoters. Subcellular grading and pull-down experiments were performed using biotinylated oligonucleotides encoding the upstream region of the AXL transcription start site. BRD4 and p-CDK9 (the p-TEFB component) in PECAPJ49 CTXR # 4 showed an increased association with the AXL promoter compared with the parental lines. Two BRD4 bands (~120 and ~100 kDa) of CTXR # 4 bind to their parental lines. The ~100 kDa band observed by this BRD4 antibody is reported to be the shortest subtype of BRD4 [19]. The shortest subtype is believed to mediate epithelial to mesenchymal cell transformation (EMT) and is associated with drug resistance [20]. In summary, these results suggest that secondary cetuximab resistance is accompanied by high BRD4 expression and recruitment to the substitute RTK promoter region and that BRD4 targeting can resensitize cetuximab-resistant cells.

Pharmacological inhibition of BET family members overcame cetuximab resistance in vitro

To investigate the sensitivity of the HNSCC model of cetuximab to BRD4 activity after pharmacological inhibition, PECAPJ49 and FADU models were treated with the BET inhibitor JQ1. Compared with BRD2 and BRD3, JQ1 preferentially binds to the bromodomain of BRD4 to block transcription [13]. Similar to the silencing of BRD4-expressed genes, the pharmacological inhibition of BET by JQ1 significantly resensitized CTXR clones to cetuximab. Another BET inhibitor, I-BET-762 [21], was also validated in eliminating the cetuximab resistance phenotype. I-BET-762 also restored the sensitivity of CTXR clones of the HNSCC model to cetuximab therapy. In addition, the combination of cetuximab and JQ1 showed similar treatment efficacy in the NSCLC model with secondary cetuximab resistance, suggesting that the BET inhibitor combined with cetuximab may be effective in several malignancies.

Consistent with BRD4 gene silencing, cetuximab and JQ1 treatment downregulated AXL expression in PECAPJ49 CTXR clones and lung cancer clones HC4 and HC8. To test the durability of treatment with cetuximab and JQ1, the study used a long-term growth trial in vitro. The cetuximab-sensitive HNSCC cell line was cultured after 3 weeks of cetuximab monotherapy or cetuximab combined with JQ1. Both treatments significantly reduced cell growth. However, after drug removal, cells treated with cetuximab showed rapid growth, whereas cells previously treated with cetuximab and the JQ1 combination showed only marginal growth.

In conclusion, bromodomain blockade can avoid cetuximab resistance and inhibit the expression of alternative RTKs, which are known to drive the cancer phenotype of cetuximab resistance.

Blocking BET can inhibit the expression of RTK in the HNSCC model of cetuximab

To validate the efficacy of the combined treatment strategy of cetuximab and JQ1 in the primary cetuximab resistance model, they first determined whether BRD4 participates in cetuximab resistance. Except for HSC6 and SCC47, the results showed no significant correlation between cetuximab resistance and BRD4 basal expression in all cell lines tested.

Next, they had primary cetuximab-resistant cell lines that accepted cetuximab monotherapy or in combination with JQ1. The results showed that all primary cetuximab-resistant cell lines were significantly sensitive to cetuximab after JQ1 combination therapy. To determine whether JQ1 sensitized drug-resistant cell lines to cetuximab by downregulating RTK, this study applied cetuximab monotherapy or combined treatment with JQ1 to the intrinsically cetuximab-resistant cell line CAL27 and performed a phosphorylated RTK array. Density analysis revealed a modest increase in RTK activation with cetuximab, while combination therapy significantly reduced the alternative activation of several RTKs, such as HER2, HER3, DTK, and MET. Western blotting confirmed that the decrease in phosphorylated RTK in CAL27 cells results from the decrease in total RTK expression and its downstream effectors phosphorylated Src family kinases (SFKs). In SCC47 and HSC6 cells, the expression of HER3 and AXL was also reduced after combination therapy. Compared to cetuximab monotherapy, BRD4 knockdown in CAL27 cells resulted in significantly improved insensitivity to cetuximab. Western blotting showed that BRD4 knockdown in CAL27 cells also led to downregulation of HER3 phosphorylation without cetuximab. These results suggest that BRD4 blockade can make cells resistant to cetuximab, resensitize to cetuximab, and downregulate RTK expression in cetuximab-resistant HNSCC models.

RTK overexpression mediates resistance to cetuximab and BET inhibition

Evaluation of the mechanism by which BET inhibition overcomes cetuximab resistance. The control vector or HER3-overexpressing vector was stably introduced into the cetuximab-sensitive cell line PECAPJ49 and treated with cetuximab monotherapy or combined with JQ1 for 72h. HER3 overexpression reduced the sensitivity of all cetuximab in this cell line, while the control group was sensitive to both therapies. Although the HER3-overexpressing cell lines were still sensitive to the combination of cetuximab and JQ1, the sensitivity of these cells was significantly lower than that of the control cell lines. In summary, cetuximab resistance can be mediated by RTK overexpression, while BET block can partially overcome cetuximab resistance by reducing RTK overexpression.

BET blocking prevents the acquisition of cetuximab resistance in the PDX model of HNSCC patients

Two patient-derived xenograft models (PDX-6282 and PDX-7157) were transplanted into mice, allowing the tumors to grow to approximately 100mm^3. The mice were then treated with cetuximab, JQ1, or a cetuximab + JQ1 combination or vehicle. These PDX models were initially cetuximab-sensitive but eventually developed a resistant phenotype. The low base level of BRD4 probably causes insensitivity to JQ1 therapy in PDX models. Notably, the cetuximab and JQ1 combination inhibited tumor growth in both PDX models. Phospho-RTK array analysis was performed on lysates obtained from two PDXs in the accepted treatment of all four groups. The results showed that PDX-6282 showed increased activation of HER2, HER3, ROR2, and TIE2 after treatment with cetuximab. PDX-7157 tumors treated with cetuximab showed upregulated activation of RyK, InSR, DDR1, and ROR2. In both models, cetuximab and JQ1 eliminated all of these RTK activation levels. In addition, the combination significantly reduced the expression of downstream signaling components of RTK, including SFKs, ERK1/2, and MYC. These results suggest that JQ1 prevents acquired resistance to cetuximab by eliminating RTK expression in preclinical models of HNSCC.

Expression of RTK and BRD4 genes in a clinical cohort of HNSCC

In a cohort of 40 HNSCC patients treated with platinum and cetuximab (NCBI GEO GSE65021) [22], in this cohort, 14 patients showed long-term progression-free survival (PFS > 12months), and 26 patients had progression-free survival (PFS < 5.6months). A sample of moderately/poorly differentiated patients from the oral and laryngeal regions ($n=25$) showed that AXL and ROR2 expression was significantly higher in patients with short PFS than in patients with long PFS. Five of eight moderately to poorly differentiated tumors from oropharynx sites were HPV positive and were excluded from the analysis. BRD4 was significantly higher in HNSCC tumor tissue than in normal tissue, but there was no difference in BRD4 gene expression between samples from short- and long-PFS patients. These results suggest that RTK upregulation is a risk factor for poor prognosis and recurrence in HNSCC patients. BRD4 is a potential therapeutic target for improving outcomes in HNSCC patients.

Conclusion

The abnormal expression and activation of substitute RTKs mediate resistance to cetuximab, but the targeting of multiple RTKs at the same time is not feasible due to the activation of alternative RTKs and the expected toxic side effects [1–4]. This study evaluated a single molecular target, the bromine domain, and the extra terminal protein BRD4, to enhance the antitumor effects of cetuximab by targeting this site to reduce the activation and expression of surrogate RTKs. Increased RTK expression and BRD4 expression were observed in the acquired cetuximab-resistant cell line model. Both genetic and pharmacological targeting of BRD4 significantly reduced RTK expression. Blocking BRD4 with the bromine domain inhibitor JQ1 delayed acquired resistance to cetuximab in the HNSCC PDX model. Both PDXs showed heterogeneous upregulation of RTK activation, while JQ1 eliminated the upregulation of RTKs. RTK expression was increased in patients with HNSCC, and the short PFS BRD4 was upregulated in HNSCC patients treated with cetuximab. These results indicated that BRD4 inhibition is a promising therapeutic target to improve the prognosis of HNSCC patients.

References

[1] Madoz-Gúrpide J, et al. Activation of MET pathway predicts poor outcome to cetuximab in patients with recurrent or metastatic head and neck cancer. J Transl Med 2015;13:282.
[2] Jiang N, et al. Combination of anti-HER3 antibody MM-121/SAR256212 and cetuximab inhibits tumor growth in preclinical models of head and neck squamous cell carcinoma. Mol Cancer Ther 2014;13:1826–36.
[3] Brand TM, et al. AXL mediates resistance to cetuximab therapy. Cancer Res 2014;74:5152–64.
[4] Wheeler DL, et al. Mechanisms of acquired resistance to cetuximab: role of HER (ErbB) family members. Oncogene 2008;27:3944–56.
[5] Hu S, et al. Broad RTK-targeted therapy overcomes molecular heterogeneity-driven resistance to cetuximab via vectored immunoprophylaxis in colorectal cancer. Cancer Lett 2016;382:32–43.
[6] Samatar AA, Poulikakos PI. Targeting RAS-ERK signalling in cancer: promises and challenges. Nat Rev Drug Discov 2014;13:928–42.
[7] Leonard B, et al. BET inhibition overcomes receptor tyrosine kinase-mediated cetuximab resistance in HNSCC. Cancer Res 2018;78:4331–43.
[8] Stuhlmiller TJ, et al. Inhibition of Lapatinib-induced kinome reprogramming in ERBB2-positive breast cancer by targeting BET family bromodomains. Cell Rep 2015;11:390–404.
[9] Stratikopoulos EE, et al. Kinase and BET inhibitors together clamp inhibition of PI3K signaling and overcome resistance to therapy. Cancer Cell 2015;27:837–51.
[10] Zeng L, Zhou MM. Bromodomain: an acetyl-lysine binding domain. FEBS Lett 2002;513:124–8.
[11] Yang Z, et al. Recruitment of P-TEFb for stimulation of transcriptional elongation by the bromodomain protein Brd4. Mol Cell 2005;19:535–45.
[12] Marshall NF, Peng J, Xie Z, Price DH. Control of RNA polymerase II elongation potential by a novel carboxyl-terminal domain kinase. J Biol Chem 1996;271:27176–83.
[13] Filippakopoulos P, et al. Selective inhibition of BET bromodomains. Nature 2010;468:1067–73.
[14] Zuber J, et al. RNAi screen identifies Brd4 as a therapeutic target in acute myeloid leukaemia. Nature 2011;478:524–8.

[15] Shi J-B, et al. Synthesis and characterization of single-crystalline zinc tin oxide nanowires. Nanoscale Res Lett 2014;9:210.

[16] Jun HJ, et al. Acquired MET expression confers resistance to EGFR inhibition in a mouse model of glioblastoma multiforme. Oncogene 2012;31:3039–50.

[17] Floyd SR, et al. The bromodomain protein Brd4 insulates chromatin from DNA damage signalling. Nature 2013;498:246–50.

[18] Gadd MS, et al. Structural basis of PROTAC cooperative recognition for selective protein degradation. Nat Chem Biol 2017;13:514–21.

[19] Conrad RJ, et al. The short isoform of BRD4 promotes HIV-1 latency by engaging repressive SWI/SNF chromatin-remodeling complexes. Mol Cell 2017;67.

[20] Alsarraj J, et al. Deletion of the proline-rich region of the murine metastasis susceptibility gene Brd4 promotes epithelial-to-mesenchymal transition- and stem cell-like conversion. Cancer Res 2011;71:3121–31.

[21] Mirguet O, et al. Discovery of epigenetic regulator I-BET762: lead optimization to afford a clinical candidate inhibitor of the BET bromodomains. J Med Chem 2013;56:7501–15.

[22] Bossi P, et al. Functional genomics uncover the biology behind the responsiveness of head and neck squamous cell cancer patients to cetuximab. Clin Cancer Res 2016;22:3961–70.

CHAPTER

13

Dual inhibition of EGFR and c-Src by cetuximab and dasatinib combined with FOLFOX chemotherapy in patients with metastatic colorectal cancer

Kewen Qian[a], *Wenyan Fu*[b], *Changhai Lei*[a], *and Shi Hu*[a]

[a]Department of Biophysics, College of Basic Medical Sciences, Naval Medical University (Second Military Medical University), Shanghai, China [b]Department of Assisted Reproduction, Shanghai Ninth People's Hospital, Shanghai Jiao Tong University School of Medicine, Shanghai, China

Abstract

Src is an intracellular tyrosine kinase. Aberrant activation of Src regulates the adhesion, migration, and invasiveness of tumor cells through a series of downstream signaling pathways. Additionally, it has been implicated in drug resistance in metastatic colorectal cancer (mCRC). Dasatinib is a multitargeted inhibitor of many kinases, and its ability to overcome drug resistance to cetuximab has been demonstrated in some preclinical studies. In this chapter, researchers have summarized the results of phase IB/II studies evaluating the efficacy and toxicity profile of the combination of cetuximab, dasatinib, and FOLFOX, which is a classical chemotherapy for mCRC. The study indicated that the addition of dasatinib did not show efficient clinical activity in EGFR resistance, which was primarily thought to be related to an incomplete inhibition of Src. Furthermore, as a phase II study found acquired resistant KRAS mutant clones as a possible limiting to reverse EGFR resistance, the research group then elucidated the role of KRAS amplification in anti-EGFR treatment. Interestingly, the clinicopathologic features of mCRC patients with KRAS amplification or KRAS mutation are differentiated.

Abbreviations

mCRC metastatic colorectal cancer
EGFR epidermal growth factor receptor

https://doi.org/10.1016/B978-0-12-821584-5.00028-6

OS overall survival
MTD maximum tolerated dose
DLT dose limiting toxicity
TTF time to treatment failure
FAK focal adhesion kinase
PFS progression-free survival
IBD inflammatory bowel disease
NGS next-generation sequencing

Conflict of interest

No potential conflicts of interest were disclosed.

Combination synergistic chemotherapy of mCRC treatment

Colorectal cancer (CRC) is the third most common malignancy in the United States, and up to 50% of all CRC patients suffer from metastasis as the disease progresses [1,2]. Notably, metastatic CRC (mCRC) is often considered clinically incurable. The combination of FOLFOX and targeted agents, such as the monoclonal antibodies cetuximab and panitumumab, that target EGFR can improve the median overall survival (OS) of mCRC patients [1]. However, resistance to these targeted treatments greatly affects the disease progression response to the regimen. In addition, the efficacy can be affected by multiple other factors, such as some molecular biomarkers. Several phase III clinical trials have demonstrated the efficacy of the combination of chemotherapy and the EGFR-targeted monoclonal antibodies cetuximab and panitumumab [3]. However, the efficacy is restricted to patients with mutational RAS mCRC. Serebriiskii et al. conducted the largest cohort of mCRCs harboring a RAS amplification that have been published in which 13,336 next-generation sequencing (NGS) databases were reviewed. In this cohort, RAS-amplified specimens made up 2% [1].

Phase IB: Safety and efficacy

There were two objectives of the Phase IB study: one was to determine the maximum tolerated dose (MTD) and dose-limiting toxicity (DLT) of the combination of dasatinib plus FOLFOX with cetuximab and whether this regimen has biological activity on c-Src activity [2]. A total of 30 patients were enrolled in this portion, of which 18 participated in the dose escalation cohort and 12 participated in the expansion cohort. Six of the 30 Phase IB patients had a reduction in the size of their measurable tumors by more than 30%, which represents a response rate of 20% [2]. Four of these six patients had ever received prior EGFR-based therapy, and three of them were treated with cetuximab. However, they had all failed to respond before. The dose of dasatinib during treatment was regulated according to adverse reactions such as myelosuppression and fatigue. Regarding the association between RAS amplification or KRAS mutation mCRC with the response to the combination regimen, a phase IB study found that five of the above six responding patients had a wild-type KRAS oncogene; moreover, in this cohort, the median progression-free survival was 4.6 months, and 23% of the patients had no progression [2].

Phase II: The role of KRAS mutation

There were also two objectives of the phase II study. One was to determine the response-rate distribution of dasatinib and FOLFOX with cetuximab. The other was to determine the time to treatment failure (TTF) and ratio of the current TTF to the TTF of the immediate prior regimen, as well as the OS, safety profile, and tolerability of the regimen [2]. Furthermore, this period of study also primarily evaluated the effects of KRAS mutation in combination with regimen treatment and disease progression. A total of 47 patients were enrolled, of whom 23 patients were KRAS c12/13 mutant and 24 patients were KRAS c12/13 wild-type. At the first restaging interval, 3 patients with wild-type KRAS who had been refractory to cetuximab had a confirmed partial response (PR). The median ratio of TTF in KRAS mutation and wild-type patients on the prior regimen (mFOLFOX6 plus dasatinib) and this study (mFOLFOX6 plus dasatinib with cetuximab) were 1.0 and 0.8, respectively, and the median OS was 5.9 and 7.7 months, respectively.

Previous studies have demonstrated that the inhibitory activity of Src can be associated with a reduction in focal adhesion kinase (FAK) and paxillin phosphorylation at Src-related sites.

The research group first conducted an in vitro experiment, treated peripheral blood mononuclear cells with 300 nM dasatinib and detected changes in the levels of FAK and paxillin. The response of the cells was detectable and confirmed the phosphorylation sites of the two proteins. However, western blot analysis of the phosphorylation of FAK and paxillin was unexpected. According to the experimental results, the correlation between the dose of dasatinib, phosphorylation level, and therapy progression was barely discernible.

RAS amplification and KRAS mutation

One phase III clinical trial (funded by Amgen and others; PRIME ClinicalTrials.gov number, NCT00364013) that involved 108 patients with RAS mutations, including KRAS and NRAS mutations located in different exons, concluded that additional RAS mutations are associated with an undesirable response to combination regimen treatment in patients who received panitumumab-FOLFOX4. Another study was conducted on 89 mCRC patients, all of whom had been treated with irinotecan-based chemotherapy but failed. After treatment with cetuximab, KRAS mutation in tumor DNA was analyzed, and it was found that resistance to cetuximab was highly associated with KRAS mutation. To correlate RAS amplification with response to anti-EGFR treatment and compare the difference with RAS mutations, we reviewed the results of molecular testing on 1286 consecutive mCRC specimens from a large cohort of 1271 mCRC patients.

RAS amplification and mutations were both detected, and we determined the exclusion between them as well as each RAS mutant gene. The 1286 specimens were submitted for targeted NGS, with a proportion of mutations in KRAS, NRAS, BRAF, and PIK3CA. For RAS amplification, a total of 22 mCRC specimens were detected, of which KRAS, HRAS, and KRAS were detected in 15, 2, and 5 cases, respectively. The conclusion was that exclusion did exist among each RAS gene, while there was no exclusive relationship between RAS amplifications and mutations.

IBD and KRAS-amplified mCRC

In the 1990s, researchers found that IBD-associated CRC was more likely to occur in young patients [4]. As genomic studies were conducted, molecular distinctions between IBD-associated CRCs with sporadic cases were found [5,6], which indicates that the disease history of IBD increased the risk of recurrent chromosomal gains and loss. Another interesting conclusion of the study was that mCRC patients with KRAS amplification were younger than those without KRAS-amplified mCRC. Additionally, 9 of the 15 KRAS-amplified mCRC patients had a disease history of inflammatory bowel disease (IBD), while 14 of the 1256 patients without KRAS amplification had no relevant medical history. However, there were no more statistically significant associations with other features. In contrast, KRAS mutation was associated with several differences in addition to the older age group.

Discussion and future directions

Multiple prospective phase II and III clinical trials have shown that repeated administration of FOLFOX alone or in combination with oxaliplatin can efficiently improve the response rates and OS of the regimens. Existing researchers have demonstrated that Src family kinases play an important role in the development of resistance to anti-EGFR therapies. Some preclinical studies have also demonstrated that inhibition of Src can alleviate chemoresistance and resistance to targeted agents to a certain extent. Furthermore, the results of this study, as well as other preclinical and clinical evidence, indicated that RAS amplification in mCRC is a potential mechanism of resistance to anti-EGFR inhibitors. It seems meaningful to consider RAS copy number analysis when determining a candidate for anti-EGFR therapy [2].

There are several limitations to this study. First, the sample size was too small and was unable to detect tiny improvements in OS and PFS. Second, molecular analysis of the tumor tissue might fail due to DNA degradation. Third, the acquisition of tumor tissue may not be representative, as the response of the tumor tissue is not universal; if the sample was posttreatment, then sampling bias may confound interpretation and result in an incorrect assessment of genetic heterogeneity. Furthermore, the sensitivity assay of mutations should be improved.

References

[1] Favazza LA, et al. KRAS amplification in metastatic colon cancer is associated with a history of inflammatory bowel disease and may confer resistance to anti-EGFR therapy. Mod Pathol 2020;33:1832–43.
[2] Parseghian CM, et al. Dual inhibition of EGFR and c-Src by cetuximab and dasatinib combined with FOLFOX chemotherapy in patients with metastatic colorectal cancer. Clin Cancer Res 2017;23:4146–54.
[3] Heskamp S, et al. Cetuximab reduces the accumulation of radiolabeled bevacizumab in cancer xenografts without decreasing VEGF expression. Mol Pharm 2014;11:4249–57.

[4] Meyers TJ, et al. Association between inflammatory bowel disease and prostate cancer: a large-scale, prospective, population-based study. Int J Cancer 2020;147:2735–42.
[5] Baker AM, et al. Evolutionary history of human colitis-associated colorectal cancer. Gut 2019;68:985–95.
[6] Lai LA, et al. Pan-colonic field defects are detected by CGH in the colons of UC patients with dysplasia/cancer. Cancer Lett 2012;320:180–8.

CHAPTER

14

A novel bispecific antibody targeting EGFR and cMet is effective against EGFR inhibitor-resistant lung tumors

Tian Li, Changhai Lei, and Shi Hu

Department of Biophysics, College of Basic Medical Sciences, Naval Medical University (Second Military Medical University), Shanghai, China

Abstract

Nonsmall cell lung cancer (NSCLC) can develop resistance to tyrosine kinase inhibitors (TKIs) through site mutations in EGFR (T790M) and activation of the cMet pathway. This chapter describes a bispecific EGFR-cMet antibody with multiple mechanisms of action. The researchers developed this antibody in NSCLC to inhibit the effects of EGFR mutations and cMet pathway activation. The developer named the antibody JNJ-61186372. This antibody blocks ligand-induced phosphorylation of EGFR and cMet, and this blockade provides more potent inhibition than the combination of single receptor-binding antibodies. In NSCLC tumor models, the therapeutic effects of JNJ-61186372 may be achieved through inhibition of signal/receptor downregulation and Fc-driven effector interactions. Antibody therapy can cause tumor regression. The therapeutic effect of concurrent use of JNJ-61186372 and a third-generation EGFR TKI were observed in a human lung xenograft tumor model. Treatment with JNJ-61186372 in mammals did not produce severe toxicity, and no common antibody drug side effects, such as skin rash, were observed. Bispecific EGFR-cMet antibodies have great potential in inhibiting EGFR mutation-induced drug resistance in NSCLC.

Abbreviations

ADCC antibody-dependent cell cytotoxicity
EGFR epidermal growth factor receptor
NSCLC nonsmall cell lung cancer
TKI tyrosine kinase inhibitor

Conflict of interest

No potential conflicts of interest were disclosed.

https://doi.org/10.1016/B978-0-12-821584-5.00005-5

There are several activating mutations in the kinase domain of EGFR, including in-frame deletions of exon 19 and L858R exon 21 mutations in humans. According to current studies, these mutations may be the main cause of nonsmall cell lung cancer (NSCLC).

In the early clinical treatment of patients, most patients with nonsmall cell lung cancer have a positive response to drugs such as gefitinib and erlotinib. These drugs are first-generation EGFR tyrosine kinase inhibitors (TKIs). However, clinical results show that the effective time of such drugs is not long-lasting, and the average time for patients to exhibit significant drug resistance is less than 1 year [1,2].

According to clinical data, approximately 50% of drug resistance cases are caused by secondary mutations of EGFR [1,3–7]. When the threonine at position T790 is mutated to methionine, the mutant retains low nanomolar affinity for small molecule tyrosine kinase inhibitors. In addition, drug-resistant tumors provide an additional alternative mechanism by activating the cMet signaling pathway, which can bypass the blockade of EGFR by TKI and help cancer cells increase the possibility of survival. There are many ways to activate the cMet signaling pathway, such as amplifying the MET gene, increasing the expression of cMet, and simultaneously or independently increasing the expression levels of cMet ligand and hepatocyte growth factor [5,6,8,9]. In patients with NSCLC who are resistant to TKIs that target EGFR, the two mechanisms of secondary mutation and cMet signal activation may also exist at the same time [3,6,10,11].

There is an interaction between the two signaling pathways related to EGFR and cMet. It is hoped that dual targeting can inhibit the key receptors in the two signaling pathways at the same time, which can successfully inhibit the activation of the compensation mechanism and enhance the ultimate antitumor effect of the treatment plan.

Researchers have developed a new bispecific antibody named JNJ-61186372, which can bind to EGFR through one Fab arm and cMet antigen through a second Fab arm [12,13]. In the follow-up research, the researchers also made optimizations to further expand the application functions of the JNJ-61186372 antibody. First, after the addition of the antibody, the ligand-induced activation in tumor cells is blocked, and the receptor is induced to degrade. These results prove that the antibody can simultaneously inhibit EGFR and cMet signals, forming a dual inhibitory effect. In addition, Fab can recognize and bind to the surface of tumor cells because EGFR and cMet are highly expressed on the surface, Fc-dependent effector mechanisms are activated, and immune cells can attack tumor cells to achieve specific anticancer purposes. Antibody-dependent cellular cytotoxicity (ADCC) is one of the most common and typical mechanisms. To further improve the anticancer effect, researchers hope to obtain nonfucosylated therapeutic antibodies [14]. They engineered the cell line to achieve this design so that the JNJ-61186372 produced can more effectively mediate ADCC functions.

Several preclinical studies have proven that the JNJ-61186372 antibody can inhibit primary EGFR mutation tumors through different mechanisms of action [15–17]. At the same time, the antibody is effective against two kinds of drug-resistant tumors: tumors with resistance mutations at the second site of T790 M in EGFR and tumors that activate the cMet pathway. In addition, the combined use of the JNJ-61186372 antibody and third-generation TKIs against EGFR can lead to complete tumor regression. From the perspective of drug safety, no drug side effects were found in the study using the JNJ-61186372 antibody to treat cynomolgus monkeys, including no rash symptoms that are common in other EGFR-targeted drug treatments. In lung cancer and some other malignant tumor models, preclinical data confirm that the JNJ-61186372 antibody is very effective against cancers related to abnormal activation of EGFR and cMet signals, and it is speculated that it will have considerable potential in clinical treatment.

According to previous studies, during the development of EGFR mutation-driven NSCLC, EGFR and cMET signaling pathways have certain mutual compensatory effects, and both participate in mediating drug resistance to inhibitors of either pathway, forming double cross-resistance [4,8,9].

For patients with EGFR mutation-driven nonsmall cell lung cancer, small molecule TKIs against EGFR will be used in the current standard treatment guidelines. The most commonly used drugs in such patients include erlotinib and gefitinib. In the initial stage of treatment, these drugs usually have a very good effect, but because the second site mutation of EGFR (e.g., T790M) or the activation of alternative pathways (e.g., cMet pathway) can mediate the emergence of drug resistance in cancer cells, such drugs will eventually lose the function of treatment, and it is difficult to inhibit the occurrence and development of cancer [4,9,10,18]. Because of this clinical data feedback, pharmaceutical companies have successively launched second-generation TKIs against EGFR, such as afatinib and dacomitinib, and third-generation TKIs against EGFR, such as osimertinib, furmonertinib, and almonertinib. The clinical effects of these drugs are ideal compared to those of first-generation TKIs, but drug resistance to treatment still lacks a way to address it, and there are great limitations in clinical practice.

Through the experimental comparison of tissue from clinical samples, acquired EGFR C797S mutations mediate the resistance of nonsmall cell lung cancer containing EGFR T790 M to third-generation TKI drugs [19].

In addition, clinical studies have also reported analyzed data of MET amplification in lung cancer cells from patients receiving AZD9291 drug treatment, which indicates that the activation of the cMet pathway in lung cancer cells is directly related to the resistance to third-generation TKIs against EGFR [20].

As shown in previous preclinical research results, to solve the problem of drug resistance in nonsmall cell lung cancers, relevant researchers developed a bispecific antibody that can recognize EGFR and the cMet signaling pathway at the same time and named the antibody JNJ-61186372 [21].

The results showed that the antibody could efficiently recognize EGFR and CMET. The novel antibody (JNJ-61186372) can inhibit the growth and progression of cancer cells through three different pathways. The first two pathways are similar, inhibiting EGFR-related signal transduction and CMET-related signal transduction. First, ligand-induced activation in the signaling pathway is inhibited by blocking the binding of the two ligands to the corresponding receptor, and then the receptor is inactivated by degradation. The third mechanism is using FC effector-mediated ADCC to kill tumor cells with high expression of both antigens [16].

Through these mechanisms, JNJ-61186372 has shown therapeutic effects in many xenotransplantation animal models, and the transplanted tumor cells in these models include a variety of EGFR mutations (e.g., exon 19 deletion, L858R, T790M). Through the phenomenon that the novel antibody can recognize the extracellular region of EGFR, it is speculated that the bispecific antibody may have the potential to function in tumors with the c797s mutation of EGFR.

In addition, when JNJ-61186372 was applied to the activation model of the CMET signaling pathway, whether the activation of the signaling pathway was through the production of autocrine ligands or the amplification of met genes, the antibody had antitumor activity. JnJ-61186372 also has antitumor activity in wild-type EGFR and autocrine HGF production models. In summary, the unique mechanism of JNJ-61186372 can effectively target all kinds of mutations that can produce resistance to EGFR and inhibit the key way that tumor cells produce acquired resistance to EGFR TKI.

According to clinical results, it may be necessary and the best choice to use multidrug combination therapy in nonsmall cell lung cancer. Combination therapy can fully inhibit epidermal growth factor receptor-related mutations and action pathways through a variety of mechanisms and effectively prevent the occurrence of drug resistance events.

The combined application of afatinib and cetuximab has been shown to be more effective in treating mice with L858R/T790M erlotinib-resistant tumors [22]. This treatment plan can nearly completely resolve the tumors, and the therapeutic effect may be achieved through complete depletion of phosphorylation and total EGFR. The therapeutic effect of the combined use of the two drugs far exceeds that of any one drug alone.

In addition to inhibiting the cMet signaling pathway at the same time, the JNJ-61186372 antibody combined with third-generation EGFR TKI therapy may have the advantage of inhibiting the EGFR signaling pathway in cancer cells more thoroughly through the mechanism of ligand inhibition and receptor degradation. JNJ-61186372 may also be able to solve tumor cells that are resistant to EGFR- or cMet-independent mechanisms. It has been proven that the combined application of JNJ-61186372 and AZD9291 can inhibit the growth of drug-resistant tumors very effectively. The relevant research is based on an animal model of xenograft tumor cells with EGFR mutations and cMet activation. In addition, in most experimental models, combination therapy always has advantages compared with single drug therapy [20,23]. Long-term observations showed that the combination therapy can prevent tumor regeneration in experimental animals. Clinical patient studies have shown that tumor heterogeneity is the main reason for tumor resistance. Two research observations in 2015 showed that after receiving the first round of EGFR TKI treatment, lung cancer patients who developed resistance to previous treatments had both T790M-positive and T790M-negative cancer cells. Compared with the existing conventional therapies, combination therapy with JNJ-61186372 has the ability to delay or prevent the emergence of drug resistance to a certain extent. Combination therapy has higher activity on subgroups of tumors with different EGFR mutations and can more completely inhibit the EGFR pathway and inhibit the activation of the alternative signaling pathway c-Met.

In recent years, the rapid development of antibody engineering has made artificially modified bispecific human antibodies (such as JNJ-61186372) have better performance and huge application potential. Bispecific antibodies have a potential advantage, usually reducing the escape of tumor cells. When there are two target antigens on the same tumor cell, the antibody significantly enhances the ultimate killing effect on cancer cells by enhancing the inhibitory effect on the downstream signal of the related antigen. It is generally believed that the affinity effect is the main reason for this characteristic.

When there is an affinity effect, tumor cells expressing two target antigens specifically bind to both arms of the bispecific antibody (cross-arm binding). Compared with cancer cells that only express one target antigen or bind to only one Fab arm by other means, bispecific antibodies undoubtedly have a higher apparent affinity. Taking JNJ-61186372 as an example, this antibody can isolate EGFR and c-MET receptors into nonsignaling heterodimer complexes, preventing their homodimerization, thereby helping JNJ-61186372 relative to monovalent targets and increasing potency to EGFR or c-MET Ab. The results of characterizing cross-arm binding efficiency show that when the bivalent antibody binds to the first receptor on the same cell and then binds to the second target, the apparent affinity increases. At the same time, the ability of antibodies to inhibit antigen phosphorylation will also increase significantly [24].

It has also been observed in a study of the H1975-HGF model that the therapeutic effect of the combined use of anti-EGFR and anti-c-Met monoclonal antibodies is not as good as that of JNJ-61186372 at the same dose level [25]. This observation is also consistent with previous theoretical assumptions.

The bispecific structure of the JNJ-61186372 antibody may also bring other unexpected properties.

According to early experimental results, drugs that target EGFR often have certain skin toxicity, and this safety risk is usually one of the key issues that limit the application of such drugs.

However, in the study of cynomolgus monkeys as experimental subjects, even if the dose reached 120 mg/kg, researchers still did not find EGFR-mediated skin toxicity of the novel antibody.

Based on the results of the experiment, an inference can be made that the weak monovalent binding of JNJ-61186372 and EGFR may not induce rash and other symptoms in cynomolgus monkeys treated with JNJ-61186372, while other antibodies that bivalently interact with EGFR (e.g., cetuximab and panitumumab)-induced adverse reactions are also related to their strong binding ability to skin and intestinal cells.

In some studies, it has been found that the binding ability of JNJ-61186372 with normal human skin fibroblasts is significantly different from that of cetuximab, and the affinity is approximately 20 times different.

Due to the low level of cMet expressed in normal tissues, the cMet binding arm of JNJ-61186372 did not affect the binding of antibodies to skin and intestinal cells.

From the overall evaluation, JNJ-61186372 has weak binding to skin and intestinal cells and has a low inhibitory effect on EGFR antigen in normal tissues.

In tumor cells, the expression levels of EGFR and cMet are both high. Both arms of the novel antibody are involved in the binding effect and achieve a stronger binding force and inhibitory function.

JNJ-61186372 has potential low toxicity characteristics, which may also become an advantage of combination therapy. In particular, the combined efficacy of afatinib and cetuximab may be limited by toxicity issues [26].

At present, three different mechanisms of JNJ-61186372 involved in the regulation of EGFR and the cMet pathway have been clearly demonstrated in preclinical experiments. Through its special structure, the JNJ-61186372 antibody can kill tumor cells through multiple mechanisms even if it is not used in combination with other drugs. The multichannel killing function produces long-lasting antitumor effects on preclinical models that have developed drug resistance due to the administration of other treatment methods. The results of these preclinical studies support the promotion of the antibody to human clinical trials. The JNJ-61186372 antibody can be used as a monotherapy for specific patients or can be combined with other drugs to treat lung cancer and other malignant tumors with related mechanisms of action.

References

[1] Kobayashi S, et al. EGFR mutation and resistance of non-small-cell lung cancer to gefitinib. N Engl J Med 2005;352:786–92.

[2] Perez-Soler R, et al. Determinants of tumor response and survival with erlotinib in patients with non-small-cell lung cancer. J Clin Oncol 2004;22:3238–47.

[3] Chen HJ, et al. Clinicopathologic and molecular features of epidermal growth factor receptor T790M mutation and c-MET amplification in tyrosine kinase inhibitor-resistant Chinese non-small cell lung cancer. Pathol Oncol Res 2009;15:651–8.

[4] Engelman JA, et al. MET amplification leads to gefitinib resistance in lung cancer by activating ERBB3 signaling. Science 2007;316:1039–43.

[5] Sequist LV, et al. Genotypic and histological evolution of lung cancers acquiring resistance to EGFR inhibitors. Sci Transl Med 2011;3.

[6] Pao W, et al. Acquired resistance of lung adenocarcinomas to gefitinib or erlotinib is associated with a second mutation in the EGFR kinase domain. PLoS Med 2005;2:225–35.

[7] Yun CH, et al. The T790M mutation in EGFR kinase causes drug resistance by increasing the affinity for ATP. Proc Natl Acad Sci U S A 2008;105:2070–5.

[8] Yano S, et al. Hepatocyte growth factor expression in EGFR mutant lung cancer with intrinsic and acquired resistance to tyrosine kinase inhibitors in a Japanese cohort. J Thorac Oncol 2011;6:2011–7.

[9] Turke AB, et al. Preexistence and clonal selection of MET amplification in EGFR mutant NSCLC. Cancer Cell 2010;17:77–88.

[10] Bean J, et al. MET amplification occurs with or without T790M mutations in EGFR mutant lung tumors with acquired resistance to gefitinib or erlotinib. Proc Natl Acad Sci U S A 2007;104:20932–7.

[11] Jiang SX, et al. EGFR genetic heterogeneity of nonsmall cell lung cancers contributing to acquired gefitinib resistance. Int J Cancer 2008;123:2480–6.

[12] Labrijn AF, et al. Efficient generation of stable bispecific IgG1 by controlled Fab-arm exchange. Proc Natl Acad Sci U S A 2013;110:5145–50.

[13] Labrijn AF, et al. Controlled Fab-arm exchange for the generation of stable bispecific IgG1. Nat Protoc 2014;9:2450–63.

[14] Satoh M, Iida S, Shitara K. Non-fucosylated therapeutic antibodies as next-generation therapeutic antibodies. Expert Opin Biol Ther 2006;6:1161–73.

[15] Grugan KD, et al. Fc-mediated activity of EGFR x c-Met bispecific antibody JNJ-61186372 enhanced killing of lung cancer cells. MAbs 2017;9:114–26.

[16] Moores SL, et al. A novel bispecific antibody targeting EGFR and cMet is effective against EGFR inhibitor-resistant lung tumors. Cancer Res 2016;76:3942–53.

[17] Haura EB, et al. JNJ-61186372 (JNJ-372), an EGFR-cMet bispecific antibody, in EGFR-driven advanced non-small cell lung cancer (NSCLC). J Clin Oncol 2019;37.

[18] Yu HA, et al. Analysis of tumor specimens at the time of acquired resistance to EGFR-TKI therapy in 155 patients with EGFR-mutant lung cancers. Clin Cancer Res 2013;19:2240–7.

[19] Thress KS, et al. Acquired EGFR C797S mutation mediates resistance to AZD9291 in non-small cell lung cancer harboring EGFR T790M. Nat Med 2015;21:560–2.

[20] Planchard D, et al. EGFR-independent mechanisms of acquired resistance to AZD9291 in EGFR T790M-positive NSCLC patients. Ann Oncol 2015;26:2073–8.

[21] Moores SL, et al. Discovery and preclinical pharmacology of JNJ-61186372: a novel bispecific antibody targeting EGFR and cMET. Cancer Res 2014;74.

[22] Regales L, et al. Dual targeting of EGFR can overcome a major drug resistance mutation in mouse models of EGFR mutant lung cancer. J Clin Investig 2009;119:3000–10.

[23] Piotrowska Z, et al. Heterogeneity underlies the emergence of EGFR(T790) wild-type clones following treatment of T790M-positive cancers with a third-generation EGFR inhibitor. Cancer Discov 2015;5:713–22.
[24] Jarantow SW, et al. Impact of cell-surface antigen expression on target engagement and function of an epidermal growth factor receptor x c-MET bispecific antibody. J Biol Chem 2015;290:24689–704.
[25] Zheng SM, et al. Cross-arm binding efficiency of an EGFR x c-Met bispecific antibody. MAbs 2016;8:551–61.
[26] Janjigian YY, et al. Dual inhibition of EGFR with Afatinib and Cetuximab in kinase inhibitor-resistant EGFR-mutant lung cancer with and without T790M mutations. Cancer Discov 2014;4:1036–45.

CHAPTER

15

Use of MET kinase inhibitors to overcome cetuximab resistance in colorectal cancer

Zetong Ma[a], Wenyan Fu[b], Changhai Lei[a], and Shi Hu[a]

[a]Department of Biophysics, College of Basic Medical Sciences, Naval Medical University (Second Military Medical University), Shanghai, China [b]Department of Assisted Reproduction, Shanghai Ninth People's Hospital, Shanghai Jiao Tong University School of Medicine, Shanghai, China

Abstract

MET amplification is considered one of the important causes of anti-EGFR resistance in many tumors. Researchers have investigated cetuximab resistance mediated by hepatocyte growth factor (HGF)-dependent mesenchymal-epithelial transition factor (MET) activation. DIFI, GEO, and LIM1215 cells were treated with EGF, HGF, cetuximab, and PHA-665752, a highly specific MET blockade, at different concentrations. Cell proliferation was measured using WST-1, and an isogram was used to study synergies. The expression and signal transduction were detected by Western blotting. Activation of EGFR and MET cooperatively accelerated cell proliferation. Cetuximab inhibits cell growth through MAPK, EGFR, and Akt dephosphorylation. HGF restored cell proliferation and saved cells from G1 arrest and apoptosis by restimulating the MAPK and Akt pathways in cetuximab-treated cells, inhibiting MET activation by PHA-665752 or downregulating the expression of MET by siRNA can counteract its effect. The activation of MET induced by HGF is a newly discovered mechanism for cetuximab resistance. Blocking HGF/MET may enhance the sensitivity of CRC patients to EGFR inhibitors and is a promising therapeutic strategy.

Abbreviations

CRCs colorectal cancers
HGF hepatocyte growth factor
MET mesenchymal-epithelial transition
NSCLC nonsmall cell lung cancer
OSCC oral squamous cell carcinoma
TKI tyrosine kinase inhibitor

https://doi.org/10.1016/B978-0-12-821584-5.00019-5

Conflict of interest

No potential conflicts of interest were disclosed.

Introduction

Hepatocyte growth factor (HGF) facilitates cell motility, growth, and morphogenesis by binding to MET (hepatocyte growth factor receptors) to activate receptor tyrosine kinase cascades and mediate cancer development and progression [1–4]. MET amplification is one of the important causes of anti-EGFR resistance in various tumors and the cause of acquired resistance to gefitinib's EGFR tyrosine kinase inhibitor (TKI) in non-small cell lung cancer (NSCLC) [5,6]. Resistance to anti-EGFR therapy is mediated by transactivation of the Met-ErBB3 and PI3K/Akt pathways, leading to signal recovery [5]. MET activation is a mechanism for the inherent resistance of NSCLC cells to gefitinib, which have active EGFR mutations but without amplified MET [7]. Similarly, MET mediates trastuzumab resistance in breast cancer cells with ERBB2 (HER2) overexpression [8]. In contrast, when cocultured with EGF or heregulin-B1, gastric cancer cells with MET amplification showed resistance to MET-specific TKI [9]. The expression of MET in the CRC microenvironment is often synergistic with HGF, whose overexpression indicates poor prognosis [10]. Malignant transformation of CRC requires autocrine or paracrine HGF [11,12]. Highly expressed in OSCC, the HGF/MET signaling pathway has been shown to induce migration, invasion, and metastasis of OSCC cells through the formation of lamellar and filamentous feet [13] and the destruction of E-cadherin [14]. Crossfire between MET and EGFR, vascular endothelial growth factor receptor, Wnt and other signaling proteins have also been found in many cell lines [5,15,16]. In addition, it has been shown that MET activation during cetuximab treatment prompts poor prognosis in head and neck squamous cell carcinoma (HNSCC) patients [17]. The combination of MET inhibitors and EGFR-targeted therapy may be a promising treatment strategy.

EGFR and MET are coexpressed in a genotypically diverse panel of cetuximab-sensitive CRC cell lines

When overexpressed in CRC, the expression intensity of EGFR and MET often indicates poor prognosis. MET and EGFR were coexpressed to varying degrees in all three cetuximab-sensitive CRC cell lines, including EGFR-amplified DIFI, KRAS mutant GRO, and wild-type LIM1215. DIFI showed that EGFR was significantly overexpressed due to gene amplification, and no cell lines showed MET amplification. Of the three cell lines sensitive to cetuximab, the DIFI cell line was the most sensitive to cetuximab, and SW620 cells (control cell line) were resistant to cetuximab. Despite KRAS mutations, GEO cell lines were inhibited by cetuximab, possibly because cetuximab blocked the growth stimulation loop formed by the autocrine EGFR ligand amphiregulin.

Combined activation of EGFR and MET enhances CRC cell proliferation by augmenting the activation of AKT and MAPK

HEF and MET are usually co-overexpressed in most CRCs, which indicates prognosis. EGF and HGF were used to investigate the activation of MET, cell proliferation, and signal transduction in EGFR-driven CRC cell lines. After treatment with EGF or HGF, LIM1215 cells showed a moderate growth response, while treatment with both EGF and HGF eventually resulted in a more robust proliferation response. A similar response has been observed in DIFI cells to dual receptor activation, while EGF stimulation did not induce significant proliferative responses in GEO cell lines. This suggests that HGF-dependent MET activation plays an essential role in EGFR-activated colorectal cancer cell proliferation. Lim1215 and DIFI cells were treated with EGF, HGF, or both. EGF and HGF treatment significantly enhanced the activation of Akt and MAPK pathways compared with either growth factor alone.

HGF induces resistance to cetuximab through MET activation

HGF enabled DIFI, GEO, and LIM1215 cells to develop cetuximab resistance in a dose-dependent manner. Moreover, the resistance of GEO and LIM1215 to cetuximab was more evident than that of DIFI, which the EGFR amplification of DIFI might cause. After treatment with HGF for 72 h, the inhibition of cetuximab on the cells disappeared, and proliferation of all three cell lines was observed. After treatment with cetuximab, HGF, and PHA-665752 (MET-specific tyrosine kinase inhibitor), cells regained sensitivity to cetuximab. HGF-induced LIM1215 resistance to cetuximab was eliminated by silencing MET expression with a specific siRNA. These results suggest that MET inhibition can eliminate HGF-induced cetuximab resistance.

HGF rescues CRC cells from cetuximab-induced G1 arrest

After treatment with cetuximab for 24 h, DIFI, GEO, and LIM1215 cells aggregated in the G1 phase. For LIM1215, GEO, and DIFI, cetuximab reduced the percentage of cells in the S phase. After HGF treatment, the proportion of S phase cells of the three kinds of cells increased to 27%, 20%, and 10%, respectively. Next, inhibition of the MET pathway using PHA-665752 eliminated the effect of HGF on cetuximab-treated cells by reducing the percentage of cells in the S phase. It was demonstrated that HGF results in cetuximab resistance by counteracting the G1 block induced by cetuximab in three types of cells. These effects can be eliminated by blocking the MET pathway.

HGF rescued the apoptosis of DiFi cells induced by cetuximab

Cetuximab inhibits, rather than kills, cell growth in most CRCs. However, the EGFR-amplified cell line DIFI underwent apoptosis under the action of cetuximab. HGF can save DIFI cells from apoptosis induced by cetuximab. After treating the mixture of HGF and PHA-665752 for 24 h, cetuximab restored the pro-apoptotic effect.

HGF rescue occurs via MET-dependent activation of AKT and MAPK

The activation of downstream signaling pathways was detected to explore the molecular mechanism of HGF leading to the resistance of CRC cells to cetuximab. PHA-665752 treatment inhibited EGFR phosphorylation and Akt and MAPK activation in DIFI and LIM1215 cells. Cells treated with cetuximab and HGF exhibited increased methylation, phosphorylation, and reactivation of Akt and MAPK. HGF-mediated resistance is not achieved by HER-3 activation. The combination of cetuximab and PHA-665752 continuously inhibits the activation of Akt and MAPK pathways in CRC expressing HGF.

Discussion

Previous studies have shown that EGFR inhibition resistance in cancer cells can be overcome by activating a shared downstream pathway via the MET receptor [5–9,18]. MET receptors are overexpressed in colorectal cancer, similar to EGFR [19]. In addition, HGF and Met are coexpressed in the CRC tumor microenvironment, and the coexpression level correlates with prognosis[10, 19]. This study hypothesizes that HGF-dependent MET activation enables CRC cells to escape inhibition from cetuximab by activating a common downstream pathway that facilitates cell survival and proliferation.

This study established the possibility of MET and EGFR interactions by recording the coexpression of MET and EGFR in CRC cell lines. These results suggest that MET/HGF protects CRC cells from cetuximab-induced apoptosis by activating the Akt and MAPK signaling pathways. In addition, in the absence of MET amplification in cell lines, these cetuximab-sensitive cells still generate resistance to EGFR inhibition after HGF stimulation. However, HGF-induced resistance to cetuximab is less evident in the EGFR-amplified cell line DIFI because it is dependent on the EGFR pathway, indicating that HGF can protect cells without MET amplification from anti-EGFR therapy.

Downregulation of MET expression with siRNA or PHA-665752 finally inhibited HGF-dependent Akt and MAPK activation and resensitized cells to the effects of cetuximab. A variety of MET-targeted therapies are currently under clinical development. However, the current xenograft model is flawed because mouse HGF cannot bind to human MET effectively. This study further developed and refined different mouse models while modifying conditions to fit the properties of CRC encountered in the clinic.

References

[1] Bottaro DP, et al. Identification of the hepatocyte growth factor receptor as the c-met proto-oncogene product. Science 1991;251:802–4.

[2] Johnson M, Koukoulis G, Matsumoto K, Nakamura T, Iyer A. Hepatocyte growth factor induces proliferation and morphogenesis in nonparenchymal epithelial liver cells. Hepatology 1993;17:1052–61.

[3] Blumenschein GR, Mills GB, Gonzalez-Angulo AM. Targeting the hepatocyte growth factor-cMET axis in cancer therapy. J Clin Oncol 2012;30:3287–96.

[4] Liska D, Chen C-T, Bachleitner-Hofmann T, Christensen JG, Weiser MR. HGF rescues colorectal cancer cells from EGFR inhibition via MET activation. Clin Cancer Res 2011;17:472–82.

[5] Engelman JA, et al. MET amplification leads to gefitinib resistance in lung cancer by activating ERBB3 signaling. Science 2007;316:1039–43.

[6] Bean J, et al. MET amplification occurs with or without T790M mutations in EGFR mutant lung tumors with acquired resistance to gefitinib or erlotinib. Proc Natl Acad Sci U S A 2007;104:20932–7.

[7] Yano S, et al. Hepatocyte growth factor induces gefitinib resistance of lung adenocarcinoma with epidermal growth factor receptor-activating mutations. Cancer Res 2008;68:9479–87.

[8] Shattuck DL, Miller JK, Carraway KL, Sweeney C. Met receptor contributes to trastuzumab resistance of Her2-overexpressing breast cancer cells. Cancer Res 2008;68:1471–7.

[9] Bachleitner-Hofmann T, et al. HER kinase activation confers resistance to MET tyrosine kinase inhibition in MET oncogene-addicted gastric cancer cells. Mol Cancer Ther 2008;7:3499–508.

[10] Kammula US, et al. Molecular co-expression of the c-met oncogene and hepatocyte growth factor in primary colon cancer predicts tumor stage and clinical outcome. Cancer Lett 2007;248:219–28.

[11] Trusolino L, Comoglio PM. Scatter-factor and semaphorin receptors: cell signalling for invasive growth. Nat Rev Cancer 2002;2:289–300.

[12] Birchmeier C, Birchmeier W, Gherardi E, Vande Woude GF. Met, metastasis, motility and more. Nat Rev Mol Cell Biol 2003;4:915–25.

[13] Yasui H, Ohnishi Y, Nakajima M, Nozaki M. Migration of oral squamous cell carcinoma cells are induced by HGF/c-met signalling via lamellipodia and filopodia formation. Oncol Rep 2017;37:3674–80.

[14] Murai M, et al. Overexpression of c-met in oral SCC promotes hepatocyte growth factor-induced disruption of cadherin junctions and invasion. Int J Oncol 2004;25:831–40.

[15] Sulpice E, et al. Cross-talk between the VEGF-A and HGF signalling pathways in endothelial cells. Biol Cell 2009;101:525–39.

[16] Vermeulen L, et al. Wnt activity defines colon cancer stem cells and is regulated by the microenvironment. Nat Cell Biol 2010;12:468–76.

[17] Madoz-Gúrpide J, et al. Activation of MET pathway predicts poor outcome to cetuximab in patients with recurrent or metastatic head and neck cancer. J Transl Med 2015;13:282.

[18] Puri N, Salgia R. Synergism of EGFR and c-met pathways, cross-talk and inhibition, in non-small cell lung cancer. J Carcinog 2008;7:9.

[19] Di Renzo MF, et al. Overexpression and amplification of the met/HGF receptor gene during the progression of colorectal cancer. Clin Cancer Res 1995;1:147–54.

CHAPTER

16

EGFR- and VEGF(R)-targeted small molecules show synergistic activity in colorectal cancer models refractory to combinations of monoclonal antibodies

Guangyao Li[a], *Tian Li*[a], *Wenyan Fu*[b], *and Shi Hu*[a]

[a]Department of Biophysics, College of Basic Medical Sciences, Naval Medical University (Second Military Medical University), Shanghai, China [b]Department of Assisted Reproduction, Shanghai Ninth People's Hospital, Shanghai Jiao Tong University School of Medicine, Shanghai, China

Abstract

The signal transduction of VEGF(R) is inseparable from the effect of EGFR, which provides theoretical possibilities for targeted drug combinations against them. However, colon cancer patients are not sensitive to the combination therapy of monoclonal antibodies (for EGFR and VEGF). The possible hypothesis is that intracellular signal transmission relies not only on surface ligands and receptors but also on small-molecule tyrosine kinase inhibitors (TKIs) that have other influence pathways.

Abbreviations

CRC colorectal cancer
EGFR epidermal growth factor receptor
mAb monoclonal antibody
RTK receptor tyrosine kinases
TKI tyrosine kinase inhibitor
VEGF vascular endothelial growth factor

https://doi.org/10.1016/B978-0-12-821584-5.00001-8

Conflict of interest

No potential conflicts of interest were disclosed.

After CRC xenotransplanted mice were exposed to two drugs (bevacizumab, cetuximab, vargatef, and afatinib) of mAb and TKI, respectively, the results showed that vargatef compared with either drug alone. The tumor growth inhibitory effect of afatinib and vargatef was significantly improved, and this result shows that the apoptosis rate of tumor cells increased by five times. In contrast, the combined application of cetuximab and bevacizumab did not show a significantly better effect than a single drug, as expected. Tumor-related intracellular phosphorylated VEGFR1 and phosphorylated EGFR were significantly reduced in the two TKI treatment groups. However, testing the two mAb exposure groups was not noticeable.

However, the status of KRAS had no effect on the excellent combined effect of the two TKI treatment groups. The signal transduction of VEGF and EGFR in the cell is consistent with the level of cytotoxic activity. This points out the direction for TKIs following in-depth clinical trials, and it is also applicable to patients with mutant KRAS.

An important oncogenic signaling pathway: Intracellular signal transduction

Targeting carcinogenic signaling pathways is an important new idea for the progress of anticancer drugs. However, part of the sudden change that occurs in the downstream part, the interference between the signal path and the feedback loop, are all factors that cannot be ignored; the current focus is on targeting multiple signal pathways simultaneously [1]. Tumor survival and growth are always accompanied by changes in the crosstalk between signaling pathways of VEGF and epidermal growth factor receptor (EGFR) [2]. Both bevacizumab (an antibody that neutralizes VEGF) and cetuximab, which targets EGFR, are already licensed for CRC [3]. The latest studies of nearly 1800 patients (PACCE and CAIRO2) showed that the inclusion of EGFR-targeted monoclonal antibodies into the treatment mix did not make a significant difference, even no exception in wild-type KRAS tumor patients [4,5].

The change in signal transduction in tumor cells is a hot spot of recent research. First, research has found that the signal transduction of receptor tyrosine kinase RTKs (such as EGFR and VEGFR) is different from the receptors bound to the plasma membrane; the intrinsic RTK can also transmit signals [6]. Second, internal autocrine (endocrine) VEGF/VEGFR1 signal transduction was found in a variety of tumor types [7,8]. This indicates that the loss of VEGF/VER endocrine signaling caused by homologous recombination lacks VEGF-A and is manifested in tumor cell growth inhibition and increased apoptosis [9].

These studies indicate that only part of oncogenic signal transduction can be achieved by inhibiting extracellular ligands or receptors, while internal RTK signal transduction and crosstalk may be affected by small-molecule TKIs.

Vargatef/BIBF 1120, VEGFR, PDGFR, and FGFR triple vascular kinase inhibitors [10] and afatinib/BIBW 2992 are two TKIs that can irreversibly inhibit EGFR and HER2 [11]. The results show that intracellular EGFR and VEGFR signal transduction and RTK are only inhibited by TKI, which induces an increase in the rate of apoptosis. However, the phosphorylation state is also related to the internalization and stability of the receptor. Excessive doses of vargatef and afatinib will lead to decreased phosphorylated VEGFR1 and phosphorylated EGFR in cells. This study clarifies the critical position of intracellular signal transduction among many factors in tumor cell survival. [6–9,12,13].

Mechanistic differences and activity comparison between TKIs and mAbs

In the HT-29 xenograft, a classic CRC model for testing anticancer drugs, the combination of vargatef and afatinib, is more effective in inhibiting tumor proliferation and viability than the combination of bevacizumab and cetuximab. More obvious synergy may be related to cell growth inhibition (related to the cell cycle) or cytotoxicity. First, we observed the changes in DNA synthesis in the body after exposure to the four reagents. Quantitative image analysis shows that any single treatment of bevacizumab, cetuximab, or vargatef inhibits approximately 40% of DNA synthesis in the body, while afatinib inhibits nearly 70% of DNA synthesis. Moreover, the combination of TKIs or mAbs did not show a therapeutic effect superior to the highly active drugs in their respective combinations.

The next step was to observe the effect of the four reagents on tumor cell apoptosis. Compared with control tumors, any single treatment of bevacizumab, cetuximab, or vargatef had no significant effect on apoptosis, while the effect of afatinib doubled and increased the apoptotic score. Compared with the control group, cell apoptosis was slightly affected by bevacizumab and cetuximab but was more affected by the combination of vargatef and afatinib, and the level of apoptotic cell score increased by approximately five times. Include a series of cells that can efficiently express EGFR and HER2 in a comparative study. The results showed that the sensitivity to afatinib was 130 times. Some research results indicate that tumor-associated VEGFR1 affects cell survival and proliferation [7,8,13]. Moreover, through Student's *t*-test, there was no correlation between vargatef and afatinib ($r2=0.14$, $P=0.22$), indicating that the sensitivity to these two drugs does not interfere with their respective mediation pathways.

To further study the effects of two TKI drugs and their combinations on the cell cycle, cell cycle analysis found that there was significant cell cycle arrest in G1 that lasted for the entire 120-h incubation time. Interestingly, the combination of the two agents caused only a slight increase in the G1 phase. Ninety-six hours after adding any drug, the apoptotic cell fraction was at least 10% in HT-29 cells. In contrast, there was no significant increase in LS513 cells.

However, simultaneous exposure of LS513 cells to vargatef and afatinib combined caused at least 20% of induced apoptosis. This indicates that the combination of the two drugs has a synergistic effect among the drug combinations that causes more than 50% of the loss of vitality.

One-time treatment of HT-29 and LS513 cells with vargatef or afatinib can lead to prolonged G1 block. However, the combination of vargatef and afatinib did not cause similar changes in the G1 phase, consistent with the in vivo findings. Therefore, the combined application effect of monoclonal antibodies has no advantage compared with its single effect, which may be because the cell cycle mediators on which the two agents are closely dependent are the same.

Final thoughts and future directions

Vargatef and afatinib are multitarget drugs, and their activity depends on their inhibitory effects on VEGFR1 and EGFR. Recently, K5-SOS mice with epidermal cancer have given answers in which the keratinocyte-specific deletion of VEGF and EGFR genes affects tumor growth compared with the pharmacological inhibitory effects of TKIs on VEGFR and EGFR signal transduction [13]. Therefore, the activity of the TKI combination may be closely related to the inhibition of signal transduction. However, VEGF and EGFR are not the only two targets that affect tumor activity. Their one-way chemical composition may affect the side effects of drugs that combine targeting VEGF and EGFR. A recent study showed that long-term oral afatinib has long-lasting antitumor activity without potential risks [14], and two phase I studies report that long-term use of vargatef is not only safe but also shows good medicine generation of dynamic characteristics as feedback of the first batch of efficacy signals [15,16]. However, the clinical results also confirmed no synergy between several monoclonal antibodies, which also indicates that signal transduction (VEGFR, EGFR, etc.) in human tumor cells may be affected by the combination of TKIs and mAbs through different mechanisms. At the same time, the mechanism of the unsatisfactory effect of the two kinds of mAbs in synergistic treatment was clarified, and a new idea was developed for the TKI trial, which is also applicable to patients with mutant KRAS.

References

[1] Gossage L, Eisen T. Targeting multiple kinase pathways: a change in paradigm. Clin Cancer Res 2010;16:1973–8.
[2] Larsen AK, Ouaret D, El Ouadrani K, Petitprez A. Targeting EGFR and VEGF(R) pathway cross-talk in tumor survival and angiogenesis. Pharmacol Ther 2011;131:80–90.
[3] Saltz LB, Lenz HJ, Kindler HL, Hochster HS, Wadler S, Hoff PM, et al. Randomized phase II trial of cetuximab, bevacizumab, and irinotecan compared with cetuximab and bevacizumab alone in irinotecan-refractory colorectal cancer: the BOND-2 study. J Clin Oncol 2007;25:4557–61.
[4] Hecht JR, Mitchell E, Chidiac T, Scroggin C, Hagenstad C, Spigel D, et al. A randomized phase IIIB trial of chemotherapy, bevacizumab, and panitumumab compared with chemotherapy and bevacizumab alone for metastatic colorectal cancer. J Clin Oncol 2009;27:672–80.
[5] Tol J, Koopman M, Cats A, Rodenburg CJ, Creemers GJ, Schrama JG, et al. Chemotherapy, bevacizumab, and cetuximab in metastatic colorectal cancer. N Engl J Med 2009;360:563–72.
[6] Lemmon MA, Schlessinger J. Cell signaling by receptor tyrosine kinases. Cell 2010;141:1117–34.
[7] Lee TH, Seng S, Sekine M, Hinton C, Fu Y, Avraham HK, et al. Vascular endothelial growth factor mediates intracrine survival in human breast carcinoma cells through internally expressed VEGFR1/FLT1. PLoS Med 2007;4:e186.
[8] Naik S, Dothager RS, Marasa J, Lewis CL, Piwnica-Worms D. Vascular endothelial growth factor receptor-1 is synthetic lethal to aberrant {beta}-catenin activation in colon cancer. Clin Cancer Res 2009;15:7529–37.
[9] Samuel S, Fan F, Dang LH, Xia L, Gaur P, Ellis LM. Intracrine vascular endothelial growth factor signaling in survival and chemoresistance of human colorectal cancer cells. Oncogene 2011;30:1205–12.
[10] Hilberg F, Roth GJ, Krssak M, Kautschitsch S, Sommergruber W, Tontsch-Grunt U, et al. BIBF 1120: triple angiokinase inhibitor with sustained receptor blockade and good antitumor efficacy. Cancer Res 2008;68:4774–82.
[11] Li D, Ambrogio L, Shimamura T, Kubo S, Takahashi M, Chirieac LR, et al. BIBW2992, an irreversible EGFR/HER2 inhibitor highly effective in preclinical lung cancer models. Oncogene 2008;27:4702–11.
[12] Bianco R, Rosa R, Damiano V, Daniele G, Gelardi T, Garofalo S, et al. Vascular endothelial growth factor receptor-1 contributes to resistance to anti-epidermal growth factor receptor drugs in human cancer cells. Clin Cancer Res 2008;14:5069–80.

[13] Lichtenberger BM, Tan PK, Niederleithner H, Ferrara N, Petzelbauer P, Sibilia M. Autocrine VEGF signaling synergizes with EGFR in tumor cells to promote epithelial cancer development. Cell 2010;140:268–79.
[14] Yap TA, Vidal L, Adam J, Stephens P, Spicer J, Shaw H, et al. Phase I trial of the irreversible EGFR and HER2 kinase inhibitor BIBW 2992 in patients with advanced solid tumors. J Clin Oncol 2010;28:3965–72.
[15] Mross K, Stefanic M, Gmehling D, Frost A, Baas F, Unger C, et al. Phase I study of the angiogenesis inhibitor BIBF 1120 in patients with advanced solid tumors. Clin Cancer Res 2010;16:311–9.
[16] Okamoto I, Kaneda H, Satoh T, Okamoto W, Miyazaki M, Morinaga R, et al. Phase I safety, pharmacokinetic, and biomarker study of BIBF 1120, an oral triple tyrosine kinase inhibitor in patients with advanced solid tumors. Mol Cancer Ther 2010;9:2825–33.

CHAPTER

17

Antitumor activity of the VEGFR inhibitor ZD6474 in human cancer cells resistant to anti-EGFR therapy

Chuqi Wang[a,b], *Wenyan Fu*[c], *Changhai Lei*[a], *and Shi Hu*[a]

[a]Department of Biophysics, College of Basic Medical Sciences, Naval Medical University (Second Military Medical University), Shanghai, China [b]Department of Pharmacy, National University of Singapore, Singapore [c]Department of Assisted Reproduction, Shanghai Ninth People's Hospital, Shanghai Jiao Tong University School of Medicine, Shanghai, China

Abstract

Tumor growth is driven by abundant signaling pathways. Tumors can circumvent one signaling blockade and generate resistance by utilizing the redundancy in growth signaling. Therefore, combining inhibitors of more than one pathway may be a rational strategy against tumor resistance. Vascular endothelial growth factor (VEGF) and EGFR pathways are interrelated. Activation of EGFR is known to induce the secretion of multiple paracrine angiogenic growth factors in cancer cells, such as VEGF, transforming growth factor alpha, basic fibroblast growth factor, and interleukin-8. Increased levels of VEGF lead to neovascularization within the tumor and acquired resistance to EGFR inhibitors. ZD6474 (vandetanib) is a small-molecule inhibitor of VEGF receptor-2 (VEGFR-2), EGFR, and RET tyrosine kinase activity. A previous study showed that ZD6474 overcomes acquired resistance to anti-EGFR therapy as monotherapy. Another study found that the combination of ZD6474 with EGFR inhibitors has synergistic antitumor activity in a series of human cancers. Collectively, double blockade of EGFR and VEGFR-2 signaling using ZD6474 presents potential and is worth further evaluation in clinical settings.

Abbreviations

VEGF vascular endothelial growth factor
EGFR epidermal growth factor receptor
VEGFR vascular endothelial growth factor receptor
MAPK mitogen-activated protein kinase
TGFα transforming growth factor alpha
EGF epidermal growth factor

https://doi.org/10.1016/B978-0-12-821584-5.00016-X

COX-2 cyclooxygenase-2
HNSCC head and neck squamous cell carcinoma
NSCLC non-small cell lung carcinoma

Conflict of interest

No potential conflicts of interest were disclosed.

Introduction

VEGF is a key player in angiogenesis (the formation of new blood vessels). It stimulates the mitogenesis and survival of endothelial cells, increases the permeability of blood vessels, and thus promotes blood vessel growth and remodeling [1]. VEGF mediates physiological processes by binding to two VEGF receptors, VEGFR-1 and VEGFR-2 [2]. VEGFR-2 is the main signaling receptor for the proliferation and differentiation of endothelial cells [3]. Given the essential role of angiogenesis in tumor development and growth, aberrant activation of the VEGF pathway has the potential to induce tumorigenesis [1,4]. In this scenario, VEGF could be a potential target for anticancer therapy [1].

Considering that VEGF and EGFR are interrelated in many tumor types [5], simultaneously targeting these two interesting therapeutic targets has a rationale in cancer treatment. EGFR activation can induce cancer cells to produce positive regulators of angiogenesis, such as VEGF and IL-8, thereby promoting endothelial cell growth and angiogenesis [6]. It has been reported that at least part of the antitumor effect of EGFR inhibitors is mediated by the inhibition of downstream proangiogenic mediators, thus reducing microvessel density and metastases [5,7]. Conversely, cancer cells can harness the VEGF paracrine mechanism to evade anti-EGFR therapies [8]. Therefore, blocking the VEGF and EGFR pathways seems to be rational and has shown efficacy in various preclinical and clinical studies [9].

ZD6474, also known as vandetanib, is an oral VEGFR2 inhibitor. Recent studies have shown that it can also inhibit EGFR and RET tyrosine kinases [10–12]. While it has been used in clinics for treating medullary thyroid cancer, ZD6474 also shows promise in other cancers. This chapter focuses on approaches to dual inhibition of the VEGF and EGFR pathways, including ZD6474 as monotherapy and combinations of ZD6474 with anti-EGFR antibody cetuximab to overcome anti-EGFR resistance.

ZD6474 efficiently inhibited the growth of EGFR-resistant tumors via inhibition of VEGFR

In their research, Ciardiello et al. investigated whether ZD6474 could inhibit the growth of tumors with acquired resistance to EGFR inhibitors and whether ZD6474 could delay or prevent the development of resistance in chronic treatment [13]. They first found that continuous administration of cetuximab and gefitinib (two anti-EGFR drugs) caused marked tumor inhibition during the first several weeks. Nevertheless, the xenografts began to grow after 11–12 weeks and eventually reached a growth rate similar to that of untreated tumors after 20–22 weeks, indicating the development of drug resistance. In contrast, chronic treatment with ZD6474 blocked GEO tumor growth for 23 weeks. Of note, reversible tumor growth was observed upon cessation of a 4-week treatment of all three drugs, and the growth rates were similar among the three groups.

They next assessed whether ZD6474 has an antitumor effect against GEO tumors after treatment with EGFR inhibitors. GEO-bearing mice were treated with either cetuximab or gefitinib for 4 weeks and allowed to grow for 4 weeks. Following the 4-week interval, the mice received continuous treatment with ZD6474, cetuximab, or gefitinib. ZD6474-treated GEO tumors, but not cetuximab and gefitinib, inhibited tumor growth for the following 9-week treatment period until the end of the experiment.

To determine the mechanisms of resistance to EGFR inhibitors, the researchers generated two cell strains, GEO-ZD1839-RES and GEO-C225-RES, that are resistant to gefitinib (ZD1839) and cetuximab (C225), respectively. Only a small reduction in EGFR levels was observed in the two resistant cell lines compared with the parental cells. Moreover, little or no changes in TGFα, Bcl-2, Bcl-xL, total mitogen-activated protein kinase (MAPK), total Akt, phospho-Akt, p53, p27, or MDM-2 were found between all three cell lines. In contrast, significant increases in COX-2, phospho-MAPK, and VEGF were observed in GEO-ZD1839-RES and GEO-C225-RES cells compared with GEO cells, according to the western blot results. Levels of secreted VEGF were also upregulated in the conditioned medium of the two resistant cell lines, implying that increased VEGF might be responsible for tumor escape from long-term EGFR inhibition.

To evaluate the in vitro sensitivity, resistant and parental GEO cells were treated with ZD6474, cetuximab, and gefitinib. The two resistant cell lines were not affected by all three drugs, whereas parental GEO cells were significantly inhibited by ZD6474, cetuximab, and gefitinib, consistent with previous results. The researchers finally tested the in vivo antitumor effect of these drugs on GEO-ZD1839-RES and GEO-C225-RES cell lines. Interestingly, while cetuximab and gefitinib did not affect the growth of either cell line, ZD6474 induced significant growth inhibition in both resistant xenografts. These results indicate that the antitumor activity of ZD6474 against anti-EGFR-resistant cells might be due to VEGF signaling inhibition in endothelial cells, but not the direct inhibition of EGFR in cancer cells. Collectively, the multitargeting ZD6474 might have broad-spectrum activity against tumors that do not respond to anti-EGFR therapy [13].

The combination of ZD6474 and cetuximab displayed a synergistic antitumor effect against multiple cancer cell lines

In addition to ZD6474 as monotherapy, ZD6474 can be combined with other drugs, such as cetuximab. According to the study of Morelli et al., ZD6474 and cetuximab showed synergistic antitumor effects and significantly prolonged median overall survival in mice as opposed to single-drug treatment in multiple cancer cell lines [4]. The researchers first assessed the synergistic effect of EGFR inhibition in GEO, MKN28, A549, ZR-75-1, KYSE-30, and MiaPaca2 cancer cell lines, which present a functional EGFR-dependent autocrine growth pathway and lack VEGFR and RET. In soft agar colony formation assays, ZD6474 and cetuximab showed strong synergism in all the cancer cell lines tested. Moreover, the induction of programmed cell death by ZD6474, cetuximab, or ZD6474 plus cetuximab also verified the synergism of these two drugs. In this assay, either single agent slightly increased apoptosis by 1.4- to 2.3-fold compared with untreated cancer cells, whereas the combination of the two drugs increased apoptosis by fivefold to fivefold compared with the untreated control. Together, these in vitro data suggested that ZD6474 could potentiate the antitumor effect of cetuximab via direct inhibition of EGFR.

The researchers next evaluated the in vivo effects of cetuximab and/or ZD6474 in mice engrafted with GEO or A549 cancer cells. Although a 4-week treatment with ZD6474 or cetuximab individually induced 70%–80% tumor inhibition, the tumors started to grow rapidly within 1–3 weeks posttreatment. However, the inhibition remained much longer in the ZD6474+ cetuximab group. In GEO-bearing mice, the average survival periods of ZD6474 plus cetuximab, ZD6474, cetuximab, and untreated control were 155, 56, 56, and 35 days, respectively. Similarly, in A549-bearing mice, the average survival periods of ZD6474 plus cetuximab, ZD6474, cetuximab, and untreated control were 175, 63, 63, and 35 days, respectively. Moreover, the combinatorial treatment eradicated tumors in 4/10 mice of the GEO group and in 3/10 mice of the A549 groups. Ki-67 and neovessel formation within the tumor were significantly reduced by ZD6474 plus cetuximab compared with either single agent. Taken together, these findings indicate that the ZD6474 and cetuximab combination is feasible and effective in human cancer cell lines.

Discussion

In this chapter, researchers reviewed two preclinical studies that indicated ZD6474's potential in combating tumors with resistance to anti-EGFR therapy. Previous investigations have found that tumors upregulate proangiogenic factors under the pressure of EGFR inhibitors to promote endothelial proliferation and permeability and eventually develop resistance to EGFR inhibitors [8]. The aforementioned article by Ciardiello et al. further confirmed that increased VEGF might be at least one of the causes of anti-EGFR resistance in a human cancer cell line [13]. Given the key role of the paracrine VEGF mechanism in anti-EGFR insensitivity, cotargeting VEGF and the EGFR pathway may be a plausible strategy for either sensitizing EGFR inhibitors or preventing the emergence of EGFR resistance. Indeed, many studies have focused on this strategy and have achieved encouraging results. For example, the efficacy of the EGFR inhibitor gefitinib in combination with bevacizumab, an anti-VEGF-A antibody, has been supported by a series of clinical trials in renal cell carcinoma [14], HNSCC [15], and NSCLC [16].

ZD6474 (vandetanib) is an inhibitor of not only VEGFR-2 but also EGFR and RET. The broad spectrum of antitumor mechanisms may presumably confer ZD6474 treatment advantages. However, although this drug has proven effective in patients with medullary thyroid carcinoma, clinical trials for other types of cancers are mixed with both dismal and encouraging results. In a randomized, double-blind phase III trial (ZEAL), vandetanib plus pemetrexed failed to prolong PFS in NSCLC patients versus placebo plus pemetrexed [17]. Another phase III trial (ZODIAC) later showed that vandetanib and docetaxel combination slightly hampered disease progression in patients with previously treated advanced NSCLC and was associated with better control of symptoms [18]. A third phase III trial (ZEPHYR) did not show an improved overall survival after vandetanib treatment in patients with NSCLC [19]. In addition to the studies in NSCLC, the assessment of ZD6474 is ongoing in various tumor types, either as monotherapy or combined with other antitumor agents. The efficacy of ZD6474 against cancers awaits confirmation in large-scale trials.

References

[1] Carmeliet P. VEGF as a key mediator of angiogenesis in cancer. Oncology 2005;69(Suppl. 3):4–10.
[2] Thomas KA. Vascular endothelial growth factor, a potent and selective angiogenic agent. J Biol Chem 1996;271:603–6.
[3] Kerbel RS. Tumor angiogenesis: past, present and the near future. Carcinogenesis 2000;21:505–15.
[4] Morelli MP, et al. Anti-tumor activity of the combination of cetuximab, an anti-EGFR blocking monoclonal antibody and ZD6474, an inhibitor of VEGFR and EGFR tyrosine kinases. J Cell Physiol 2006;208:344–53.
[5] Tabernero J. The role of VEGF and EGFR inhibition: implications for combining anti-VEGF and anti-EGFR agents. Mol Cancer Res 2007;5:203–20.
[6] Ciardiello F, et al. Interaction between the epidermal growth factor receptor (EGFR) and the vascular endothelial growth factor (VEGF) pathways: a rational approach for multi-target anticancer therapy. Ann Oncol 2006;17 (Suppl. 7), vii109-114.
[7] Ellis LM. Epidermal growth factor receptor in tumor angiogenesis. Hematol Oncol Clin North Am 2004;18:1007–21. viii.
[8] Viloria-Petit A, et al. Acquired resistance to the antitumor effect of epidermal growth factor receptor-blocking antibodies in vivo: a role for altered tumor angiogenesis. Cancer Res 2001;61:5090–101.
[9] Byers LA, Heymach JV. Dual targeting of the vascular endothelial growth factor and epidermal growth factor receptor pathways: rationale and clinical applications for non-small-cell lung cancer. Clin Lung Cancer 2007;8 (Suppl. 2):S79–85.
[10] Ciardiello F, et al. Antitumor effects of ZD6474, a small molecule vascular endothelial growth factor receptor tyrosine kinase inhibitor, with additional activity against epidermal growth factor receptor tyrosine kinase. Clin Cancer Res 2003;9:1546–56.
[11] Vitagliano D, et al. The tyrosine kinase inhibitor ZD6474 blocks proliferation of RET mutant medullary thyroid carcinoma cells. Endocr Relat Cancer 2011;18:1–11.
[12] Ryan AJ, Wedge SR. ZD6474—a novel inhibitor of VEGFR and EGFR tyrosine kinase activity. Br J Cancer 2005;92: S6–S13.
[13] Ciardiello F, et al. Antitumor activity of ZD6474, a vascular endothelial growth factor receptor tyrosine kinase inhibitor, in human cancer cells with acquired resistance to antiepidermal growth factor receptor therapy. Clin Cancer Res 2004;10:784–93.
[14] Hainsworth JD, Sosman JA, Spigel DR, Edwards DL, Baughman C, Greco A. Treatment of metastatic renal cell carcinoma with a combination of bevacizumab and erlotinib. J Clin Oncol 2005;23:7889–96.
[15] Vokes EE, et al. A phase I study of erlotinib and bevacizumab for recurrent or metastatic squamous cell carcinoma of the head and neck (HNC). J Clin Oncol 2005;23:5504.
[16] Herbst RS, et al. Phase I/II trial evaluating the anti-vascular endothelial growth factor monoclonal antibody bevacizumab in combination with the HER-1/epidermal growth factor receptor tyrosine kinase inhibitor erlotinib for patients with recurrent non-small-cell lung cancer. J Clin Oncol 2005;23:2544–55.
[17] Felip E, Rosell R. Pemetrexed as second-line therapy for advanced non-small-cell lung cancer (NSCLC). Ther Clin Risk Manag 2008;4:579–85.
[18] Herbst RS, et al. Vandetanib plus docetaxel versus docetaxel as second-line treatment for patients with advanced non-small-cell lung cancer (ZODIAC): a double-blind, randomised, phase 3 trial. Lancet Oncol 2010;11:619–26.
[19] Lee JS, et al. Vandetanib versus placebo in patients with advanced non-small-cell lung cancer after prior therapy with an epidermal growth factor receptor tyrosine kinase inhibitor: a randomized, double-blind phase III trial (ZEPHYR). J Clin Oncol 2012;30:1114–21.

CHAPTER

18

Antibody-mediated delivery of anti-KRAS-siRNA overcomes therapy resistance in colon cancer

Chuqi Wang[a,b], *Changhai Lei*[a], *and Shi Hu*[a]

[a]**Department of Biophysics, College of Basic Medical Sciences, Naval Medical University (Second Military Medical University), Shanghai, China** [b]**Department of Pharmacy, National University of Singapore, Singapore**

Abstract

KRAS mutation causes resistance to anti-EGFR antibodies in a variety of tumor types. Despite huge efforts in finding drugs targeting KRAS, little progress has been achieved due to the unusual structure of KRAS, which is often referred to as undruggable. Recently, siRNA has been proposed as a promising approach to silence the expression of disease-causing genes. In this chapter, researchers describe an innovative study using KRAS-specific siRNA coupled to the anti-EGFR antibody cetuximab to treat KRAS-mutated cancers. The siRNA-antibody complex was demonstrated to be effective in finding, binding, and internalizing into EGFR-positive cells and inhibited the growth of anti-EGFR-resistant tumors both in vitro and in vivo. Collectively, the siRNA-antibody complex is a promising candidate to address the unmet clinical need in combating KRAS-driven human cancers.

Abbreviations

EGFR epidermal growth factor receptor
GTP guanosine triphosphate
GDP guanosine diphosphate
GEFs guanine nucleotide exchange factors
GAPs GTPase-activating proteins
siRNA small interfering RNA

Conflict of interest

No potential conflicts of interest were disclosed.

https://doi.org/10.1016/B978-0-12-821584-5.00003-1

Introduction

The KRAS protein is a key component that relays signaling from EGFR to instruct cell proliferation and differentiation [1]. Acting like a switch, KRAS can be turned on when bound to guanosine triphosphate (GTP) and deactivate itself by transforming GTP into GDP via its intrinsic GTPase activity [2]. In an abnormal scenario, however, mutations in KARS transform it into a constitutively active conformation [3]. It is not surprising that KRAS mutations could lead to resistance to anti-EGFR antibodies, considering that KRAS is downstream of EGFR and activates both PI3K/AKT and MAPK independent of EGFR. In fact, mutant KRAS has been found to cause resistance to anti-EGFR monoclonal antibodies and has been identified as a predictive marker for EGFR inhibitors [4]. KRAS mutations are not rare in human cancers. In contrast, KRAS mutation isoforms occur in over 20% of lung cancer [5], colon cancer [6], and pancreatic cancer patients [7], highlighting the necessity of finding effective KRAS inhibitors. Unfortunately, despite huge efforts to find KRAS inhibitors, nearly all attempts failed. KRAS is often considered an undruggable target because the KRAS protein does not seem to have suitable pockets that inhibitors can bind to [7].

Small interfering RNAs (siRNAs) are a class of nucleic acids that can downregulate gene expression by interfering with mRNA [8]. Gene silencing by siRNA has emerged as a powerful treatment approach across a potentially broad range of diseases. The dominant limitation of siRNA drugs is their instability, which weakens the activity of siRNA in vivo. Therefore, an effective and specific delivery approach is critical for preventing siRNA degradation and for siRNA entry into target cells. Previously reported siRNA delivery platforms include but are not limited to cationic lipids, nanoparticles, exosomes, and peptides [9–12]. In fact, various anti-KRAS siRNA drugs have shown encouraging outcomes in preclinical studies. For example, siRNA specific to the KRAS G12D mutant showed an antitumor effect against pancreatic ductal adenocarcinoma [13]. Codelivery of siRNA targeting KRAS and miR-34a, a p53-regulated tumor suppressor miRNA, demonstrated efficacy in lung cancer models [14]. These encouraging results suggested that siRNA may be a promising approach for KRAS targeting.

KRAS has proven to be responsible for resistance to anti-EGFR antibodies and is extremely difficult to target with small molecules. To tackle these problems, Bäumer et al. coupled anti-KRAS siRNA to anti-EGFR antibodies to treat KRAS-mutated EGFR-positive cancer [3]. In this chapter, researchers recapitulate the generation, characterization, and functional validation of this novel therapeutic strategy.

Generation and characterization of KRAS-siRNA-anti-EGFR antibody complexes

To deliver siRNA to target cells, Bäumer et al. chose the anti-EGFR antibody cetuximab to form complexes with anti-KRAS siRNA because this antibody could bind EGFR and provoke internalization into the cell [3,15]. To load siRNA into the antibody, they first coupled protamine and cetuximab by the crosslinker sulfo-SMCC. Protamine is a highly cationic charged protein with good biocompatibility that can bind with negatively charged siRNA through electrostatic interactions, thereby forming antibody-siRNA complexes. The specific siRNA payload was assessed by electrophoresis. According to the results, one molecule of cetuximab-protamine could bind with eight molecules of siRNA, indicating a significant siRNA load for the antibody-protamine complex.

One limitation of siRNA drugs is the short half-life time and vulnerability to enzymatic digestion. The stability of cetuximab-protamine-siRNA was measured. After incubation with cell culture medium for 1–22h, no marked degradation was observed at any time point, suggesting the protective effect of the antibody-protamine complex for siRNA against enzymatic digestion.

Binding to the cell receptor and subsequent internalization of the antibody-siRNA complex are critical for intracellular siRNA functions. Hence, the binding and internalization of the antibody and antibody complex with protamine and siRNA were evaluated with flow cytometry. Flow cytometry results suggested no interference of protamine and siRNA on the binding and internalization of EGFR. The internalization was further verified with a fluorescence microscope, and fluorescence-labeled siRNA was observed in EGFR-positive cells after incubation.

In vitro validation of KRAS-siRNA-anti-EGFR antibody complexes: KRAS silencing and tumor inhibitory activity

Enzymatically produced (e)siRNA was used to target KRAS. esiRNA is composed of a complex pool of active inhibitory RNA fragments and has high specificity and efficiency [3,16]. When added to KRAS-mutant HCT116 and LoVo adenocarcinoma cells, the cetuximab-KRAS-esiRNA complex efficiently reduced the expression of KRAS and subsequently the level of ERK1/2 phosphorylation and c-Myc protein expression. In contrast to KRAS-esiRNA, control-esiRNA did not influence KRAS expression levels or downstream signaling.

The researchers next assessed tumor growth inhibition of the cetuximab-siRNA complex in a panel of cancer cell lines through the soft agar colony formation assay. As expected, the cetuximab-protamine-KRAS-esiRNA complex, but not cetuximab and cetuximab coupled to control-esiRNA, caused significant tumor inhibition in KRAS-mutant EGFR-positive colon cancer cell lines HCT116, LoVo, HCT15, SW480, and DLD1. However, the cetuximab-protamine-KRAS-esiRNA complex did not inhibit the BRAF-mutant cell line HT29. In KRAS wild-type MDA cells, cetuximab with or without esiRNA showed a similar antitumor effect, suggesting that KRAS-siRNA did not enhance the response to KRAS wild-type cells. In an MTS viability assay, cetuximab coupled to KRAS-siRNA, but not control-siRNA, killed HCT116, LoVo, DLD1, and SW480 cells in a dose-dependent manner. Not surprisingly, BRAF-mutant HT29 cells remained resistant to cetuximab-KRAS-esiRNA treatment. Consistently, BrdUrd results showed reduced proliferation in the same KRAS-mutant cancer cell lines after treatment with the KRAS-esiRNA complex compared with the control-siRNA complex, and the Annexin-V staining assay also validated the antitumor effect of the cetuximab-KRAS-esiRNA complex. Taken together, the cetuximab-protamine-KRAS-esiRNA complex has good specificity in silencing KRAS and has a pronounced antitumor effect against KRAS-driven cancer cells in vitro.

In vivo antitumor effect of the cetuximab-KRAS-esiRNA complex

The effect of the cetuximab-protamine-KRAS-esiRNA complex was analyzed in mouse tumor xenograft models. The researchers first investigated the bioavailability of the complex. Accumulation of antibody-siRNA conjugates was found in tumors following intraperitoneal or subcutaneous administration, whereas naked siRNA did not show enrichment in tumors. They next assessed the therapeutic effect of the drug. As expected, mAb-protamine-KRAS-esiRNA complexes inhibited the growth of SW480, DLD1, and HCT116 cell lines. In contrast, BRAF-mutated HT29 cells did not respond to the treatment. The KRAS mRNA levels in HCT116 tumors were consistently suppressed in the cetuximab-KRAS-esiRNA group compared with the control groups. Of note, HCT116 cells contain a PIK3CA activating in addition to the KRAS mutation. Combined KRAS and PIK3CA esiRNAs showed a synergistic antitumor effect against HCT116 cells.

Final thoughts and future directions

In this chapter, researchers introduce a novel KRAS-targeting strategy using the KRAS-siRNA-anti-EGFR antibody complex. The complex effectively inhibited the growth of KRAS-mutant tumor cells by downregulating KRAS expression and subsequently suppressed ERK and the MAPK pathway. This approach showed promise in overcoming resistance to anti-EGFR antibodies caused by KRAS mutation. Two problems with siRNA-based therapy are siRNA degradation and off-target effects. By coupling siRNA to the antibody-protamine complex, the siRNA could be protected from enzymatic digestion and be taken up by cells via internalization of the antibody. Moreover, the dual specificity of both the antibody and siRNA decreases the possibility of off-target effects. One can therefore choose antibodies and siRNAs with more flexibility. The more specific the antibody is, the less specific the siRNA needs to be, or vice versa [15].

In addition to targeting tumors that have mutated KRAS and EGFR expression, the use of antibody-siRNA complexes can potentially be broadened to many tumor types and almost any siRNA. In tumors bearing both KRAS and PIK3CA mutations, combined KRAS and PIK3CA esiRNAs showed a syngenetic effect in reducing colony growth, indicating that combinatorial RNA interference toward more than one aberrant pathway might achieve synergistic efficacy. Even personalized treatment that targets different tumor drivers according to each individual's tumor specification could become possible with this approach. In conclusion, the KRAS-siRNA-anti-EGFR antibody complex is an attractive strategy for the treatment of KRAS-driven human cancers and warrants further evaluation.

References

[1] Brand TM, Wheeler DL. KRAS mutant colorectal tumors: past and present. Small GTPases 2012;3:34–9.

[2] Boguski MS, McCormick F. Proteins regulating Ras and its relatives. Nature 1993;366:643–54.

[3] Bäumer S, et al. Antibody-mediated delivery of anti-KRAS-siRNA in vivo overcomes therapy resistance in colon cancer. Clin Cancer Res 2015;21:1383–94.

[4] Normanno N, Tejpar S, Morgillo F, De Luca A, Van Cutsem E, Ciardiello F. Implications for KRAS status and EGFR-targeted therapies in metastatic CRC. Nat Rev Clin Oncol 2009;6:519–27.

[5] Adderley H, Blackhall FH, Lindsay CR. KRAS-mutant non-small cell lung cancer: converging small molecules and immune checkpoint inhibition. EBioMedicine 2019;41:711–6.

[6] Dinu D, et al. Prognostic significance of KRAS gene mutations in colorectal cancer—preliminary study. J Med Life 2014;7:581–7.

[7] McCormick F. KRAS as a therapeutic target. Clin Cancer Res 2015;21:1797–801.

[8] Lamberti G, Barba AA. Drug delivery of siRNA therapeutics. Pharmaceutics 2020;12.

[9] Choi YS, et al. The systemic delivery of siRNAs by a cell penetrating peptide, low molecular weight protamine. Biomaterials 2010;31:1429–43.

[10] El-Andaloussi S, et al. Exosome-mediated delivery of siRNA in vitro and in vivo. Nat Protoc 2012;7:2112–26.

[11] Liu B. Exploring cell type-specific internalizing antibodies for targeted delivery of siRNA. Brief Funct Genomic Proteomic 2007;6:112–9.

[12] Hsu CY, Uludağ H. A simple and rapid nonviral approach to efficiently transfect primary tissue-derived cells using polyethylenimine. Nat Protoc 2012;7:935–45.

[13] Zorde Khvalevsky E, et al. Mutant KRAS is a druggable target for pancreatic cancer. Proc Natl Acad Sci U S A 2013;110:20723–8.

[14] Xue W, et al. Small RNA combination therapy for lung cancer. Proc Natl Acad Sci U S A 2014;111:E3553–61.

[15] Bäumer N, et al. Antibody-coupled siRNA as an efficient method for in vivo mRNA knockdown. Nat Protoc 2016;11:22–36.

[16] Surendranath V, Theis M, Habermann BH, Buchholz F. Designing efficient and specific endoribonuclease-prepared siRNAs. Methods Mol Biol 2013;942:193–204.

CHAPTER

19

Primary and acquired resistance of colorectal cancer cells to anti-EGFR antibodies converge on MEK/ERK pathway activation and can be overcome by combined MEK/EGFR inhibition

Tian Li and Shi Hu

Department of Biophysics, College of Basic Medical Sciences, Naval Medical University (Second Military Medical University), Shanghai, China

Abstract

This chapter briefly introduces the pathogenesis of colorectal cancer, treatment options, the reasons why some patients with colorectal cancer develop drug resistance, and potential solutions. Relevant studies have found that EGFR-independent RAS-RAF-MEK-MAPK pathway activation is one of the important mechanisms by which tumor cells develop resistance to cetuximab. Activation of MEK is involved in primary and acquired resistance to cetuximab, and inhibition of EGFR and MEK may be a strategy to overcome anti-EGFR resistance in colorectal cancer patients.

The researchers used a selective MEK1/2 inhibitor (BAY 86-9766) in preclinical experiments. According to reports, the inhibitor showed significant benefit in the treatment of multidrug-resistant human colorectal cancer cell lines. Furthermore, combination therapy with cetuximab and BAY 86-9766 induced synergistic antiproliferative and apoptotic effects in cetuximab-resistant colorectal cancer cells by blocking the MAPK and AKT pathways. In a model of xenograft tumors, combination therapy with cetuximab and BAY 86-9766 significantly inhibited tumor growth and improved survival.

https://doi.org/10.1016/B978-0-12-821584-5.00014-6

Abbreviations

EGFR epidermal growth factor receptor
mAbs monoclonal antibodies
MTD maximum tolerated dose
TKIs tyrosine kinase inhibitors

Conflict of interest

No potential conflicts of interest were disclosed.

Colorectal cancer is one of the major diseases with extremely high morbidity and mortality worldwide [1]. Through the development of new drugs and the improvement of treatment plans, the data show that the evaluation of clinical data after treatment of patients with metastatic colorectal cancer has increased significantly in the past 10 years, and the survival period has also been greatly extended. In the past, treatment options were limited; even with the most ideal treatment plan, the median survival time of colorectal cancer patients was only 6 months, but now patients can choose to use active chemotherapeutic drugs (e.g., fluoropyrimidines, oxaliplatin, and irinotecan) and molecularly targeted tumor therapy drugs such as cetuximab, panitumumab, and regorafenib. The patients' median survival time has also been extended to more than 2 years [2,3].

Researchers have developed several monoclonal antibodies (mAbs) for treatment. The two most widely used drugs are cetuximab and panitumumab. Both molecular targets of these antibodies are epidermal growth factor receptors (EGFRs) and can inhibit EGFR by blocking EGFR and two downstream signals. After a long period of research by researchers, it was discovered that RAS-RAF-MEK-MAPK and PTEN-PIK3CA-AKT are two key ways for EGFR-related signals to function after activation [4–6]. According to clinical trial data from patients with metastatic colorectal cancer, cetuximab alone or cetuximab combined with standard chemotherapy for the treatment is effective and recommended by the guidelines [7]. However, after carrying out anti-EGFR colorectal cancer treatment for a period of time, patients may become resistant to drugs that target EGFR. According to the researchers' analysis and speculation, this resistance may be due to the simultaneous activation of multiple signaling pathways in tumor cells and then cooperation for downstream signals or acting on the EGFR signaling pathway alone. Based on sequencing and experimental analysis, the researchers proposed that the point mutation at the 12th or 13th codon in exon 2 of the KRAS gene is the main negative predictor of cetuximab efficacy [8,9]. Therefore, in clinical treatment, doctors only recommend the treatment of cetuximab or combination chemotherapy with cetuximab to metastatic colon cancer patients whose KRAS gene phenotypes of the tumors are wild-type.

In many retrospective studies, it has been proven that the primary resistance of EGFR inhibitors in patients with colorectal cancer after treatment is not only directly related to KRAS gene mutations but may also be related to other downstream effector mutations of EGFR in cells. Multigene mutations, including those in BRAF, NRAS, and PIK3CA (exon 20), are all related to drug resistance [10]. However, even colorectal cancer patients (subgroups of molecular enriched patients) have not detected mutations that can cause primary drug resistance in KRAS, BRAF, NRAS, and PIK3CA (exon 20) genes. The initial effective rate of anti-EGFR in the clinic is not close to 100%. This result suggests that there are other undiscovered mechanisms in anti-EGFR therapy that lead to primary drug resistance [11].

In the current clinical treatment, patients with colorectal cancer will first be evaluated to determine whether there are gene mutations in the EGFR-related signaling pathway and to determine whether treatment with cetuximab is appropriate. However, all patients who initially obtained the treatment effect from cetuximab eventually developed resistance, which is also known as acquired resistance to cetuximab [12]. Furthermore, the primary resistance mechanism of anti-EGFR drugs discovered by researchers may also contribute to some effects in this process.

Some researchers hold the view and believe that the acquired resistance of patients with metastatic colorectal cancer to targeted EGFR therapy may be due to the mutation of the KRAS gene in cancer cell subtypes and its dominant expression during treatment [13,14]. In addition, among metastatic colorectal cancer patients who have not undergone gene sequencing and evaluation, there is a 3% probability that HER2 gene amplification will be found in the patient, while HER2 gene amplification will occur in KRAS wild-type tumor patients [15]. The frequency is significantly higher, and amplification of the HER2 gene will have a negative impact on the use of anti-EGFR drugs to treat patients with colorectal cancer.

In addition to HER2 gene amplification, elevated lectin levels may be one of the crucial factors for the acquired resistance of a subgroup of metastatic colorectal cancer patients resistant to cetuximab [16]. Heregulin is a ligand that can bind to other HER members (HER3 and HER4). The amplification of the HER2 gene and lectin expression levels in tumors may lead to acquired resistance to anti-EGFR monoclonal antibody therapy, and the related effect may be caused by the continuous activation of the RAS-RAF-MEK-MAPK signaling pathway.

In this regard, the intracellular RAS-RAF-MEK-MAPK signaling pathway has a great influence and effect on the transmission of proliferation signals that transmit the proliferation signal among activated growth factor receptors and the nucleus of normal cells and tumor cells. RAF protein kinase is a kinase that phosphorylates MEK1 and MEK2 and activates downstream signals through increased phosphorylation levels, which in turn phosphorylates and activates MAPK (or ERK1 and ERK2). MAPK can increase the phosphorylation level of many cell substrates [17]. This reaction plays an essential role in the proliferation and survival of cells. Therefore, using MEK inhibitors may become a promising tumor treatment strategy that can control tumor growth based on an abnormal MEK signaling pathway. There is also a potential advantage for MEK targeting. Multiple pathways, including RAS-RAF-MEK-MAPK, require MEK to assist in the transmission of signals. At MEK targets, many correlative signaling pathways can be blocked by directly inhibiting MEK. In current development, most MEK inhibitors specifically target MEK1 and/or MEK2, and no off-target effects have been found in clinical trials [18].

BAY 86-9766 is a small molecule drug and a non-ATP-competitive MEK1/2 inhibitor with high selectivity and strong effect and is suitable for oral use [19]. This drug prevents MEK from binding to ERK from the substrate and then blocks the phosphorylation of ERK [20]. The way the drug works is to bind to the allosteric site adjacent to the ATP-binding region of MEK and then interact with ATP, the activation loop, and surrounding residues following the principle of structure-determined function. According to in vitro experiments, BAY 86-9766 can inhibit the proliferation of human tumor cell lines, including cell lines that carry the BRAF V600E mutation [19]. Moreover, BAY 86-9766 can also exert certain antitumor effects in animal models of human xenotransplantation. Researchers have completed a multicenter clinical trial and evaluated the safety and tolerability of BAY 86-9766. The relevant tests showed that BAY 86-9766 has good tolerability, good oral absorption, dose ratio pharmacokinetics, and maximum tolerated dose (MTD)-targeted inhibition, and the evidence is far more than that [20,21]. BAY 86-9766 showed unexpected clinical benefits in this type of tumor. According to a phase 2 clinical study, the therapeutic effect of the combination drug (BAY 86-9766 and gemcitabine) surpassed the existing treatment for advanced pancreatic cancer, and the safety was within the license [22]. Trends in improving ORR, PFS, and OS can be observed in patients whose KRAS gene was wild-type, but the link between KRAS mutations and clinical benefits remains to be confirmed.

In tumor cells, the RAS-RAF-MEK-MAPK pathway is the most important downstream pathway of EGFR to assist tumor proliferation. The specific manifestation is that cancer patients whose RAS-RAF-MEK-MAPK signal is continuously activated often show resistance to targeted EGFR drugs. Drug resistance may appear in the initial stage of treatment, or it may manifest as the therapeutic effect of the drug on the tumor becoming negligible after the tumor recurs. To study related drug resistance mechanisms in depth, researchers used a human colorectal cancer model to evaluate the effect of BAY 86-9766 on drug-resistant tumors, in which the tumors are not sensitive to existing anti-EGFR drugs. Different human colorectal cancer cell lines were used to establish eight models, and the antiproliferative sensitivity of cetuximab and KRAS, BRAF, PIK3CA, and NRAS gene mutation profiles of these cell lines were characterized [23]. Based on the previous identification results, two human colorectal cancer cell lines resistant to cetuximab were selected for research at the same time. The cell lines resistant to cetuximab were derived from cancer cells that had been continuously treated with cetuximab [24,25]. The above studies further clarified the mechanisms of patient resistance to cetuximab and tried to help patients solve the problem of drug resistance.

A complete understanding of the resistance mechanism of tumor cells to anticancer drugs is the key factor for the more effective development of anticarcinogen and treatment methods. In recent years, researchers have focused on the mechanism of tumor drug resistance, especially resistance to EGFR inhibitors.

As the retrospective studies mentioned at the beginning showed that KRAS mutations in the hot spots of exons 2–4, BRAF, NRAS, and PIK3CA exon 20 genes are the main reasons for patients' resistance to either cetuximab or panitumumab therapy, these patients all have chemotherapy-refractory metastatic colorectal cancer [8,9,26–29]. The mechanism of action of this factor is due to the activation of downstream signals in EGFR-independent cells.

In follow-up studies, some researchers found that in a group of human colorectal cancer cell lines. Cell lines resistant to cetuximab are often due to genetic mutations in the RAS-RAF-MEK-MAPK system [23]. It should be pointed out that cells often used in experiments are activated by mutations in the KRAS gene in codons 12 or 13 of the second exon, including LOVO, HCT116, HCT15, SW620, and SW480 cells. In addition to the above cell lines, HT29 cells have mutations in the BRAF gene.

It is currently recognized that MAPK activation is the key influencing factor of the EGFR pathway. MAPK activation may also be a potential mechanism that makes tumors resistant to anti-EGFR drugs. The clinician discovers cetuximab does not work in some colorectal cancer cell lines while treating. It may not be effective at the beginning, or it may fail after a period of treatment. These two situations correspond to primary resistance and acquired resistance, respectively. Although cetuximab may be able to inhibit EGFR, it cannot completely block the activation of MAPK and AKT. The phosphorylation level of MAPK protein and the activation of signaling pathways have been proven to be important mechanisms for targeted EGFR drugs to produce acquired resistance. Many laboratories have reached similar conclusions and reported experimental results. In particular, Yonesaka's team has developed a GEO cell clone resistant to cetuximab, in which ERBB2 amplification or the upregulation of the heregulin protein activates the ERBB2 signal, which will continue to activate constitutive ERK1/2 [16].

Normanno's team proved that the continuous activation of MEK/MAPK signaling is the main reason that causes the human breast cancer cell line to become an acquired drug-resistant cell line model [30]. This resistance is mainly for the small molecule drug gefitinib that targets EGFR, which is a tyrosine kinase inhibitor. In follow-up studies, researchers have obtained two cell lines that are not sensitive to TKIs (erlotinib and gefitinib) through experimental separation [31]. To make cancer cells resistant to drugs, lung adenocarcinoma cells (CALU-3) and colorectal cancer cells (HCT116) have been exposed to the corresponding therapeutic drugs for a long time. Studies have found that the acquired resistance of EGFR inhibitors in these cells is mainly attributed to the activation of the RAS-RAF-MEK-MAPK signaling system. Sorafenib can block RAF signal transduction by administering sorafenib. Drug resistance can be overcome by sorafenib treatment.

In addition, related studies have also demonstrated that MEK-dependent epithelial-to-mesenchymal phenotype transition is a common phenotype of tumor cells acquired resistance to different types of TKIs. The drug resistance of such cancer cells can be solved by using selective MEK inhibitors [32]. Meanwhile, clones of GEO cell lines with acquired resistance to cetuximab can be inhibited by selumetinib (a MEK1/2 inhibitor), and the growth inhibitory effect of this drug is relatively sensitive [16]. Based on the above experimental results, it can be concluded that inhibition of MEK may be a potential and effective advanced therapeutic strategy to prevent and/or overcome the resistance of cancer cells to various molecular targeted drugs that block the intracellular signal transduction driven by growth factor receptor.

Combination therapy targeting EGFR antibody and MEK1/2 inhibitor can produce a very strong inhibitory effect on colorectal cancer cell lines with cetuximab resistance. This enhanced inhibition of proliferation was only observed in cell lines resistant to cetuximab. The results of this experiment suggest that there may be an antagonistic effect between different drugs when combined with medication, and the therapeutic effect of the combined medication needs to be evaluated in advance. According to the experimental results, the combined use of the two drugs can increase the lysis of PARP and increase the activation level of caspase-3. These indicators all show that the level of apoptosis is increased at this time. Therefore, the experimental results suggest that the drug's inhibitory effect on tumors may be achieved by inhibiting the activation of pMAPK and pAKT.

Activation of the RAS-RAF-MEK-MAPK signaling system may have an impact on determining cetuximab resistance. To verify this inference and whether the abovementioned resistance is related to MEK activation, the researchers also verified the results with selective MEK1/2 siRNA experiments. The recovery of the sensitivity of anti-EGFR inhibitor-resistant colorectal cancer cells to cetuximab is partly related to the inhibition of MEK1/2. MEK silencing can reverse drug resistance to a certain extent, enabling cetuximab to inhibit MAPK and inhibit tumor cell proliferation. In vivo experiments of tumor xenograft models showed that the combination of cetuximab and BAY 86-9766 can inhibit tumor growth, thereby improving the survival rate of mice. According to the experimental design, the tumors in this model acquired resistance to cetuximab.

MEK inhibitors are one of the hot spots in clinical drug research, and there are currently more than 10 related drugs under development [20,33–37]. Early-stage MEK inhibitors alone have demonstrated clinical efficacy in metastatic melanoma patients with BRAF-mutant cancers [38]. However, it is rarely observed in patients with nonselective chemically refractory metastases of other cancer types [39,40].

According to a phase I study report on selumetinib combined with cetuximab in the treatment of chemotherapy-refractory solid tumors, the combination drug is well tolerated, and relevant evidence shows that the drug has antitumor effects in patients with corresponding conditions. Among them, two patients had partial relief of symptoms, and two patients were in stable condition [41]. In a phase Ib clinical trial, patients with metastatic colorectal cancer who had the BRAF mutation participated in treatment and evaluation [42]. The MEK inhibitor LGX818 was evaluated in combination with cetuximab or the PIK3CA inhibitor BYL719, and the related treatment plan also had a certain effect.

In conclusion, the current experimental evidence shows that cetuximab combined with selective MEK inhibitors can play a synergistic antitumor role in colorectal cancer cells with primary or acquired resistance to cetuximab. Cetuximab alone could not inhibit RAS-RAF-MEK-MAPK signal transduction. In contrast, after treatment with BAY 86-9766 in the experiment, selective MEK inhibitors significantly prevented MAPK activation, inhibited tumor cell proliferation, and played a role in inducing apoptosis in almost all colorectal cancer cell line models resistant to cetuximab. Finally, treating the tumor cells with BAY 86-9766 can restore the sensitivity of cetuximab, indicating that the combined inhibition of EGFR and MEK can prevent and/or overcome the resistance of metastatic colorectal cancer patients to cetuximab, which is a reasonable treatment strategy. In this regard, targeting the MEK pathway or eventually combining MEK1/2 inhibitors with cetuximab or other EGFR-targeted drugs to delay the emergence of drug resistance may have clinical significance for patients who are not sensitive to anti-EGFR drugs.

References

[1] Malvezzi M, Arfe A, Bertuccio P, Levi F, La Vecchia C, Negri E. European cancer mortality predictions for the year 2011. Ann Oncol 2011;22:947–56.

[2] Schmoll HJ, et al. ESMO consensus guidelines for management of patients with colon and rectal cancer. A personalized approach to clinical decision making. Ann Oncol 2012;23:2479–516.

[3] Martinelli E, et al. Emerging VEGF-receptor inhibitors for colorectal cancer. Expert Opin Emerg Drugs 2013;18:25–37.

[4] Ciardiello F, Tortora G. Drug therapy: EGFR antagonists in cancer treatment. N Engl J Med 2008;358:1160–74.

[5] Mendelsohn J, Baselga J. The EGF receptor family as targets for cancer therapy. Oncogene 2000;19:6550–65.

[6] Galizia G, et al. Cetuximab, a chimeric human mouse anti-epidermal growth factor receptor monoclonal antibody, in the treatment of human colorectal cancer. Oncogene 2007;26:3654–60.

[7] Martinelli E, De Palma R, Orditura M, De Vita F, Ciardiello F. Anti-epidermal growth factor receptor monoclonal antibodies in cancer therapy. Clin Exp Immunol 2009;158:1–9.

[8] Normanno N, Tejpar S, Morgillo F, De Luca A, Van Cutsem E, Ciardiello F. Implications for KRAS status and EGFR-targeted therapies in metastatic CRC. Nat Rev Clin Oncol 2009;6:519–27.
[9] De Roock W, et al. Association of KRAS p.G13D mutation with outcome in patients with chemotherapy-refractory metastatic colorectal cancer treated with Cetuximab. JAMA 2010;304:1812–20.
[10] Van Cutsem E, et al. Cetuximab plus irinotecan, fluorouracil, and leucovorin as first-line treatment for metastatic colorectal cancer: updated analysis of overall survival according to tumor KRAS and BRAF mutation status. J Clin Oncol 2011;29:2011–9.
[11] Bokemeyer C, et al. Fluorouracil, leucovorin, and oxaliplatin with and without cetuximab in the first-line treatment of metastatic colorectal cancer. J Clin Oncol 2009;27:663–71.
[12] Trusolino L, Comoglio PM. Scatter-factor and semaphorin receptors: cell signalling for invasive growth. Nat Rev Cancer 2002;2:289–300.
[13] Misale S, et al. Emergence of KRAS mutations and acquired resistance to anti-EGFR therapy in colorectal cancer. Nature 2012;486:532–U131.
[14] Diaz LA, et al. The molecular evolution of acquired resistance to targeted EGFR blockade in colorectal cancers. Nature 2012;486:537–40.
[15] Bertotti A, et al. A molecularly annotated platform of patient-derived xenografts ("Xenopatients") identifies HER2 as an effective therapeutic target in cetuximab-resistant colorectal cancer. Cancer Discov 2011;1:508–23.
[16] Yonesaka K, et al. Activation of ERBB2 signaling causes resistance to the EGFR-directed therapeutic antibody cetuximab. Sci Transl Med 2011;3.
[17] Fremin C, Meloche S. From basic research to clinical development of MEK1/2 inhibitors for cancer therapy. J Hematol Oncol 2010;3.
[18] Yeh TC, et al. Biological characterization of ARRY-142886 (AZD6244), a potent, highly selective mitogen-activated protein kinase kinase 1/2 inhibitor. Clin Cancer Res 2007;13:1576–83.
[19] Iverson C, et al. RDEA119/BAY 869766: a potent, selective, allosteric inhibitor of MEK1/2 for the treatment of cancer. Cancer Res 2009;69:6839–47.
[20] Weekes CD, et al. Multicenter phase I trial of the mitogen-activated protein kinase 1/2 inhibitor BAY 86-9766 in patients with advanced cancer. Clin Cancer Res 2013;19:1232–43.
[21] Chang Q, Chapman MS, Miner JN, Hedley DW. Antitumour activity of a potent MEK inhibitor RDEA119/BAY 869766 combined with rapamycin in human orthotopic primary pancreatic cancer xenografts. BMC Cancer 2010;10.
[22] Van Laethem JL, et al. Phase I/II study of Refametinib (BAY 86-9766) in combination with gemcitabine in advanced pancreatic cancer. Target Oncol 2017;12:97–109.
[23] Troiani T, et al. Primary and acquired resistance of colorectal cancer cells to anti-EGFR antibodies converge on MEK/ERK pathway activation and can be overcome by combined MEK/EGFR inhibition. Clin Cancer Res 2014;20:3775–86.
[24] Troiani T, et al. Increased TGF-alpha as a mechanism of acquired resistance to the anti-EGFR inhibitor cetuximab through EGFR-MET interaction and activation of MET signaling in colon cancer cells. Clin Cancer Res 2013;19:6751–65.
[25] Ciardiello F, et al. Antitumor activity of ZD6474, a vascular endothelial growth factor receptor tyrosine kinase inhibitor, in human cancer cells with acquired resistance to antiepidermal growth factor receptor therapy. Clin Cancer Res 2004;10:784–93.
[26] De Roock W, et al. Effects of KRAS, BRAF, NRAS, and PIK3CA mutations on the efficacy of cetuximab plus chemotherapy in chemotherapy-refractory metastatic colorectal cancer: a retrospective consortium analysis. Lancet Oncol 2010;11:753–62.
[27] Douillard JY, et al. Panitumumab-FOLFOX4 treatment and RAS mutations in colorectal cancer. N Engl J Med 2013;369:1023–34.
[28] Ciardiello F, et al. Molecular profiling of the CAPRI GOIM trial in KRAS wild type (wt) metastatic colorectal cancer (mCRC) patients (pts): Cetuximab plus FOLFIRI followed by FOLFOX4 +/− cetuximab. Eur J Cancer 2013;49:S14.
[29] Stintzing S, et al. Analysis of KRAS/NRAS and BRAF mutations in FIRE-3: a randomized phase III study of FOLFIRI plus cetuximab or bevacizumab as first-line treatment for wild-type (WT) KRAS (exon 2) metastatic colorectal cancer (mCRC) patients. Eur J Cancer 2013;49:S8–9.
[30] Normanno N, et al. Breast cancer cells with acquired resistance to the EGFR tyrosine kinase inhibitor gefitinib show persistent activation of MAPK signaling. Breast Cancer Res Treat 2008;112:25–33.

[31] Morgillo F, Martinelli E, Troiani T, Orditura M, De Vita F, Ciardiello F. Antitumor activity of sorafenib in human cancer cell lines with acquired resistance to EGFR and VEGFR tyrosine kinase inhibitors. PLoS One 2011;6.
[32] Morgillo F, et al. Antitumour efficacy of MEK inhibitors in human lung cancer cells and their derivatives with acquired resistance to different tyrosine kinase inhibitors. Br J Cancer 2011;105:382–92.
[33] Martinez-Garcia M, et al. First-in-human, phase I dose-escalation study of the safety, pharmacokinetics, and pharmacodynamics of RO5126766, a first-in-class dual MEK/RAF inhibitor in patients with solid tumors. Clin Cancer Res 2012;18:4806–19.
[34] Houede N, et al. Safety and evidence of activity of MSC1936369, an oral MEK1/2 inhibitor, in patients with advanced malignancies. J Clin Oncol 2011;29.
[35] Leijen S, et al. Phase I (Ph) safety, pharmacodynamic (PD), and pharmacokinetic (PK) trial of a pure MEK inhibitor (i), RO4987655, in patients with advanced/metastatic solid tumor. J Clin Oncol 2011;29.
[36] Shapiro G, et al. Clinical combination of the MEK inhibitor GDC-0973 and the PI3K inhibitor GDC-0941: a first-in-human phase Ib study testing daily and intermittent dosing schedules in patients with advanced solid tumors. J Clin Oncol 2011;29.
[37] Tolcher AW, et al. A phase I study of RO4929097, a novel gamma secretase inhibitor, in patients with advanced solid tumors. J Clin Oncol 2010;28.
[38] Robert C, et al. Selumetinib plus dacarbazine versus placebo plus dacarbazine as first-line treatment for BRAF-mutant metastatic melanoma: a phase 2 double-blind randomised study. Lancet Oncol 2013;14:733–40.
[39] Hainsworth JD, et al. A phase II, open-label, randomized study to assess the efficacy and safety of AZD6244 (ARRY-142886) versus pemetrexed in patients with non-small cell lung cancer who have failed one or two prior chemotherapeutic regimens. J Thorac Oncol 2010;5:1630–6.
[40] Bennouna J, et al. A phase II, open-label, randomised study to assess the efficacy and safety of the MEK1/2 inhibitor AZD6244 (ARRY-142886) versus capecitabine monotherapy in patients with colorectal cancer who have failed one or two prior chemotherapeutic regimens. Invest New Drugs 2011;29:1021–8.
[41] Deming DA, et al. A phase I study of selumetinib (AZD6244/ARRY-142866) in combination with cetuximab (cet) in refractory solid tumors and KRAS mutant colorectal cancer (CRC). J Clin Oncol 2012;30.
[42] van Geel R, et al. A phase Ib dose-escalation study of Encorafenib and Cetuximab with or without Alpelisib in metastatic BRAF-mutant colorectal cancer. Cancer Discov 2017;7:610–9.

CHAPTER

20

MEK1/2 inhibitors AS703026 and AZD6244 may be potential therapies for KRAS-mutated colorectal cancer that is resistant to EGFR monoclonal antibody therapy

Guangyao Li, Changhai Lei, and Shi Hu

Department of Biophysics, College of Basic Medical Sciences, Naval Medical University (Second Military Medical University), Shanghai, China

Abstract

EGFR monoclonal antibodies are a new popular targeted drug for the recent clinical treatment of CRC patients. Nevertheless, patients with K-RAS mutations are resistant to it, and the therapeutic effect is significantly lower than expected, similar to cetuximab (Erbitux) and panitumumab (Vectibix). In this chapter, two MEK inhibitors (AZD6244 and AS703026) were tested in both cells and xenotransplantation to study the ability of the two drugs to respond to EGFR inhibitor resistance in patients with K-RAS mutations. Research has focused on human colorectal cancer cells, including several types that express mutant K-RAS or WT isotopes. The results showed that D-WT cells were significantly inhibited regardless of exposure to cetuximab in vivo or in vitro, including the RAS–ERK signaling pathway. Nevertheless, D-MUT cells had no noticeable abnormal changes. In contrast, the key to the effects of the two MEK inhibitors, AS703026 and AZD6244, lies in the downstream target kinase ERK of MEK, so in vivo and in vitro have a significant inhibitory effect on the proliferation of D-MUT cells. These research results pave the way for MEK inhibitors in K-RAS mutant CRC to become a mechanical preclinical concept application.

https://doi.org/10.1016/B978-0-12-821584-5.00024-9

Abbreviations

EGFR epidermal growth factor receptor
mAb monoclonal antibody
mCRC metastatic colorectal cancer
VEGF vascular endothelial growth factor

Conflict of interest

No potential conflicts of interest were disclosed.

The status of K-RAS mutations is the key influencing factor in determining cetuximab resistance sensitivity

Among the leading cancers causing death, colorectal cancer is a member not to be ignored [1,2]. In recent years, due to the gradual application of oxaliplatin and irinotecan chemotherapy, clinical treatment has made significant progress, and the average overall survival time of patients has increased by nearly 9 months [3]. Recently, vascular endothelial growth factor (VEGF) and epidermal growth factor (EGF) have been the most popular therapeutic targets, including popular molecularly targeted drugs such as cetuximab and bevacizumab, which also provides new ideas for the treatment of CRC patients [4–6].

The outcome and prognosis of many cancer diseases, including colorectal cancer, are affected by a combination of many factors. EGFR is one of the critical receptors, so it has attracted the attention of researchers [7].

However, some studies have found that mCRC patients with K-RAS-activating mutations are not sensitive to molecular antibody drugs targeting EGFR, and molecularly targeted drug clinical efficacy is greatly reduced [8–12]. Therefore, in July 2009, the FDA proposed the necessity of screening the K-RAS gene mutation degree of mCRC patients before using EGFR mAb treatment to avoid ineffective treatment and waste of medical resources [13].

RAS activation in cancer cells can activate their proliferation and development through pathways such as PI3 kinase-Akt and Raf–MEK–ERK [14–16].

Cetuximab and other monoclonal antibodies that target EGFR inhibit the signal cascade by recognizing the extracellular site binding of EGFR to inhibit cancer cell proliferation [17]. Since RAS–Raf–MEK–ERK belongs to the upstream signaling pathway, it is speculated that resistance to cetuximab due to K-RAS mutation in CRC may play a key role in MEK inhibition.

Colorectal cancer cell lines (DLD-1) with the same genetic background and genes were selected, and alleles were inactivated through homologous recombination [18]. The use of these cell lines avoids the possible effects of mutations of K-RAS [19–22].

These inhibitors, AS703026 and AZD6244, are MEK1/2 inhibitors that have progressed to the initial stage of clinical trials [23,24]. The two types of DLD-1 cells were treated with the above two reagents to observe their status. The results showed that MEK inhibitors have significantly better effects on K-RAS mutation types in vivo and in vitro than cetuximab. There is a more noticeable effect of inhibiting proliferation. The results show that the paired cell line is an excellent drug screening system.

Efficacy in the ERK pathway, proliferation, transformation, and tumor growth in cetuximab-resistant tumor cells caused by K-RAS mutation

RAS and ERK activity are inextricably linked, and cetuximab exerts its effect by inhibiting its GTP load, but it has no similar effect in wild-type cells. The decrease in ERK activity levels in D-MUT cells was consistent with the use of AS703026 or AZD6244, but no combined effect was observed, and AS703026 had a better inhibitory effect than AZD6244. No significant changes in Akt activity were observed during the entire experiment.

Subsequently, MTT and BrdU incorporation tests were used to measure the effects of the above two drugs on cell proliferation. The combination of cetuximab and the two drugs was applied to the DLD cell line to maintain a 1:3 action time. The proliferation of D-WT cells exposed to cetuximab alone was inhibited, and BrdU incorporation was reduced by 54%–65%. The D-MUT cells were not affected. However, adding two MEK inhibitors has inhibitory effects on both types of DLD cells. In both cell lines, BrdU incorporation was reduced by approximately 40%. Although the incorporation of BrdU in D-WT cells proved the existence of the combined effect, it was not reflected in the activation, proliferation, or transformation of ERK in cells.

The next step is to use a mouse tumor xenograft model to measure the antitumor effects of the three drugs. Throughout the experiment, the tumor volume was measured with calipers every week. The results showed no significant change in the size of D-MUT xenograft tumors exposed to cetuximab. However, AS703026 and AZD6244 showed similar apparent growth inhibitory effects, reducing the tumor volume and weight due to D-MUT cells by nearly 65%, which is related to cell-based in vitro and is consistent with cell studies. Unfortunately, the combined effect of AS703026 and AZD6244 was not observed.

Finally, we performed immunohistochemical analysis to clarify the mechanism of action of the three drugs on D-MUT xenograft tumors. An anti-pan-RAS antibody was used to detect GTP-loaded pan-RAS protein (RAS-GTP). However, a significant decrease in p-ERK staining appeared in D-MUT tumors administered AS703026 and AZD6244. After AS703026 or AZD6244 treatment of D-MUT tumors, a decrease in ATF2 activity monitored by the corresponding antibody T71 was observed, proving that the regulation of target genes is inhibited by ERK activity. In D-MUT tumors, Akt activity was not affected by exposure to cetuximab or MEK inhibitor therapy. No obvious relationship between the β-catenin level and cetuximab treatment was observed.

The development of MEK inhibitors and the discovery of their mechanism of action

Since the development of the first MEK inhibitor, PD98059, a variety of MEK inhibitors have been developed. However, due to side effects, early MEK inhibitors did not reach the clinical trial stage [25].

However, through continuous screening or modification and improvement of the original MEK inhibitors, nearly 10 MEK inhibitors have been included in the clinical trial treatment plan, the leading one is CI-1040, developed by Pfizer [24]. However, its activity, efficacy, and other aspects have not reached a satisfactory level, published after Phase II [24]. AS703026 is a particular ATP allosteric MEK1/2 inhibitor; its tumor-suppressive effect is being tested [26]. AS703026 can inhibit multiple myeloma (MM) cells when administered in combination with a variety of anti-MM therapies, and the mutation of RAS and BRAF has no effect on it [27]. The experimental results proved the excellent efficacy of MEK inhibitor drugs for patients with CRC (with K-RAS mutation). As an upper pathway, EGFR activation is a prerequisite for downstream activation [17]. The role of the RAS protein in the activation of Akt is crucial [28]. Two downstream pathways are closely involved in cell proliferation [29].

This study shows that tumor growth in xenotransplantation trials is not affected by the inhibitory effects of MEK1/2 inhibitor drugs. The results showed that neither cetuximab nor MEK inhibitor inhibited Akt activity, indicating that cancer-causing tumors with K-RAS mutations can be effectively resolved by simultaneously blocking the PI3K-Akt and RAS–ERK pathways.

In the experiment, it was observed that AS703026 and AZD6244 both stimulated an increase in the GTP load of RAS. This may be due to the desensitization of RAS activation by dissociating the Grb2-Sos complex by phosphorylation of Sos [30].

This also confirms the previous report that MEK inhibition makes RAS activation longer by preventing the phosphorylation and dissociation of Sos from Grb2 [31]. Interestingly, the BrdU incorporation experiment discovered that MEK inhibitor and cetuximab have a synergistic effect on the new DNA synthesis of wild-type cells. This indicates that the pathway that transmits proliferation signals is only substituted in wild-type DLD-1 cells. The pathway of PI3K to downstream Akt signals in parental DLD-1 cells may be blocked [32]. Therefore, the PI3K-Akt pathway may not participate in this synergy. It may be an approach involving RALGDS [33]. This may be a potential way to cause synergy.

Final thoughts and future directions

This study used DLD-1-cell syngeneic colorectal cancer cell line pairing experiments. Compared with previous similar experiments, the improvement lies in that using these cell lines will control the potentially irrelevant influence of the interaction between K-RAS mutation and wild type [19–22]. The experimental results show that MUT has apparent advantages over a variety of colon cancer cells with K-RAS mutations in that it has apparent resistance to EGFR mAb. This study proves the possibility of using a CRC xenograft tumor model with syngeneic K-RAS mutation to overcome K-RAS mutation cetuximab resistance. This shows that in the era of personalized medicine, paired syngeneic colorectal cancer cells have the potential to be used as a screening combination to determine whether they can overcome the K-RAS mutation associated with cetuximab resistance.

References

[1] Sung JJ, Lau JY, Goh KL, Leung WK. Increasing incidence of colorectal cancer in Asia: implications for screening. Lancet Oncol 2005;6:871–6.
[2] Yee YK, Tan VP, Chan P, Hung IF, Pang R, Wong BC. Epidemiology of colorectal cancer in Asia. J Gastroenterol Hepatol 2009;24:1810–6.
[3] Jemal A, Siegel R, Ward E, Hao Y, Xu J, Thun MJ. Cancer statistics, 2009. CA Cancer J Clin 2009;59:225–49.

[4] Cunningham D, Humblet Y, Siena S, Khayat D, Bleiberg H, Santoro A, et al. Cetuximab monotherapy and cetuximab plus irinotecan in irinotecan-refractory metastatic colorectal cancer. N Engl J Med 2004;351:337–45.
[5] Hurwitz H, Fehrenbacher L, Novotny W, Cartwright T, Hainsworth J, Heim W, et al. Bevacizumab plus irinotecan, fluorouracil, and leucovorin for metastatic colorectal cancer. N Engl J Med 2004;350:2335–42.
[6] Mayer RJ. Targeted therapy for advanced colorectal cancer–more is not always better. N Engl J Med 2009;360:623–5.
[7] Scartozzi M, Bearzi I, Berardi R, Mandolesi A, Fabris G, Cascinu S. Epidermal growth factor receptor (EGFR) status in primary colorectal tumors does not correlate with EGFR expression in related metastatic sites: implications for treatment with EGFR-targeted monoclonal antibodies. J Clin Oncol 2004;22:4772–8.
[8] Amado RG, Wolf M, Peeters M, Van Cutsem E, Siena S, Freeman DJ, et al. Wild-type KRAS is required for panitumumab efficacy in patients with metastatic colorectal cancer. J Clin Oncol 2008;26:1626–34.
[9] Di Fiore F, Blanchard F, Charbonnier F, Le Pessot F, Lamy A, Galais MP, et al. Clinical relevance of KRAS mutation detection in metastatic colorectal cancer treated by Cetuximab plus chemotherapy. Br J Cancer 2007;96:1166–9.
[10] Lievre A, Bachet JB, Boige V, Cayre A, Le Corre D, Buc E, et al. KRAS mutations as an independent prognostic factor in patients with advanced colorectal cancer treated with cetuximab. J Clin Oncol 2008;26:374–9.
[11] Lievre A, Bachet JB, Le Corre D, Boige V, Landi B, Emile JF, et al. KRAS mutation status is predictive of response to cetuximab therapy in colorectal cancer. Cancer Res 2006;66:3992–5.
[12] Karapetis CS, Khambata-Ford S, Jonker DJ, O'Callaghan CJ, Tu D, Tebbutt NC, et al. K-ras mutations and benefit from cetuximab in advanced colorectal cancer. N Engl J Med 2008;359:1757–65.
[13] August J. Market watch: emerging companion diagnostics for cancer drugs. Nat Rev Drug Discov 2010;9:351.
[14] Hancock JF. Ras proteins: different signals from different locations. Nat Rev Mol Cell Biol 2003;4:373–84.
[15] Downward J. Targeting RAS signalling pathways in cancer therapy. Nat Rev Cancer 2003;3:11–22.
[16] Downward J. Targeting RAS and PI3K in lung cancer. Nat Med 2008;14:1315–6.
[17] Giehl K. Oncogenic Ras in tumour progression and metastasis. Biol Chem 2005;386:193–205.
[18] Yun J, Rago C, Cheong I, Pagliarini R, Angenendt P, Rajagopalan H, et al. Glucose deprivation contributes to the development of KRAS pathway mutations in tumor cells. Science 2009;325:1555–9.
[19] Benvenuti S, Sartore-Bianchi A, Di Nicolantonio F, Zanon C, Moroni M, Veronese S, et al. Oncogenic activation of the RAS/RAF signaling pathway impairs the response of metastatic colorectal cancers to anti-epidermal growth factor receptor antibody therapies. Cancer Res 2007;67:2643–8.
[20] Yeh JJ, Routh ED, Rubinas T, Peacock J, Martin TD, Shen XJ, et al. KRAS/BRAF mutation status and ERK1/2 activation as biomarkers for MEK1/2 inhibitor therapy in colorectal cancer. Mol Cancer Ther 2009;8:834–43.
[21] Mahoney CL, Choudhury B, Davies H, Edkins S, Greenman C, Haaften G, et al. LKB1/KRAS mutant lung cancers constitute a genetic subset of NSCLC with increased sensitivity to MAPK and mTOR signalling inhibition. Br J Cancer 2009;100:370–5.
[22] Zoppoli G, Moran E, Soncini D, Cea M, Garuti A, Rocco I, et al. Ras-induced resistance to lapatinib is overcome by MEK inhibition. Curr Cancer Drug Targets 2010;10:168–75.
[23] Bennouna J, Lang I, Valladares-Ayerbes M, et al. A Phase II, open-label, randomised study to assess the efficacy and safety of the MEK1/2 inhibitor AZD6244 (ARRY-142886) versus capecitabine monotherapy in patients with colorectal cancer who have failed one or two prior chemotherapeutic regimens. Invest New Drugs Epub 2010. Google Scholar.
[24] Fremin C, Meloche S. From basic research to clinical development of MEK1/2 inhibitors for cancer therapy. J Hematol Oncol 2010;3:8.
[25] Friday BB, Adjei AA. Advances in targeting the Ras/Raf/MEK/Erk mitogen-activated protein kinase cascade with MEK inhibitors for cancer therapy. Clin Cancer Res 2008;14:342–6.
[26] Price S. Putative allosteric MEK1 and MEK2 inhibitors. Expert Opin Ther Patents 2008;18:603–27.
[27] Kim K, Kong SY, Fulciniti M, Li X, Song W, Nahar S, et al. Blockade of the MEK/ERK signalling cascade by AS703026, a novel selective MEK1/2 inhibitor, induces pleiotropic anti-myeloma activity in vitro and in vivo. Br J Haematol 2010;149:537–49.
[28] Rodriguez-Viciana P, Warne PH, Dhand R, Vanhaesebroeck B, Gout I, Fry MJ, et al. Phosphatidylinositol-3-OH kinase as a direct target of Ras. Nature 1994;370:527–32.
[29] Yarden Y, Sliwkowski MX. Untangling the ErbB signalling network. Nat Rev Mol Cell Biol 2001;2:127–37.
[30] Waters SB, Holt KH, Ross SE, Syu LJ, Guan KL, Saltiel AR, et al. Desensitization of Ras activation by a feedback disassociation of the SOS-Grb2 complex. J Biol Chem 1995;270:20883–6.

[31] Langlois WJ, Sasaoka T, Saltiel AR, Olefsky JM. Negative feedback regulation and desensitization of insulin- and epidermal growth factor-stimulated p21ras activation. J Biol Chem 1995;270:25320–3.
[32] Wee S, Jagani Z, Xiang KX, Loo A, Dorsch M, Yao YM, et al. PI3K pathway activation mediates resistance to MEK inhibitors in KRAS mutant cancers. Cancer Res 2009;69:4286–93.
[33] Matsubara K, Kishida S, Matsuura Y, Kitayama H, Noda M, Kikuchi A. Plasma membrane recruitment of RalGDS is critical for Ras-dependent Ral activation. Oncogene 1999;18:1303–12.

CHAPTER

21

EGFR/JIP-4/JNK2 signaling attenuates cetuximab-mediated radiosensitization of squamous cell carcinoma cells

Kewen Qian[a], *Tian Li*[a], *Wenyan Fu*[b], *Changhai Lei*[a], *and Shi Hu*[a]

[a]Department of Biophysics, College of Basic Medical Sciences, Naval Medical University (Second Military Medical University), Shanghai, China [b]Department of Assisted Reproduction, Shanghai Ninth People's Hospital, Shanghai Jiao Tong University School of Medicine, Shanghai, China

Abstract

EGF receptor (EGFR) can promote the growth of malignant tumors and enhance the tolerance of malignant tumor cells to radiation and chemotherapy, including squamous cell carcinomas (SCCs). In addition to inhibiting pro-survival signals, cetuximab, a targeted drug for EGFR, may also induce signal bypass related to cell apoptosis. Therefore, researchers need to confirm these bypasses to improve the effectiveness of EGFR-targeted therapy. In this chapter, the researchers constructed three cell culture models and found that when cetuximab-treated SCC cells were cultured in a three-dimensional cell matrix (lrECM), the EGFR signals in these cells and radiation tolerance of the cells were similar to tumor xenografts but different from the traditional two-dimensional cultured SCC cells. According to phosphorylation analysis, the phosphorylation signal of JNK2 was significantly enhanced with cetuximab tolerance. Using immunoprecipitation mass spectrometry, it was found that the downstream JNK/JIP-4 pathway was activated after cetuximab-inhibited EGFR. When JIP-4 or JNK2 was knocked out or JNK2 inhibitory drugs were used, the therapeutic effect of cetuximab and the radiosensitivity of tumors were obviously enhanced. Therefore, it is possible to reduce cetuximab tolerance in tumor cells by inhibiting JIP-4/JNK2 signal bypass. In addition, through the construction of three cell culture models, it was found that the growth status of three SCC cell lines under three-dimensional matrix culture was close to that of tumor cells in tumor transplants, which is constructive for studying the resistance mechanism of cetuximab and the intracellular phosphorylation signals in the treatment of HNSCC.

https://doi.org/10.1016/B978-0-12-821584-5.00002-X

Abbreviations

CREB	cAMP response element-binding protein
Elk1	E twenty-six-like transcription factor 1
HNSCC	head and neck squamous cell carcinomas
JIP-4	c-jun-NH2-kinase-interacting protein-4
JNK	c-jun-NH2-kinase
lrECM	Cultrex 3D Culture Matrix
MEK	MAP–ERK kinase
MNK1	MAPK-interacting kinase 1
PI3K	phosphoinositide 3-kinase
SCC	squamous cell carcinomas

Conflict of interest

No potential conflicts of interest were disclosed.

3-Dimensional (3D) extracellular matrix (ECM) for studying how HNSCC cells grow in vivo

Although cetuximab has been widely used in the clinic, the drug resistance of this tumor-targeted therapy has always been a shortcoming. However, research on its mechanism is not in depth. Existing studies suggest that the EGFR-Src protein kinase family, ErbB2 signaling, and p53 signaling are related to the decreased efficacy of cetuximab in the treatment of colon cancer and nonsmall cell lung cancer [1–3]. In fact, the study of EGFR signaling in tumor tissues is still marginal, and the differences in different tumors also determine the different mechanisms related to the EGFR pathway. Therefore, researchers need to study the mechanisms of drug resistance by determining the growth status of HNSCC cells in vivo. Previous studies have confirmed that protein expression, protein–protein interactions, intracellular signals, and sensitivity to radiotherapy and chemotherapy under 3D growth conditions are more similar to the physiological state than in the traditional two-position culture mode [4–6].

Growth conditions play an important role in EGFR phosphorylation and signaling

First, we compared the EGFR signal expression of SCC cell lines and the phosphorylation of downstream-related signaling pathways and molecules in cells cultured in traditional monolayer culture medium, 3D lrECM-cultured cells and tumor transplants. Comparing the expression levels of EGFR, ERK1/2 and Akt and the phosphorylation of corresponding phosphorylation sites in three SCC cell lines, FaDu, UTSCC14 and A431, under the above three culture conditions, it was found that except for A431, the expression of EGFR and the activation of related pathways in FaDu and UTSCC14 cells were similar to those in tumor transplants and significantly different from those under 2D culture conditions. This result supported the in vivo physical similarity of 3D lrECM-cultured cells.

Cytotoxicity and radiosensitizing potential of cetuximab

The tumor cells were stimulated with EGF under serum-free culture conditions to test the inhibitory effect of cetuximab. It was found that under 2D and 3D lrECM culture conditions, cetuximab treatment alone had the same effect on SCC cells, but when combined with radiotherapy, the radiosensitivity of the cells was cetuximab concentration dependent, and the higher the concentration was, the higher the cell death rate. This result indicated that the effect of cetuximab alone is not closely related to the cell culture method.

Phosphoproteome modifications after cetuximab-mediated EGFR inhibition

The phosphoproteome array technique was used to detect the changes in the intracellular signal network under cetuximab treatment. The test contained 1318 phosphorylation-specific antibodies. The proteins whose phosphorylation level was between 65% and 150% in FaDu cells cultured with 3D lrECM were selected, which represented the decrease in kinase phosphorylation levels and the increase in kinase inhibitory protein phosphorylation levels under cetuximab treatment, respectively. The results showed that EGFR-specific phosphorylation sites were reduced by approximately 70% under cetuximab treatment, which was significantly different and more noticeably decreased under 3D lrECM culture conditions. Following EGFR inhibition, the phosphorylation levels of MAPK, ERK1/2, Elk1, MNK1, CREB, and other proteins in the downstream MAPK signaling pathway are reduced, and proteins such as PI3K and Akt are dephosphorylated. In addition to the dephosphorylation of kinases, hyperphosphorylation of Rac-MEK4/7-JNK-c-Jun signals was also observed in 3D lrECM and tumor grafts, in which the excessive phosphorylated activation levels of JNK1/2 and c-Jun S63 reached 600% and 250%, respectively. These results indicated that the dephosphorylation of EGFR-related pathways under the action of cetuximab is accompanied by selective inhibition of MEK1/2-ERK1/2 and PI3K/Akt pathway-related proteins.

JNK2 depletion enhances cetuximab-induced radiosensitization

The JNK family members function in a multifunctional, environment-driven, and cell line/type-dependent manner [7]. Among many factors, X-ray radiation is a well-known JNK activator [7,8]. To determine whether JNK2 hyperphosphorylation changes the efficacy of cetuximab, researchers used siRNA to knock out JNK2 or used the inhibitory drug SP600125 to target JNK2 to explore the radiosensitization effect on FaDu cells. It was found that under the action of cetuximab, the lack of JNK2 is beneficial to improve the radiosensitivity of FaDu cells, and EGFR knockout inhibited the induction of cetuximab. JNK2 is highly phosphorylated, and JNK2 knockout affected the activation of c-Jun protein. This result confirms the above conclusion of the phosphoproteome array technique test.

JIP-4 connects EGFR to JNK2

After cetuximab-treated FaDu cells were cultured with 3D lrECM, mass spectrometry analysis of the EGFR immunoprecipitation complex was performed to identify a variety of interacting proteins, and the link molecule JIP-4 between EGFR and JNK signaling was found. JIP-4 knockout also radiosensitized FaDu cells to cetuximab treatment, and inhibition of the hyperphosphorylation of JNK2 induced by cetuximab was observed. The results of immunofluorescence staining and colocalization detection showed that the decrease in EGFR signaling under the action of cetuximab was accompanied by a decrease in JIP-4 signaling.

Discussion

In conclusion, the above results showed that JNK2 affects the radiosensitivity of 3D lrECM-cultured SCC cells under cetuximab treatment through the interaction of JIP-4 and EGFR. The significance of this research is mainly manifested in two aspects. First, a three-dimensional cell culture method was constructed and characterized. Under this culture condition, the growth state of tumor cells is similar to that in tumor tissue mass, which is helpful for cell growth in vivo. It is meaningful for the study of intracellular signaling pathways. Second, a bypass signaling pathway related to the drug resistance mechanism of EGFR-targeted therapy was found, which is the EGFR/JIP-4/JNK2 pathway. JNK2 participates in the cetuximab antagonistic signaling pathway through the interaction between EGFR and JIP-4. Interfering with JNK2 signaling can enhance the radiosensitivity of tumor cells. Therefore, the results of this study provide new ideas for the drug resistance mechanism of cetuximab-targeted therapy of tumors.

References

[1] Wheeler DL, et al. Epidermal growth factor receptor cooperates with Src family kinases in acquired resistance to cetuximab. Cancer Biol Ther 2009;8:696–703.

[2] Huang S, Benavente S, Armstrong EA, Li C, Wheeler DL, Harari PM. p53 modulates acquired resistance to EGFR inhibitors and radiation. Cancer Res 2011;71:7071.

[3] Lu Y, et al. Epidermal growth factor receptor (EGFR) ubiquitination as a mechanism of acquired resistance escaping treatment by the anti-EGFR monoclonal antibody cetuximab. Cancer Res 2007;67:8240.

[4] Kenny P. Three-dimensional extracellular matrix culture models of EGFR signalling and drug response. Biochem Soc Trans 2007;35:665–8.

[5] Nelson CM, Inman JL, Bissell MJ. Three-dimensional lithographically defined organotypic tissue arrays for quantitative analysis of morphogenesis and neoplastic progression. Nat Protoc 2008;3:674–8.

[6] Weaver VM, et al. beta4 integrin-dependent formation of polarized three-dimensional architecture confers resistance to apoptosis in normal and malignant mammary epithelium. Cancer Cell 2002;2:205–16.

[7] Eke I, Schneider L, Förster C, Zips D, Kunz-Schughart LA, Cordes N. EGFR/JIP-4/JNK2 signaling attenuates cetuximab-mediated radiosensitization of squamous cell carcinoma cells. Cancer Res 2013;73:297–306.

[8] Hochedlinger K, Wagner EF, Sabapathy K. Differential effects of JNK1 and JNK2 on signal specific induction of apoptosis. Oncogene 2002;21:2441–5.

CHAPTER

22

EGFR/Notch antagonists enhance the response to inhibitors of the PI3K-Akt pathway by decreasing tumor-initiating cell frequency

Tian Li, Changhai Lei, and Shi Hu

Department of Biophysics, College of Basic Medical Sciences, Naval Medical University (Second Military Medical University), Shanghai, China

Abstract

EGFR and the PI3K-Akt signaling pathway are common targets in cancer therapy and are often reported to develop associated drug resistance. This chapter describes the development of bispecific antibodies against the EGFR and Notch signaling pathways and appears to be able to overcome the problems caused by these resistances.

Bispecific antibodies targeting EGFR/Notch have been found to exhibit significant antitumor stem cell effects in both in vitro and in vivo trials. Bispecific antibodies delayed the onset of resistance to EGFR inhibitors in triple-negative breast cancer cell lines and showed efficacy in xenograft tumor experiments. Enrichment of cancer stem cell (CSC) subsets and dysregulation of Notch signaling may be challenges for treatment. Some studies have reported that bispecific antibody therapy inhibits stem cell-like subsets, reduces the frequency of tumor stem cells, and downregulates mesenchymal gene expression. What was initially confirmed is that the joint blocking of EGFR and Notch signals has the potential to increase the response to PI3K inhibition. Incorporating bispecific antibodies into combination therapy for cancer may result in additional clinical benefits.

Abbreviations

CSC cancer stem cell
DLL Delta-like ligand
EGFR epidermal growth factor receptor
EMT epithelial-mesenchymal transition
ER estrogen receptor

Novel Sensitizing Agents for Therapeutic Anti-EGFR Antibodies
https://doi.org/10.1016/B978-0-12-821584-5.00010-9

HER human epidermal growth factor receptor
PDX patient-derived xenograft
PR progesterone receptor
TIC tumor-initiating cell
TKI tyrosine kinase inhibitor

Conflict of interest

No potential conflicts of interest were disclosed.

Triple-negative breast cancer (TNBC) means that the patient's tissue tests fail to find three routine positive detections. The three routine detection indicators are estrogen receptor, progesterone receptor, and human epidermal growth factor receptor 2. Doctors and researchers pay attention to these indicators because, in most cases, they can be targeted as therapeutic targets for cancer patients, and the treatment effect is generally optimistic in the early stage. However, the prognosis of triple-negative breast cancer is poor and has a higher recurrence rate than other breast cancers. Cancer recurrence has little relationship with tumor size and lymph node status. According to the latest data, the diagnosis rate of triple-negative breast cancer accounts for more than 15% of the total number of patients with breast cancer. This also means that the healthcare system faces enormous challenges. At present, even with the best treatment regimen prescribed by the doctor, patients still have a high risk of recurrence [1,2]. Statistics show that the 5-year survival rate of triple-negative breast cancer patients after metastasis is only 12%, while the local recurrence rate is as high as 72% [3]. Surprisingly, no targeted therapy or drug is yet FDA-recommended to improve the outcomes for TNBC patients [4]. Are there any promising targets in clinical research that could be developed into specific targeted therapies for TNBC? EGFR is a common overexpressed receptor in cancer, and the probability of overexpression of the gene encoding this receptor in TNBC patients is approximately 50% [5]. According to the literature reports, a missing phosphatase on chromosome 10 was found in TNBC [2,6]. Genes with tensin homologs are frequently inactivated or have reduced expression. These research results suggest that EGFR may be an ideal target for the treatment of TNBC, and further clinical data also show that the use of EGFR inhibitors and pan-PI3K inhibitors has significant benefits for the treatment of TNBC patients [7,8]. However, according to further data, the benefits of such treatment regimens are not durable, and similar to numerous reports, these patients who achieve response with EGFR inhibitors often become resistant to the treatment drugs quickly [9,10], which cast a shadow on the clinical prospects of EGFR-targeted drugs that have been widely optimistic before. A representative head and neck cancer-related clinical trial reported in 2016 showed that the addition of cetuximab to radiotherapy-cisplatin did not improve prognosis and should therefore not be routinely prescribed [11]. If this result is extended to TNBC, then therapeutic strategies targeting EGFR also need to be reconsidered.

In those clinical experiments and preclinical experiments that failed to achieve the expected goals, the researchers also found some regularities by summarizing, such as the theory of cancer stem cells (CSCs) and tumor-initiating cells (TICs) obtained after more than 10 years of research [12]. A relatively solid theoretical foundation is used to guide further research work. Chemotherapy and radiotherapy, which are the most widely used in cancer treatment, also stimulate CSC turnover through cytokine production and DNA-autonomous repair mechanisms while killing a large number of cancer cell subpopulations in unanticipated circumstances [13]. This stimulation of CSCs may ultimately lead to some degree of resistance of CSCs to radiotherapy and chemotherapy. This is the most important cause of cancer recurrence. After drug treatment, scientists find more direct results of tumor resistance. A few tumor cells of different origins that are resistant to EGFR-blocking drugs also show resistance to other treatment options, such as paclitaxel [14]. There are also studies that suggest that CSCs are inherently resistant to EGFR inhibitor therapy [15]. Several research groups have reported that patients who received EGFR blockade therapy detected an enrichment of CSCs in the subsequent lungs [16,17]. These studies also found that enriched CSCs in lung cancer showed high resistance to both the most-established EGFR-blocking regimens (antibodies and TKIs).

At present, researchers generally agree that CSCs and their drug resistance are highly related to the Notch gene, which encodes a class of highly conserved cell surface receptors [18]. The Notch signaling pathway is a conserved cell fate-determining signal whose function is involved in many aspects of cancer research, including cancer cell phenotype, angiogenesis around cancer cells, tumor microenvironment, cancer metastasis, and tumor immune escape. The Notch signaling pathway acts by linking a transmembrane ligand on a cell to a Notch receptor paralog [19]. The key genes of the Notch signaling pathway include the HES gene and related genes, whose effect is related to the CSC phenotype and is a transcriptional repressor [20]. Researchers have found abnormal expression and modification of Notch and related proteins in a variety of cancers, and these changes prevent the pathway from functioning properly. Genetic studies have also shown that there is a cross-regulation between the two signal pathways, EGFR and Notch, during embryonic development, but whether such cross-influence hinders each other or serves similar functions depends largely on the type of animal or tissue model used in the study [21,22]. It is widely believed that Notch-EGFR acts synergistically in breast cancer, making patients resistant to targeted therapy by inducing epithelial-mesenchymal transition (EMT) and increasing the invasion of cancer [23]. The study found a kinase-dependent link between the EGFR receptor and the tyrosine phosphorylation of Notch3. Inhibition of EGFR kinase activity initiates activation of the Notch transcription site and subsequently fully activates the complete Notch signaling pathway [17]. Studies have shown that cetuximab (blocking EGFR signaling) has a strong antitumor effect, but it also promotes EMT and activates Notch, which induces significant tumor changes that promote subsequent tumor growth and aggressiveness [24]. In 2017, Hu et al. found that combined blockade of bispecific antibodies in a preclinical model could reduce EGFR and Notch signaling while reducing the frequency of CSCs and enhancing the response of cancer cells to radiotherapy [16]. In addition, Notch is also associated with PI3K, PTEN, AKT, and other signals. Abnormalities in Notch-related signaling pathways may also be the key to inducing tumorigenesis [25,26]. Clinical data suggest that such abnormalities are frequently detected in aggressive human cancers [27]. A clear demonstration is that the use of - PI3K-AKT inhibitors together with bispecific antibodies targeting EGFR and HER3 can significantly inhibit the growth of TNBC in a xenograft model [28].

Many data and inferences made by researchers have shown the close interplay between EGFR signaling, PI3K signaling, and Notch signaling in cancer and the mechanism for cotargeted therapeutic strategies. Therefore, the researchers hypothesized that coblocking EGFR signaling and Notch signaling combined with the therapeutic strategy of inhibiting PI3K can effectively improve the therapeutic effect of EGFR-positive TNBC, reduce the proportion of CSCs, and reduce the probability of recurrence and metastasis.

Preclinical experimental results also confirmed this hypothesis [29]. First, drugs that inhibit EGFR and PI3K-Akt signaling in a variety of tumor cells show that the dynamics of stem-like properties in TNBC are thus affected. Second, the study elucidated that Notch receptors and related genes were significantly altered after the use of drugs to inhibit EGFR and PI3K-Akt signaling. At the same time, enrichment of TNBC stem-like cells was observed in the study, indicating that the two are closely related, and the study also verified this connection in reverse. The design and characterization of IgG-like antibodies can act as HER/Notch antagonists to exert antitumor effects similar to those reported earlier. Regarding drug resistance, which is of most concern to researchers and patients, the results show that simultaneous targeting of multiple targets, such as HER and Notch2/3, can largely delay the occurrence of drug resistance. More surprising results are the results of animal experiments. The therapeutic strategy targeting both HER and Notch2/3 has significant antitumor activity not only at the cellular level in vitro but also in vivo.

Further analyzing the multitargeted therapy strategy, the researchers found that blocking both the HER signaling pathway and the Notch signaling pathway could attenuate the increase in the proportion of stem-like cells in tumors, which was closely related to PI3K-Akt signaling pathway inhibitors. This fully shows that multitargeted therapy can enhance the effect of PI3K-Akt signaling inhibitors and increase the sensitivity of tumors to therapy. Following the discovery that this treatment strategy could be a groundbreaking advance, the treatment regimen has been further promoted. Simultaneous HER/Notch-targeting antibody therapy can yield benefits for different kinds of solid tumors, such as pancreatic cancer, ovarian cancer, and nonsmall cell lung cancer. Efficacy treatment of many solid tumors in turn supports the hypothesis that coblocking EGFR signaling and Notch signaling combined with a therapeutic strategy to inhibit PI3K can effectively improve tumor therapy and reduce the proportion of CSCs.

Existing clinical and preclinical research data suggest that tumors can develop resistance to targeted drugs through multiple pathways. In tumors sensitive to EGFR-targeting drugs, both EMT and a higher proportion of cancer stem cells contribute to resistance to therapy [30]. Preliminary studies have shown that therapeutic strategies targeting EGFR alone or HER3 alone ultimately lead to an increase in the relative proportion of ALDH-positive cells. In patient-derived xenograft (PDX) studies, simultaneous targeting of EGFR and PI3K highly increases the proportion of CSCs and ultimately leads to the possibility of earlier emergence of resistance. Previous studies have identified a radio- and chemotherapy-resistant phenotype in CSCs in tumors [31]. Such treatment regimens would be of considerable importance if the elimination of such CSC subsets could reduce the likelihood of relapse after EGFR-targeted therapy.

A number of research reports have shown that EGFR signaling and Notch signaling are key factors that are difficult to solve in CSC-related tumor research and the researchers mentioned above developed multispecific antibodies to solve one of the problems [29]. The researchers named the engineered antibody MTJ16. It is worth emphasizing again that the MTJ16 antibody binds to EGFR and Notch2/3, which not only significantly inhibits tumor growth in animal models but also reduces the increase in the proportion of CSCs and tumor recurrence that are easily mediated by many antitumor drugs. MTJ1G was engineered as an IgG-based bivalent tetraspecific antibody similar to the previously reported tetrabody FL518 (recognizing EGFR, HER2, HER3, and VCGF simultaneously) [32]. However, the treatment regimen using this antibody drug is not foolproof, and MTJ16 does not seem to show additional therapeutic benefit compared to another dual-targeted antibody targeting EGFR and Notch. For example, in the treatment of drug-resistant tumor cells, neither antibody showed a significant therapeutic effect. This result, which is similar to other previous studies [31–33], suggests that there are other mechanisms of tumor resistance and there is complex tumor cell heterogeneity in resistance to targeted blockade of EGFR therapy. The Notch antagonist tarextumab was also used in the study, and the application of this drug demonstrated stronger tumor growth inhibition in the PDX model, suggesting that the Notch signaling pathway and the corresponding enrichment of CSCs have a crucial effect on tumor growth in vivo, which cannot be observed at the cellular level. In addition, the findings suggest that the regulation of tumor heterogeneity largely affects the expression of EMT genes. EGFR-blocking treatment of different cell lines and different cell subsets was accompanied by enrichment of cell subsets with a high expression of mesenchymal markers, and such treatments also positively affected EMT gene upregulation in tumors.

In conclusion, the unremitting efforts of researchers have preliminarily confirmed that blocking the two targets of EGFR and Notch receptors can produce better tumor treatment effects and can also exert an inhibitory effect on CSCs. This positive effect is closely related to promoting the blocking response of PI3K, which has been demonstrated in various cancer cell lines and PDX animal models. Some therapeutic strategies that adhere to specific principles have already had a large number of practical cases and proofs in clinical practice, so the combination of molecular-targeted drugs and CSC-targeted drugs may obtain further therapeutic benefits. The findings presented in this chapter also support the conclusion from the preclinical level that combination immunotherapy (drug therapy targeting both EGFR and Notch) as a recently developed treatment strategy is expected to improve the prognosis of TNBC.

References

[1] Foulkes WD, Smith IE, Reis-Filho JS. Triple-negative breast cancer. N Engl J Med 2010;363:1938–48.

[2] Shah SP, et al. The clonal and mutational evolution spectrum of primary triple-negative breast cancers. Nature 2012;486:395–9.

[3] Berger ER, Park T, Saridakis A, Golshan M, Greenup RA, Ahuja N. Immunotherapy treatment for triple negative breast cancer. Pharmaceuticals (Basel, Switzerland) 2021;14.

[4] Gradishar WJ, et al. Breast cancer, version 3.2020, NCCN clinical practice guidelines in oncology. J Natl Compr Canc Netw 2020;18:452–78.

[5] Carey L, Winer E, Viale G, Cameron D, Gianni L. Triple-negative breast cancer: disease entity or title of convenience? Nat Rev Clin Oncol 2010;7:683–92.

[6] Ibrahim YH, et al. PI3K inhibition impairs BRCA1/2 expression and sensitizes BRCA-proficient triple-negative breast cancer to PARP inhibition. Cancer Discov 2012;2:1036–47.

[7] Carey LA, et al. TBCRC 001: randomized phase II study of cetuximab in combination with carboplatin in stage IV triple-negative breast cancer. J Clin Oncol Off J Am Soc Clin Oncol 2012;30:2615–23.

[8] Ellis H, Ma CX. PI3K inhibitors in breast cancer therapy. Curr Oncol Rep 2019;21:110.

[9] Chong CR, Jänne PA. The quest to overcome resistance to EGFR-targeted therapies in cancer. Nat Med 2013;19:1389–400.

[10] Suda K, Mizuuchi H, Maehara Y, Mitsudomi T. Acquired resistance mechanisms to tyrosine kinase inhibitors in lung cancer with activating epidermal growth factor receptor mutation—diversity, ductility, and destiny. Cancer Metastasis Rev 2012;31:807–14.

[11] Truong MT, et al. Quality of life and performance status from a substudy conducted within a prospective phase 3 randomized trial of concurrent accelerated radiation plus cisplatin with or without Cetuximab for locally advanced head and neck carcinoma: NRG oncology radiation therapy oncology group 0522. Int J Radiat Oncol Biol Phys 2017;97:687–99.
[12] Batlle E, Clevers H. Cancer stem cells revisited. Nat Med 2017;23:1124–34.
[13] Li X, et al. Intrinsic resistance of tumorigenic breast cancer cells to chemotherapy. J Natl Cancer Inst 2008;100:672–9.
[14] Shien K, et al. Acquired resistance to EGFR inhibitors is associated with a manifestation of stem cell-like properties in cancer cells. Cancer Res 2013;73:3051–61.
[15] Corominas-Faja B, et al. Stem cell-like ALDH(bright) cellular states in EGFR-mutant non-small cell lung cancer: a novel mechanism of acquired resistance to erlotinib targetable with the natural polyphenol silibinin. Cell Cycle (Georgetown, Tex) 2013;12:3390–404.
[16] Hu S, et al. Antagonism of EGFR and notch limits resistance to EGFR inhibitors and radiation by decreasing tumor-initiating cell frequency. Sci Transl Med 2017;9.
[17] Arasada RR, Amann JM, Rahman MA, Huppert SS, Carbone DP. EGFR blockade enriches for lung cancer stem-like cells through Notch3-dependent signaling. Cancer Res 2014;74:5572–84.
[18] Locatelli M, Curigliano G. Notch inhibitors and their role in the treatment of triple negative breast cancer: promises and failures. Curr Opin Oncol 2017;29:411–27.
[19] Gu JW, Rizzo P, Pannuti A, Golde T, Osborne B, Miele L. Notch signals in the endothelium and cancer "stem-like" cells: opportunities for cancer therapy. Vasc Cell 2012;4:7.
[20] Chen J, et al. 3,6-dihydroxyflavone suppresses the epithelial-mesenchymal transition in breast cancer cells by inhibiting the notch signaling pathway. Sci Rep 2016;6:28858.
[21] Kumar JP, Moses K. EGF receptor and notch signaling act upstream of eyeless/Pax6 to control eye specification. Cell 2001;104:687–97.
[22] Price JV, Savenye ED, Lum D, Breitkreutz A. Dominant enhancers of Egfr in Drosophila melanogaster: genetic links between the Notch and Egfr signaling pathways. Genetics 1997;147:1139–53.
[23] Baker AT, Zlobin A, Osipo C. Notch-EGFR/HER2 bidirectional crosstalk in breast cancer. Front Oncol 2014;4:360.
[24] Schmitz S, Bindea G, Albu RI, Mlecnik B, Machiels JP. Cetuximab promotes epithelial to mesenchymal transition and cancer associated fibroblasts in patients with head and neck cancer. Oncotarget 2015;6:34288–99.
[25] Palomero T, et al. Mutational loss of PTEN induces resistance to NOTCH1 inhibition in T-cell leukemia. Nat Med 2007;13:1203–10.
[26] Kwon OJ, et al. Notch promotes tumor metastasis in a prostate-specific Pten-null mouse model. J Clin Invest 2016;126:2626–41.
[27] Muellner MK, et al. A chemical-genetic screen reveals a mechanism of resistance to PI3K inhibitors in cancer. Nat Chem Biol 2011;7:787–93.
[28] Tao JJ, et al. Antagonism of EGFR and HER3 enhances the response to inhibitors of the PI3K-Akt pathway in triple-negative breast cancer. Sci Signal 2014;7:ra29.
[29] Fu W, et al. EGFR/notch antagonists enhance the response to inhibitors of the PI3K-Akt pathway by decreasing tumor-initiating cell frequency. Clin Cancer Res 2019;25:2835–47.
[30] Yauch RL, et al. Epithelial versus mesenchymal phenotype determines in vitro sensitivity and predicts clinical activity of erlotinib in lung cancer patients. Clin Cancer Res 2005;11:8686–98.
[31] Hu S, et al. Broad RTK-targeted therapy overcomes molecular heterogeneity-driven resistance to cetuximab via vectored immunoprophylaxis in colorectal cancer. Cancer Lett 2016;382:32–43.
[32] Hu S, et al. Four-in-one antibodies have superior cancer inhibitory activity against EGFR, HER2, HER3, and VEGF through disruption of HER/MET crosstalk. Cancer Res 2015;75:159–70.
[33] Rho JK, et al. MET and AXL inhibitor NPS-1034 exerts efficacy against lung cancer cells resistant to EGFR kinase inhibitors because of MET or AXL activation. Cancer Res 2014;74:253–62.

CHAPTER

23

Inhibition of SLC1A5 sensitizes colorectal cancer to cetuximab

Zetong Ma[a], *Tian Li*[a], *Wenyan Fu*[b], *Changhai Lei*[a], *and Shi Hu*[a]

[a]Department of Biophysics, College of Basic Medical Sciences, Naval Medical University (Second Military Medical University), Shanghai, China [b]Department of Assisted Reproduction, Shanghai Ninth People's Hospital, Shanghai Jiao Tong University School of Medicine, Shanghai, China

Abstract

Targeting metabolic resource inputs is a tragedy to overcome cetuximab resistance and resensitize cancer cells to antitumor therapies in metastatic colorectal cancer (mCRC). Huanrong Ma and colleagues demonstrated that the glutamine transporter solute carrier family 5 (SLC1A5) was significantly overexpressed in patients resistant to cetuximab compared to those who responded to cetuximab [1]. SLC1A5 can significantly inhibit cell proliferation and growth in mCRC. In addition, inhibition of SLC1A5 can significantly enhance cetuximab's effect on inhibiting CRC proliferation. Mechanistically, the inhibition of SLC1A5 promotes EGFR degradation through the ubiquitin–proteasome pathway and decreases nuclear EGFR expression. This study provides a potential therapeutic target for increasing the efficacy of cetuximab in colorectal cancer with SLC1A5.

Abbreviations

DSBS	double-stranded DNA breaks
GLUT1	glucose transporter-1
mCRC	metastatic colorectal cancer
mEGFR	membrane EGFR
MET	mesenchymal epithelial transition
nEGFR	nuclear EGFR
NSDHL	NADP-dependent steroid dehydrogenase-like
PTEN	phosphatase and tensin homolog
SLC1A5	solute carrier family 5
TCGA	The Cancer Genome Atlas

https://doi.org/10.1016/B978-0-12-821584-5.00023-7

Conflict of interest

No potential conflicts of interest were disclosed.

Introduction

RAS status and other molecular changes after treatment with cetuximab, such as increased ERBB2, ERBB3, or MET levels, may contribute to cetuximab resistance [2,3]. Therefore, it is clinically important to investigate potential strategies to enhance the sensitivity of mCRC to cetuximab. Disordered cell metabolism has been regarded as a feature of cancer [4]. Two basic metabolic resources, glucose and glutamine, facilitate cancer cell importation to promote cell growth and survival. Blocking cancer cells from taking metabolic resources can effectively increase the response of cancer cells to antitumor drugs [5]. In this paper, the researchers investigate whether targeting glucose and glutamine transporters improves cancer response to cetuximab. The results showed that GLUT1 (glucose transporter-1) was upregulated in mCRC cells with KRAS or BRAF mutation [6]. SLC1A5 (a glutamine transporter) can be cotargeted with EGFR by cetuximab in HNSCC cells [7], and it has been found that the expression of SLC1A5, but not GLUT1, in colorectal cancer with cetuximab resistance is high.

SLC1A5 is a sodium-dependent transporter responsible for neutral amino acid exchange between the membrane and peripheral tissues [8]. In addition to mCRC, SLC1A5 is highly expressed in various cancers and promotes tumor growth [9–14]. The well-known mechanism by which SLC1A5 promotes tumor growth is increased by glutamine supplementation, which facilitates cellular energy production, mTOR signaling activation, macromolecular synthesis, and redox homeostasis [15]. However, the role of SLC1A5 in treating CRC with cetuximab and its molecular mechanism is still largely unknown. This study elucidates the role of SLC1A5 in the treatment of colorectal cancer with cetuximab and evaluates SLC1A5 as a potential sensitizer for cetuximab to enhance its antitumor effect.

SLC1A5 is overexpressed in cancer tissues of cetuximab-resistant CRC patients

Immunohistochemical detection of the expression status of GLUT1 and SLC1A5 in tumor cells from CRC patients treated with cetuximab revealed that the expression of GLUT1 was correlated with age. However, there was no significant difference among patients sensitive to or resistant to cetuximab treatment. The expression of SLC1A5 in cancer tissues of patients resistant to cetuximab treatment was significantly higher than that of patients sensitive to cetuximab treatment. This suggests that SLC1A5 may mediate cetuximab resistance in CRC patients.

SLC1A5 promotes the proliferation of CRC both in vitro and in vivo

They examined the expression and biological function of SLC1A5 and further analyzed the role of SLC1A5 in the treatment of cetuximab. SLC1A5 expression was positive in the colon cancer cell lines CACO2 and SW480. Treatment with the SLC1A5 inhibitor GPNA16 significantly decreased glutamine uptake and cell viability of CACO2 and SW480 cells. WB showed that SLC1A5 knockdown cells grew more slowly than those with regular SLC1A5 expression, and

GPNA inhibition of SLC1A5 could lead to G1 phase cell accumulation and a slight decrease in S-phase cells and increase the production of the autophagy-related protein LC3-II. According to the results in a xenograft mouse model, SLC1A5 inhibition with GPNA can inhibit tumor growth and reduce tumor weight. Targeting SLC1A5 can inhibit CRC proliferation in vivo and in vitro.

SLC1A5 was overexpressed in colorectal tumor tissues compared with normal colorectal tissues after TCGA CRC dataset analysis. Meanwhile, patients with higher SLC1A5 expression in stage III CRC had a worse prognosis than patients in another stage. TCGA CRC data analysis showed that patients with high SLC1A5 expression in stage III colon cancer had a poor prognosis, suggesting that SLC1A5 may play an essential role in the progression of CRC, but further verification is needed.

Targeting SLC1A5 significantly enhanced the inhibitory effect of cetuximab on CRC proliferation

The authors found that inhibition of SLC1A5 results in resensitization to cetuximab in CRC. The MTT test showed that downregulation of SLC1A5 or inhibition of SLC1A5 with GPNA increased the response of the CRC cell lines CaCO2 and SW480 to cetuximab. The same results were further confirmed in SLC1A5-positive cell lines HCT8, HCT116, SW48, and SW620. The colony formation assay showed that the proliferation of CRC cells was effectively inhibited after treatment with a combination of cetuximab and GPNA. Meanwhile, cetuximab plus GPNA treatment also increased the expression of γH2AX, a DNA damage marker that responds to the efficacy of the combinatorial therapeutic strategy [16], and LC3-II, which is essential in autophagy. In conclusion, SLC1A5 inhibition can significantly enhance the effect of cetuximab on CRC.

Targeting SLC1A5 increases ubiquitin–proteasome pathway-mediated EGFR degradation

Previous studies have shown that EGFR degradation, including degradation, transport to the nucleus, or recycling back to the membrane [17,18], can enhance the therapeutic effect of EGFR-targeted drugs [19,20]. Using TCGA CRC datasets, EGFR degradation-related genes were downregulated in tumor tissues [21–25], and these genes were associated with different levels of SLC1A5 expression in normal and tumor tissues. EGFR levels were significantly reduced by SLC1A5 blockade with GPNA or gene silencing. Moreover, the GPNA-induced EGFR reduction can be reversed by the proteasome inhibitor MG132 rather than the lysosomal inhibitor chloroquine. These results suggest that targeting SLC1A5 can improve the efficacy of cetuximab by increasing EGFR degradation mediated by the ubiquitin–proteasome pathway.

Inhibition of SLC1A5 induces DNA damage in CRC cells

After GPNA treatment, both membrane EGFR and nuclear EGFR were significantly decreased, confirming that SLC1A5 inhibition can promote EGFR degradation and reduce the ability of endocytosis and EGFR transport to the cell surface or nucleus. A previous study showed that nEGFR facilitates DNA replication and mediates drug resistance to antitumor therapies [26].

Therefore, SLC1A5 may have the ability to promote DNA repair by regulating the expression of nEGFR. The production of γH2AX was markedly increased after 48 h of SLC1A5 inhibition by GPNA. Similarly, γH2AX expression was increased in CACO2 cells and SW480 cells with SLC1A5-SH3 silenced. ATR, ATR, and DNAPKCs are the three major kinases responsible for H2AX phosphorylation and triggering different DNA damage responses [16]. The ATR inhibitor VE821 inhibited the increase in H2AX phosphorylation induced by GPNA. This is due to the deletion or suppression of SLC1A5, which increases the phosphorylation of ATR. These results suggest that inhibition of the SLC1A5-induced DNA damage response is caused by activation of ATR. Further observation indicated that after 48 h of GPNA stimulation, the bromodeoxyuridine (BrdU) immunofluorescence staining of γH2X-positive cells was positive, indicating the accumulation of single-stranded DNA (ssDNA) and double-stranded DNA breaks (DSBS) [27]. SLC1A5 plays an essential role in the DNA damage response.

Discussion

Overactivation of the EGFR downstream signaling pathway is considered to be the leading cause of resistance to cetuximab during treatment [28]. However, strategies targeting signal transduction pathways often fail with drug resistance. In contrast, strategies that inhibit key metabolic molecules are more conducive to preventing drug resistance because cancer cells are less able to compensate for the loss of essential metabolic pathways than the signal transduction system [5,29]. The study identified an essential glutamine transporter, SLC1A5, as an important therapeutic target for patients with colorectal cancer. SLC1A5 is highly expressed in cetuximab-resistant CRC patients. Multiple models showed that inhibiting SLC1A5 significantly increased the sensitivity to cetuximab in CRC. The use of cetuximab plus GPNA, a SLC1A5 inhibitor, improved the response to treatment in both RAS-wild-type CRC cell lines and RAS-mutated CRC cell lines and even RAS-mutated CRC xenograft models.

This study also reveals the mechanism by which SLC1A5 affects the efficacy of cetuximab. The results showed that SLC1A5 inhibition facilitates EGFR degradation. Inactivation of PTEN has been demonstrated to cause resistance to anti-EGFR therapies by disrupting EGFR degradation [19]. Another study showed that two cholesterol biosynthesis pathway enzymes [NADP-dependent steroid dehydrogenase-like (NSDHL) and sterol C4-methyl oxidase-like SC4mol] could be used as sensitizers for cetuximab by promoting EGFR degradation [20]. Combined targeting of SLC1A5 can improve the efficacy of cetuximab.

Another barrier to the efficacy of cetuximab in the nuclear transport of EGFR, which acts as a transcription factor regulating gene expression, leads to resistance. In this process, SLC1A5 mediated the expression of nEGFR, and the enhanced efficacy of cetuximab induced by inhibition of SLC1A5 may be related to the increased degradation of EGFR and the decreased expression of NEGFR. In addition, SLC1A5 inhibition also decreased the expression of mEGFR. SLC1A5 plays a vital role in regulating EGFR degradation or internalization and the DNA damage response and has become an important target for improving the therapeutic effect of EGFR inhibitors in CRC patients.

References

[1] Ma H, et al. Inhibition of SLC1A5 sensitizes colorectal cancer to cetuximab. Int J Cancer 2018;142:2578–88.

[2] Vlacich G, Coffey RJ. Resistance to EGFR-targeted therapy: a family affair. Cancer Cell 2011;20:423–5.

[3] Bronte G, et al. New findings on primary and acquired resistance to anti-EGFR therapy in metastatic colorectal cancer: do all roads lead to RAS? Oncotarget 2015;6:24780–96.

[4] Butler EB, Zhao Y, Muñoz-Pinedo C, Lu J, Tan M. Stalling the engine of resistance: targeting cancer metabolism to overcome therapeutic resistance. Cancer Res 2013;73:2709–17.

[5] Yun J, et al. Glucose deprivation contributes to the development of KRAS pathway mutations in tumor cells. Science 2009;325:1555–9.

[6] Lu H, Li X, Lu Y, Qiu S, Fan Z. ASCT2 (SLC1A5) is an EGFR-associated protein that can be co-targeted by cetuximab to sensitize cancer cells to ROS-induced apoptosis. Cancer Lett 2016;381:23–30.

[7] Kanai Y, et al. The SLC1 high-affinity glutamate and neutral amino acid transporter family. Mol Aspects Med 2013;34:108–20.

[8] Wang Q, et al. Targeting ASCT2-mediated glutamine uptake blocks prostate cancer growth and tumour development. J Pathol 2015;236:278–89.

[9] van Geldermalsen M, et al. ASCT2/SLC1A5 controls glutamine uptake and tumour growth in triple-negative basal-like breast cancer. Oncogene 2016;35:3201–8.

[10] Huang F, et al. Upregulated SLC1A5 promotes cell growth and survival in colorectal cancer. Int J Clin Exp Pathol 2014;7:6006–14.

[11] Hassanein M, et al. SLC1A5 mediates glutamine transport required for lung cancer cell growth and survival. Clin Cancer Res 2013;19:560–70.

[12] Willems L, et al. Inhibiting glutamine uptake represents an attractive new strategy for treating acute myeloid leukemia. Blood 2013;122:3521–32.

[13] Witte D, Ali N, Carlson N, Younes M. Overexpression of the neutral amino acid transporter ASCT2 in human colorectal adenocarcinoma. Anticancer Res 2002;22:2555–7.

[14] Hensley CT, Wasti AT, DeBerardinis RJ. Glutamine and cancer: cell biology, physiology, and clinical opportunities. J Clin Invest 2013;123:3678–84.

[15] Marzi L, et al. FOXO3a and the MAPK p38 are activated by cetuximab to induce cell death and inhibit cell proliferation and their expression predicts cetuximab efficacy in colorectal cancer. Br J Cancer 2016;115:1223–33.

[16] Ciardiello F, Tortora G. EGFR antagonists in cancer treatment. N Engl J Med 2008;358:1160–74.

[17] Wang YN, Yamaguchi H, Hsu JM, Hung MC. Nuclear trafficking of the epidermal growth factor receptor family membrane proteins. Oncogene 2010;29:3997–4006.

[18] Vivanco I, et al. The phosphatase and tensin homolog regulates epidermal growth factor receptor (EGFR) inhibitor response by targeting EGFR for degradation. Proc Natl Acad Sci U S A 2010;107:6459–64.

[19] Sukhanova A, et al. Targeting C4-demethylating genes in the cholesterol pathway sensitizes cancer cells to EGF receptor inhibitors via increased EGF receptor degradation. Cancer Discov 2013;3.

[20] Lee H-H, Wang Y-N, Hung M-C. Non-canonical signaling mode of the epidermal growth factor receptor family. Am J Cancer Res 2015;5:2944–58.

[21] Dobbelstein M, Sørensen CS. Exploiting replicative stress to treat cancer. Nat Rev Drug Discov 2015;14:405–23.

[22] McCormick F. Mutant onco-proteins as drug targets: successes, failures, and future prospects. Curr Opin Genet Dev 2011;21:29–33.

[23] Xia W, et al. Nuclear expression of epidermal growth factor receptor is a novel prognostic value in patients with ovarian cancer. Mol Carcinog 2009;48:610–7.

[24] Iida M, Brand TM, Campbell DA, Li C, Wheeler DL. Yes and Lyn play a role in nuclear translocation of the epidermal growth factor receptor. Oncogene 2013;32:759–67.

[25] Brand TM, et al. Nuclear epidermal growth factor receptor is a functional molecular target in triple-negative breast cancer. Mol Cancer Ther 2014;13:1356–68.

[26] Zhang T, et al. Metformin sensitizes prostate cancer cells to radiation through EGFR/p-DNA-PKCS in vitro and in vivo. Radiat Res 2014;181:641–9.

[27] Lin MZ, Marzec KA, Martin JL, Baxter RC. The role of insulin-like growth factor binding protein-3 in the breast cancer cell response to DNA-damaging agents. Oncogene 2014;33:85–96.

[28] Wang D, et al. HER3 targeting sensitizes HNSCC to cetuximab by reducing HER3 activity and HER2/HER3 dimerization: evidence from cell line and patient-derived xenograft models. Clin Cancer Res 2017;23:677–86.

[29] Silva VAO, et al. Rapid diminution in the level and activity of DNA-dependent protein kinase in cancer cells by a reactive nitro-Benzoxadiazole compound. Int J Mol Sci 2016;17.

Further reading

Nicklin P, et al. Bidirectional transport of amino acids regulates mTOR and autophagy. Cell 2009;136:521–34.

CHAPTER

24

Targeting of PYK2 synergizes with EGFR antagonists in basal-like TNBC and circumvents HER3-associated resistance via the NEDD4–NDRG1 axis

Tian Li, Changhai Lei, and Shi Hu

Department of Biophysics, College of Basic Medical Sciences, Naval Medical University (Second Military Medical University), Shanghai, China

Abstract

This chapter mainly introduces the reasons why EGFR is highly expressed in TNBC and is considered a potential therapeutic target, but the clinical benefit is negligible and improved therapeutic strategies. Triple-negative breast cancer (TNBC) is a highly aggressive and heterogeneous disease with poor efficacy of various current treatment options. The unsatisfactory therapeutic effect of targeting EGFR may be due to the activation of compensatory signaling pathways. Some studies suggest that this result is related to the upregulation of HER3. Targeting EGFR and PYK2/FAK synergistically inhibited basal-like TNBC cell proliferation and attenuated tumor growth in a xenograft model in vitro. Dual targeting of EGFR and PYK2/FAK inhibited complementary critical growth and survival signaling pathways activated in tumors. Inhibition of PYK2 may also abolish the upregulation of HER3 in response to EGFR antagonists, thereby circumventing the associated resistance. Some experimental results have provided preclinical evidence for a therapeutic strategy targeting both EGFR and PYK2/FAK, which may have the potential to improve the treatment status of TNBC.

Abbreviations

EGFR	epidermal growth factor receptor
FAK	focal adhesion kinase
HER	human epidermal growth factor receptor
KD	Knockdown

Novel Sensitizing Agents for Therapeutic Anti-EGFR Antibodies
https://doi.org/10.1016/B978-0-12-821584-5.00018-3

NDRG1 N-myc downstream regulatory gene 1
RTKs receptor tyrosine kinases
TNBC triple-negative breast cancer

Conflict of interest

No potential conflicts of interest were disclosed.

Triple-negative breast cancer (TNBC) refers to a special type of breast cancer. The biopsy results of this type of breast cancer patient showed that the patient's estrogen receptor and progesterone receptor were negative and that there was no amplification of HER2 [1]. At the same time, the prognosis of triple-negative breast cancer is not ideal in all types of cancer. The tumor is very aggressive and highly heterogeneous. There is no ideal targeted therapy drug or treatment method [2]. Approximately 50% of TNBC patients show highly abnormal expression of EGFR [3]. EGFR has always been regarded as a target that plays a key role in the progression and metastasis of cancer. Therefore, EGFR is also regarded by researchers as a potential target for the treatment of patients with triple-negative breast cancer. However, the use of a single drug to target EGFR cannot improves the prognosis of TNBC patients. Generally, the failure of drug efficacy to achieve the expected target is attributed to the activation of the EGFR compensation signaling pathway. Some subsequent studies have shown that signals generated by the c-Met receptor, HER3, and AXL-related signaling pathways can compensate for EGFR inhibition [2,4,5], and the upregulation of HER3 is related to the reduced effect of EGFR antagonists on TNBC patients [6]. According to the data of HER2-positive breast cancer patients, antagonists targeting HER2 therapy may cause upregulation of HER3. The initiation of this mechanism is usually attributed to the emergence of resistance to therapies targeting EGFR and HER2 [7].

Both the upregulation and degradation of receptors can be regulated at the transcriptional and posttranscriptional levels [8]. For example, knocking down NEDD4 expression in breast cancer cells can induce HER3 upregulation and promote breast cancer cell proliferation in in vitro experiments and tumor growth in vivo in animal experiments [9], while AXL degradation can overcome the resistance of nonsmall cell lung cancer to EGFR antagonists [10]. Therefore, drugs that prevent the upregulation of receptor tyrosine kinase in response to EGFR antagonists may provide an effective treatment strategy to improve the curative effect on tumors and overcome EGFR-related drug resistance problems. Previous studies have also shown consistent conclusions with this concept. Dual-targeting EGFR and one of other RTKs (receptor tyrosine kinases) that heterodimerize or interfere with EGFR signaling can enhance drug effects and/or reverse the subsequent development of drug resistance. For example, the use of drugs that target EGFR and cMet receptors in combination is considered an effective antineoplastic protocol for the treatment of TNBC [2,11].

In subsequent studies, researchers suggested that the nonreceptor tyrosine kinase PYK2 is an important downstream factor of EGFR, cMet, and their cross-correlated signaling pathway in some TNBC cell lines [12]. PYK2 and related focal adhesion kinase (FAK) are important mediators of cell mitosis and migration. Such mediation is regulated by the aggregation of RTKs, cytokine receptors, and integrins [13].

Studies have shown that both PYK2 and FAK are related to the progression and invasion of a variety of human cancer diseases, including breast cancer [14,15]. Inhibitors of FAK/PYK2 have been shown to inhibit tumor growth and metastasis in some cell and animal model studies. Usually, these inhibitors are small-molecule drugs. A small number of studies have also given such drugs to patients in the clinic [14]. The drugs show initial clinical activity and have relatively few side effects.

Some subsequent studies have shown that the expression of FAK is often higher in tumors of TNBC patients, which may become an important basis for the treatment of TNBC to develop targeted drugs [16]. In addition, proteomic analysis showed that large amounts of PYK2 are expressed in basal-like breast cancer and become active. This result is closely related to the expression level of EGFR and the activation of ERK1/2 [17].

After in-depth exploration of clinical specimens, database analysis, TNBC cell line function research, and animal xenotransplantation models, researchers found that EGFR and PYK2 are highly expressed in TNBC and are closely related to the poor prognosis of patients [18]. Further studies have found that after inhibiting PYK2 and FAK, the sensitivity of TNBC to EGFR antagonist treatment is significantly increased, and the dual-targeting EGFR and PYK2/FAK treatment strategy could synergistically inhibit the growth of tumors in vitro and in vivo [19].

In addition, after inhibiting PYK2, it was found that the degradation of the HER3 proteasome was increased, the upregulation of HER3 usually caused by the use of EGFR antagonists was eliminated, and the feedback resistance to the therapeutic effect mediated by the upregulation of HER3 was avoided [20]. PYK2 interacts with HER3 and NEDD4 and interferes with the formation of the HER3-NEDD4 complex. Knockdown (KD) of PYK2 induces the upregulation of NDRG1 (N-myc downstream regulatory gene 1). The protein expressed by this gene is a protein closely related to tumor suppression and resistance to anticancer drugs. In addition, NDRG1 can significantly enhance the binding of NEDD4 and HER3, thereby increasing the degradation of HER3. Therefore, based on the above experimental data and theories, researchers proposed that the circuit composed of PYK2, NEDD4, and NDRG1 played an important role in the degradation process of HER3. The dual-targeted drug therapy of EGFR and PYK2/FAK could not only inhibit tumor growth but also largely avoid the occurrence of drug resistance. Therefore, drugs that dually target EGFR and PYK2/FAK provide a potential and effective ideal treatment strategy for basal-like TNBC.

According to clinical statistics, TNBC cases account for approximately 20% of all breast cancer cases [21]. TNBC usually shows a high mitotic index, high metastasis rate, and unsatisfactory prognosis. TNBC is a highly heterogeneous cancer that is generally classified by molecular subtypes during diagnosis and treatment. There are differences in clinicopathological characteristics, metastasis patterns, and treatment options between different molecular subtypes. At present, 6 different TNBC subtypes have been identified and distinguished, and the differences are reflected in gene expression patterns and gene ontologies [22]. In basal-like TNBC, there are usually specific marker proteins that are highly expressed, including cytokeratin 5/6, P-cadherin, p63, and EGFR [23]. The high level of EGFR expression and corresponding carcinogenic effects indicates that targeting EGFR may be an ideal treatment strategy for patients [24]. However, analysis of the results of some clinical trials found that treatment targeting EGFR alone failed to improve patient survival. Therefore, researchers have proposed some multitargeted combination drug therapies based on clinical feedback and preliminary animal experiment research results [25].

In subsequent research, researchers proposed an effective and promising combination therapy for patients with basal-like TNBC [18]. The main change in therapy was the use of drugs to simultaneously target EGFR and Pyk2/FAK kinase. The results of cell and animal experiments suggest that this therapy has considerable potential and value for further research. This project once again verified that the high levels of PYK2 and EGFR expression were related to the unsatisfactory clinical outcome of TNBC patients. The researchers verified this theory by establishing multiple basal-like TNBC cell line models, and the results showed that the combination targeting both EGFR and PYK2/FAK can induce cell death in TNBC cell lines in vitro and significantly affect the growth of transplanted tumors in animal models. The researchers analyzed multiple signaling pathways, and the results of the analysis showed that targeting both EGFR and PYK2/FAK had complementary effects, which can inhibit key growth and survival signaling pathways in TNBC cells, thereby inducing cell death. What needs the most attention was that the steady-state expression levels of EGFR, cMet, and HER3 would be interfered by knocking down PYK2 and inhibiting PYK2/FAK. Positive effects of treating HER3 have been found in multiple basal-like TNBC cell lines. The interesting discovery and the involvement of HER3 in the development of tumor resistance to EGFR antagonists have promoted research on the mechanism by which PYK2 regulates the expression level of HER3.

A systematic analysis of the expression levels of HER3 in control cells and the PYK2 depletion group showed that PYK2 depletion was accompanied by increased proteasome degradation of HER3 and possibly induced upregulation of NDRG1. In previous studies, NDRG1 has been identified as a key factor in inhibiting the metastasis of prostate cancer, colon cancer, and breast cancer [26], but the specific mechanism by which NDRG1 exerts the corresponding physiological functions is still unclear. Earlier research results also indicated that NDRG1 was involved in the regulation of the circulation and degradation of multiple receptors, such as LDL receptor and E-cadherin [27,28]. It plays a role in the operation of vesicles after interacting with related proteins.

Since HER3 undergoes endocytosis in a clathrin-dependent manner and then circulates to the plasma membrane, researchers speculated that PYK2 affects HER3 circulation and inhibits its degradation [29]. Subsequent research found that pPYK2 is located in early and recycled endosomes, while HER3 is basically located in recycled endosomes [18]. At the same time, HER3 and pPYK2 are co-localized on the vesicle-like structure, which is most likely to be the recirculating endosome. In addition, the results of the study also found that there is a co-localization between pPYK2 and NEDD4, and there is also a co-localization between HER3 and NEDD4. Further research results prove that PYK2, HER3, and NEDD4–1/2 directly interact with each other in pairs. In addition, experiments confirmed that PYK2 can interfere with the binding of NEDD4 and HER3. The above research results revealed a new protein–protein interaction network and further showed that PYK2 can inhibit receptor degradation by isolating its ubiquitin ligase NEDD4 and/or affecting the expression of NDRG1. The experimental results of the effect of NDRG1 on NEDD4-HER3 binding supported this theory and strongly suggested that the PYK2-NDRG1-NEDD4 axis played an important role in regulating the expression level of HER3, and this effect will have a further effect on drug response and resistance mechanisms.

Related researchers speculated that the PYK2-NDRG1-NEDD4 axis not only regulates the transport and degradation of HER3 but also probably affects the degradation of other cell surface receptors and/or transporters that use similar pathways. EGFR, ErbB4, and some other membrane proteins have been shown to bind to NEDD4–1/2 or be ubiquitinated by NEDD4–1/2 in certain signaling pathways. Researchers also found an interaction between PYK2 and NEDD4–1/2 [30]. This result suggested that PYK2 can participate in the regulation of the expression and modification of some membrane proteins. Through this mechanism, Pyk2 can participate in the regulation of various characteristics of cancer cells and cancer, such as the emergence of drug resistance in cancer patients.

Through experiments, researchers also observed an important experimental phenomenon in which PYK2 depletion affects the expression levels of EGFR and cMet in TNBC-related cell lines [31]. These two kinds of RTK will affect the therapeutic effect and make the patient develop resistance to the drug. At the same time, the two-way compensatory response mediated by the drug may be the main reason why the efficacy of the drug cannot reach the expected level. This suggested that inhibiting the activity/expression of PYK2 can avoid HER3-related drug resistance and overcome the problem of specific TNBC molecular subtypes or EGFR- or cMet-related drug resistance in response to different targeted therapies.

Research on PYK2/FAK as a therapeutic target for TNBC (especially PYK2) has not received attention and has not been extensively studied. The research basis of PYK2/FAK as a therapeutic target of TNBC is consistent with the high expression of PYK2 in TNBC and its correlation with lymph node metastasis. In addition, recent research results clearly indicate that inhibiting the activity/expression of PYK2 is likely to overcome the problem of drug resistance. The related mechanism may be because it has a significant impact on the steady-state level of important intermediate signal components such as RTKs and S6K. Relevant researchers have put forward the inference that NEDD4 and NDRG1 may participate in the regulation and mediation of part of the mechanism in this process. Therefore, drugs that can effectively inhibit PYK2 and FAK may show significant benefits in the clinical treatment of TNBC.

In studying drugs that target both PYK2 and EGFR, researchers found that the synergy of the drugs in basal-like TNBC may avoid HER3-related resistance through the NEDD4-NDRG1 axis. Researchers have proposed a combined targeted therapy that can effectively treat basal-like TNBC and described the new drug regulation circuit in more detail and accurately, which plays a key role in pharmacological effects and resistance mechanisms [19]. In view of the fact that there is no ideal method to treat TNBC patients in the clinic and according to previous clinical experience, drug resistance often develops into a major problem in the late stage of targeted therapy, such research provides a new and potential treatment strategy for TNBC, which is worthy of in-depth study and is expected to play a role in clinical treatment.

References

[1] Peddi PF, Ellis MJ, Ma C. Molecular basis of triple negative breast cancer and implications for therapy. Int J Breast Cancer 2012;2012.

[2] Sohn J, et al. cMET activation and EGFR-directed therapy resistance in triple-negative breast cancer. J Cancer 2014;5:745–53.

[3] Burness ML, Grushko TA, Olopade OI. Epidermal growth factor receptor in triple-negative and basal-like breast cancer promising clinical target or only a marker? Cancer J 2010;16:23–32.

[4] Sergina NV, et al. Escape from HER-family tyrosine kinase inhibitor therapy by the kinase-inactive HER3. Nature 2007;445:437–41.

[5] Meyer AS, Miller MA, Gertler FB, Lauffenburger DA. The receptor AXL diversifies EGFR signaling and limits the response to EGFR-targeted inhibitors in triple-negative breast cancer cells. Sci Signal 2013;6.

[6] Tao JJ, et al. Antagonism of EGFR and HER3 enhances the response to inhibitors of the PI3K-Akt pathway in triple-negative breast cancer. Sci Signal 2014;7.
[7] Jiang N, Saba NF, Chen ZG. Advances in targeting HER3 as an anticancer therapy. Chemother Res Pract 2012;2012, 817304.
[8] Garrett JT, et al. Transcriptional and posttranslational up-regulation of HER3 (ErbB3) compensates for inhibition of the HER2 tyrosine kinase. Proc Natl Acad Sci U S A 2011;108:5021–6.
[9] Huang Z, et al. The E3 ubiquitin ligase NEDD4 negatively regulates HER3/ErbB3 level and signaling. Oncogene 2015;34:1105–15.
[10] Bae SY, Hong JY, Lee HJ, Park HJ, Lee SK. Targeting the degradation of AXL receptor tyrosine kinase to overcome resistance in gefitinib-resistant non-small cell lung cancer. Oncotarget 2015;6:10146–60.
[11] Kim YJ, et al. MET is a potential target for use in combination therapy with EGFR inhibition in triple-negative/basal-like breast cancer. Int J Cancer 2014;134:2424–36.
[12] Verma N, Keinan O, Selitrennik M, Karn T, Filipits M, Lev S. PYK2 sustains endosomal-derived receptor signalling and enhances epithelial-to-mesenchymal transition. Nat Commun 2015;6.
[13] Mitra SK, Hanson DA, Schlaepfer DD. Focal adhesion kinase: in command and control of cell motility. Nat Rev Mol Cell Biol 2005;6:56–68.
[14] Sulzmaier FJ, Jean C, Schlaepfer DD. FAK in cancer: mechanistic findings and clinical applications. Nat Rev Cancer 2014;14:598–610.
[15] Lipinski CA, Loftus JC. Targeting Pyk2 for therapeutic intervention. Expert Opin Ther Targets 2010;14:95–108.
[16] Yom CK, Noh DY, Kim WH, Kim HS. Clinical significance of high focal adhesion kinase gene copy number and overexpression in invasive breast cancer. Breast Cancer Res Treat 2011;128:647–55.
[17] Boyd ZS, et al. Proteomic analysis of breast cancer molecular subtypes and biomarkers of response to targeted kinase inhibitors using reverse-phase protein microarrays. Mol Cancer Ther 2008;7:3695–706.
[18] Verma N, et al. Targeting of PYK2 synergizes with EGFR antagonists in basal-like TNBC and circumvents HER3-associated resistance via the NEDD4-NDRG1 Axis. Cancer Res 2017;77:86–99.
[19] Shen T, Guo Q. EGFR signaling pathway occupies an important position in cancer-related downstream signaling pathways of Pyk2. Cell Biol Int 2020;44:2–13.
[20] Sun J, et al. Metastasis suppressor, NDRG1, mediates its activity through signaling pathways and molecular motors. Carcinogenesis 2013;34:1943–54.
[21] Foulkes WD, Smith IE, Reis-Filho JS. Triple-negative breast cancer. N Engl J Med 2010;363:1938–48.
[22] Lehmann BD, et al. Identification of human triple-negative breast cancer subtypes and preclinical models for selection of targeted therapies. J Clin Investig 2011;121:2750–67.
[23] Paredes J, Lopes N, Milanezi F, Schmitt FC. P-cadherin and cytokeratin 5: useful adjunct markers to distinguish basal-like ductal carcinomas in situ. Virchows Arch 2007;450:73–80.
[24] Corkery B, Crown J, Clynes M, O'Donovan N. Epidermal growth factor receptor as a potential therapeutic target in triple-negative breast cancer. Ann Oncol 2009;20:862–7.
[25] Lee MJ, et al. Sequential application of anticancer drugs enhances cell death by rewiring apoptotic signaling networks. Cell 2012;149:780–94.
[26] Kovacevic Z, Richardson DR. The metastasis suppressor, Ndrg-1: a new ally in the fight against cancer. Carcinogenesis 2006;27:2355–66.
[27] Pietiainen V, et al. NDRG1 functions in LDL receptor trafficking by regulating endosomal recycling and degradation. J Cell Sci 2013;126:3961–71.
[28] Kachhap SK, et al. The N-Myc down regulated Gene1 (NDRG1) is a Rab4a effector involved in vesicular recycling of E-cadherin. PLoS One 2007;2.
[29] Sak MM, et al. The oncoprotein ErbB3 is endocytosed in the absence of added ligand in a clathrin-dependent manner. Carcinogenesis 2012;33:1031–9.
[30] Goh LK, Sorkin A. Endocytosis of receptor tyrosine kinases. Cold Spring Harb Perspect Biol 2013;5.
[31] Logue JS, Morrison DK. Complexity in the signaling network: insights from the use of targeted inhibitors in cancer therapy. Genes Dev 2012;26:641–50.

CHAPTER

25

Overcoming acquired resistance to EGFR-targeted therapy by regulating autophagy

Kewen Qian[a], *Wenyan Fu*[b], *and Shi Hu*[a]

[a]Department of Biophysics, College of Basic Medical Sciences, Naval Medical University (Second Military Medical University), Shanghai, China [b]Department of Assisted Reproduction, Shanghai Ninth People's Hospital, Shanghai Jiao Tong University School of Medicine, Shanghai, China

Abstract

Autophagy is a highly conserved metabolic activity that participates in various biological activities, such as biological development and growth. However, autophagy is also a double-edged sword in the process of tumor occurrence and development. It can both promote and inhibit tumor development, which mainly depends on the type of tumor tissue and cells, tumor microenvironment, and tumor stage. Autophagy is mainly divided into two stages: the formation of autophagosomes and the fusion of autophagosomes and lysosomes. The formation of autophagosomes depends on phosphorylated Beclin1 and activated PI3K (VSP34) to form the Beclin1-VPS34 complex. The EGFR-PI3K/AKT/mTOR signaling pathway can inhibit Beclin1 phosphorylation and block autophagy, while the EGFR-PI3K/AKT/mTOR pathway can promote the phosphorylation of Beclin1 and activate autophagy. The relationship between autophagy and EGFR signaling pathways in tumor cells with drug resistance suggests modulation of autophagy for anti-EGFR treatments. In fact, autophagy inhibitors either target autophagosome formation or fusion of autophagosomes and lysosomes work well in some cancers, such as enhancing sensitivity to EGFR tyrosine kinase inhibitors (EGFR-TKIs). In addition, antigens such as CAGE, which are closely related to autophagy, have been identified as potential targets. Researchers can also combine extracellular vesicles (EVs) and miRNAs to explore candidates for combination therapies that both modulate autophagy and target EGFR.

Abbreviations

TKI	tyrosine kinase inhibitors
EVs	extracellular vesicles
PI3K	phosphatidylinositol-3-kinase
EGFR	epidermal growth factor receptor
ULK	unc-51-like kinase

https://doi.org/10.1016/B978-0-12-821584-5.00020-1

Bcl-2	B-cell lymphoma 2
3-MA	3-methyladenine
NSCLC	nonsmall cell lung cancer
HNSCCs	head and neck squamous cell carcinomas
MVB	multivesicular body
LAPTM4B	lysosomal-associated protein transmembrane 4 beta
GSK3β	glycogen synthase kinase 3 beta
DIRAS	distinct subgroup of the Ras family

Conflict of interest

No potential conflicts of interest were disclosed.

Overlap between EGFR and autophagy regulatory signaling

When autophagy occurs, cells undergo self-digestion and remove nonfunctional proteins and organelles. Many factors induce autophagy, such as reactive oxygen species, hypoxia, and unfolded proteins [1–3]. Autophagy is actually a process by which cells avoid the adverse effects of harmful stimuli for self-protection, so autophagy may be one of the mechanisms by which tumor cells survive. Autophagy is mainly divided into three types: macroautophagy, microautophagy, and chaperone-mediated autophagy [4]. For instance, macroautophagy refers to the process in which autophagosomes first form with a double-membrane structure to wrap intracellular substances and then fuse with lysosomes. The formation of autophagosomes mainly depends on the phosphorylation of Beclin1 induced by unc-51-like kinase (ULK), which then activates class III PI3-kinase (VPS34) to form a complex [5]. EGFR tyrosine kinase inhibitors (EGFR-TKIs), on the one hand, induce tyrosine phosphorylation of Beclin1, decompose the Beclin1-VPS34 complex, and form Beclin1 homodimers. On the other hand, autophagy inhibitors such as Rubicon and B-cell lymphoma 2 (Bcl-2) bind to Beclin1 and inhibit autophagy. Regarding the downstream signaling pathways of EGFR, the EGFR-PI3K/AKT/mTOR pathway inhibits autophagy by inhibiting Beclin1 phosphorylation, and the EGFR-RAF/MEK/ERK pathway promotes autophagy by activating Beclin1 phosphorylation.

Therefore, the coincidence of EGFR signaling pathways and the key molecules of autophagy suggests a possible mutually beneficial relationship between anti-EGFR therapies and autophagy inhibition.

Autophagy facilitates tumor resistance to targeted biological therapies

EGFR-TKIs such as erlotinib and gefitinib are relatively ideal treatment options for nonsmall cell lung cancer (NSCLC) and other malignancies; however, the EGFR T790 M mutation accounts for approximately 50% of tumor resistance to EGFR-TKIs [6]. Interestingly, when inhibitors are used to inhibit the downstream signaling pathways of EGFR, the sensitivity of tumor cells to EGFR-TKIs cannot be increased, so there may be other mechanisms leading to such drug resistance [7].

Tumor cells may avoid the adverse stimulation caused by antitumor drugs through the self-recycling of antigenic proteins and organelles. Studies have found that the enhancement of autophagy is related to the intensity of anticancer therapies [8]. Studies have found that the resistance of various cancers to antibody drugs is related to the enhancement of autophagy, such as resistance to EGFR-TKIs AG1478 of ovarian cancer cells [9] and resistance of head and neck squamous cell carcinomas (HNSCCs) to erlotinib [10]. Osimertinib is a third-generation EGFR-TKI for the treatment of NSCLC with EGFR mutations (EGFR T790M, L858R and exon 19). Both in vitro and in vivo experiments have shown that resistance to osimertinib is accompanied by enhanced autophagy. Osimertinib can induce autophagy and reduce apoptosis of cancer cells by increasing the level of reactive oxygen species (ROS) [11,12].

Targeting autophagy helps solve the problem of resistance to anti-EGFR treatments

Autophagy inhibitors inhibit autophagy primarily by inhibiting the formation of autophagosomes or their subsequent fusion with lysosomes.

For example, LY294002 and wortmannin inhibit the process by inhibiting the formation of Beclin1-VPS34, while spautin-1 promotes the degradation of the complex. Wortmannin is an AKT inhibitor that reduces Rad51 expression and enhances sensitivity of erlotinib-resistant NSCLC cells to erlotinib [13]. Another AKT inhibitor, LY294002, is used as an adjuvant of salinomycin for NSCLC, which inhibits the activity of AKT, thereby enhancing the efficacy of erlotinib in NSCLC [14].

Several drugs that have been widely used have been shown to inhibit the fusion of autophagosomes and lysosomes, such as hydroxychloroquine (HCQ) and chloroquine (CQ). Some antibiotics, such as cepharanthine, are also effective autophagy inhibitors that enhance the sensitivity of several cancer cells to EGFR-TKIs by blocking the fusion of autophagosomes and lysosomes [15]. EGFR-TKIs, such as lapatinib and gefitinib, also elicit potent autophagy inhibitory effects, whereas the combination of some autophagy inhibitors, such as cotreatment with CQ and bafilomycin A1, has a good anticancer effect on bladder cancer [16].

miRNAs and EVs regulate both autophagy and the response of cancer cells to anti-EGFR treatments

miRNA (micro-RNA) is a short RNA composed of 21–23 nucleotides that can regulate the expression of target genes by inhibiting the function of mRNA or using the instability of mRNA. Multiple reports indicate that numerous miRNAs play an important role in resistance to anticancer therapies and promote autophagy. Some kinds of miRNAs can enhance the sensitivity or alleviate drug resistance of tumor cells and are called tumor suppressor miRNAs. For example, let-7 and miR-34 have been proven to enhance the sensitivity of NSCLC cells to erlotinib [17]. miR-217 can enhance the killing effect of gefitinib and trastuzumab on melanoma cells by blocking the binding of EGFR and antigen CAGE [18]. In terms of autophagy, miR-123 inhibitors can upregulate the expression of phosphatase and PTEN, among which PTEN acts as an inhibitory regulator of the PI3K/AKT pathway [19]. By distinguishing the difference in miRNAs in drug-sensitive and drug-resistant NSCLC cells during antitumor therapy, researchers can find potential key miRNAs.

Extracellular vesicles (EVs) are a kind of tumor-targeted therapy delivery media and bioactive components that have been studied more. Multivesicular bodies (MVBs) are endosomal vesicles that bud intracellularly and enter the endosomal lumen. After MVBs fuse with the plasma membrane, they can be released in the form of extracellular vesicles. Research has shown that some inactive EGFRs can interact with lysosomal-associated protein transmembrane 4 beta (LAPTM4B) upon stimulation by EGFR-TKIs, leading to the accumulation of endosomal EGFR. These EGFRs can bind to autophagy inhibitors such as Rubicon, induce Rubicon to dissociate from Beclin1, and then continue to activate autophagy [20]. Tumor cells can release EVs, often with signature antigens such as miRNAs and cytokines. Second-generation EGFR-TKIs, such as PF-00299804 and CI-1033, can trigger the release of EVs containing EGFR, phospho-EGFR (pEGFR), and DNA wrapped in exosomes, and there have been some advances in targeted therapy for this type of tumor-promoting EV [21]. EVs containing long noncoding RNAs (lncRNAs), such as RP11 838N2.4, can also slow resistance to erlotinib [22].

Antiautophagic peptides that modulate the effect of anti-EGFR therapy on tumor cells

CAGE is a cancer/testis antigen. On one hand, it can directly regulate the expression of EGFR and promote the resistance of melanoma cells to targeted treatments. On the other hand, it can also enhance autophagy in tumor cells. CAGE can bind to Beclin1 and modulate the sensitivity of erlotinib-resistant NSCLC cells to erlotinib. A peptide derived from CAGE (^{266}AQTGTGKT^{273}) can bind to CAGE, block its binding with Beclin1, inhibit autophagy of NSCLC cells PC-9/ER, and ultimately enhance their sensitivity to erlotinib and osimertinib [23]. Another CAGE-derived peptide (^{269}GTGKT^{273}) can block the interaction of CAGE with glycogen synthase kinase 3 beta (GSK3β), which also increases the killing effect of drug-resistant melanoma cells [24]. Combining these derivative peptides with EGFR-targeted therapy can solve some drug resistance problems. The distinct subgroup of the Ras family (DIRAS) also has a tumor suppressor effect. DIRAS-derived peptides can bind to Beclin1 to form an autophagy-inhibitory complex [25].

Discussion

Autophagy induced by ROS, DNA damage, or metabolic stress can promote the survival of tumor cells during anti-EGFR therapy. Therefore, reducing the autophagy response is expected to solve some drug resistance problems. Studies have reported that peptide derivatives and miRNAs targeting the autophagy response of tumor cells can be used as treatment options for drug-resistant cancer patients, which need a large number of clinical tests before being put into clinical treatment.

References

[1] Klionsky DJ, et al. Guidelines for the use and interpretation of assays for monitoring autophagy (3rd edition). Autophagy 2016;12:1–222.

[2] Saint-Martin A, et al. Functional interaction of hypoxia-inducible factor 2-alpha and autophagy mediates drug resistance in colon cancer cells. Cancer 2019;11.

[3] Ye J, et al. Diosbulbin B-induced mitochondria-dependent apoptosis in L-02 hepatocytes is regulated by reactive oxygen species-mediated autophagy. Front Pharmacol 2019;10:676.
[4] Yu L, Chen Y, Tooze SA. Autophagy pathway: cellular and molecular mechanisms. Autophagy 2018;14:207–15.
[5] Russell RC, et al. ULK1 induces autophagy by phosphorylating Beclin-1 and activating VPS34 lipid kinase. Nat Cell Biol 2013;15:741–50.
[6] Xie M, He C-S, Wei S-H, Zhang L. Notch-1 contributes to epidermal growth factor receptor tyrosine kinase inhibitor acquired resistance in non-small cell lung cancer in vitro and in vivo. Eur J Cancer 2013;49:3559–72.
[7] Moreira-Leite FF, Harrison LR, Mironov A, Roberts RA, Dive C. Inducible EGFR T790M-mediated gefitinib resistance in non-small cell lung cancer cells does not modulate sensitivity to PI103 provoked autophagy. J Thorac Oncol 2010;5:765–77.
[8] Hu Y-L, Jahangiri A, Delay M, Aghi MK. Tumor cell autophagy as an adaptive response mediating resistance to treatments such as antiangiogenic therapy. Cancer Res 2012;72:4294–9.
[9] Yang Z, et al. Co-targeting EGFR and autophagy impairs ovarian cancer cell survival during detachment from the ECM. Curr Cancer Drug Targets 2015;15:215–26.
[10] Sobhakumari A, et al. NOX4 mediates cytoprotective autophagy induced by the EGFR inhibitor erlotinib in head and neck cancer cells. Toxicol Appl Pharmacol 2013;272:736–45.
[11] Li L, et al. Protective autophagy decreases osimertinib cytotoxicity through regulation of stem cell-like properties in lung cancer. Cancer Lett 2019;452:191–202.
[12] Tang ZH, Cao WX, Su MX, Chen X, Lu JJ. Osimertinib induces autophagy and apoptosis via reactive oxygen species generation in non-small cell lung cancer cells. Toxicol Appl Pharmacol 2017;321:18–26.
[13] Ko JC, et al. Roles of MKK1/2-ERK1/2 and phosphoinositide 3-kinase-AKT signaling pathways in erlotinib-induced Rad51 suppression and cytotoxicity in human non-small cell lung cancer cells. Mol Cancer Res 2009;7:1378–89.
[14] Kim KY, et al. Inhibition of autophagy promotes salinomycin-induced apoptosis via reactive oxygen species-mediated PI3K/AKT/mTOR and ERK/p38 MAPK-dependent signaling in human prostate cancer cells. Int J Mol Sci 2017;18.
[15] Tang ZH, et al. Identification of a novel autophagic inhibitor cepharanthine to enhance the anti-cancer property of dacomitinib in non-small cell lung cancer. Cancer Lett 2018;412:1–9.
[16] Hu X, et al. Blocking autophagy improves the anti-tumor activity of afatinib in lung adenocarcinoma with activating EGFR mutations in vitro and in vivo. Sci Rep 2017;7:4559.
[17] Stahlhut C, Slack FJ. Combinatorial action of microRNAs let-7 and miR-34 effectively synergizes with erlotinib to suppress non-small cell lung cancer cell proliferation. Cell Cycle 2015;14:2171–80.
[18] Kim Y, et al. miR-217 and CAGE form feedback loop and regulates the response to anti-cancer drugs through EGFR and HER2. Oncotarget 2016;7:10297–321.
[19] Han Z, Zhou X, Li S, Qin Y, Chen Y, Liu H. Inhibition of miR-23a increases the sensitivity of lung cancer stem cells to erlotinib through PTEN/PI3K/Akt pathway. Oncol Rep 2017;38:3064–70.
[20] Tan X, Thapa N, Sun Y, Anderson RA. A kinase-independent role for EGF receptor in autophagy initiation. Cell 2015;160:145–60.
[21] Montermini L, et al. Inhibition of oncogenic epidermal growth factor receptor kinase triggers release of exosome-like extracellular vesicles and impacts their phosphoprotein and DNA content. J Biol Chem 2015;290:24534–46.
[22] Zhang W, Cai X, Yu J, Lu X, Qian Q, Qian W. Exosome-mediated transfer of lncRNA RP11-838N2.4 promotes erlotinib resistance in non-small cell lung cancer. Int J Oncol 2018;53:527–38.
[23] Yeon M, et al. CAGE binds to Beclin1, regulates autophagic flux and CAGE-derived peptide confers sensitivity to anti-cancer drugs in non-small cell lung cancer cells. Front Oncol 2018;8:599.
[24] Kim Y, et al. The pentapeptide Gly-Thr-Gly-Lys-Thr confers sensitivity to anti-cancer drugs by inhibition of CAGE binding to GSK3β and decreasing the expression of cyclinD1. Oncotarget 2017;8:13632–51.
[25] Sutton MN, et al. DIRAS3-derived peptide inhibits autophagy in ovarian cancer cells by binding to Beclin1. Cancer 2019;11.

CHAPTER

26

Hedgehog signaling alters reliance on EGF receptor signaling and mediates anti-EGFR therapeutic resistance in head and neck cancer

Chuqi Wang[a,b], *Changhai Lei*[a], *and Shi Hu*[a]

[a]**Department of Biophysics, College of Basic Medical Sciences, Naval Medical University (Second Military Medical University), Shanghai, China** [b]**Department of Pharmacy, National University of Singapore, Singapore**

Abstract

Head and neck squamous cell carcinoma (HNSCC) is an aggressive, recurrent neoplasm that is characterized by poor prognosis and low overall survival. EGFR has been found to be associated with HNSCC, and cetuximab is the first anti-EGFR antibody approved for use in HNSCC. Despite the efficacy of cetuximab, most patients will relapse due to acquired resistance. The Hedgehog signaling pathway (HhP) is also frequently active in HNSCC and has been reported to integrate with EGFR signaling to induce tumorigenesis and drug resistance. In this chapter, researchers describe a study that found that long-term exposure to EGFR inhibition upregulated the HhP transcription factor GLI1 in HNSCC cells as they developed resistance to cetuximab. Conversely, inhibition of HhP enhanced the dependence of EGFR in HNSCC cells and therefore delayed or blocked tumor recurrence when used together with cetuximab. Moreover, the combination of cetuximab and the HhP inhibitor IPI-296 has a more pronounced antitumor effect than cetuximab monotherapy in HNSCC patient-derived xenografts. Together, these findings suggest that HhP inhibitors have the potential to sensitize HNSCC to anti-EGFR therapy. A phase I clinical trial (NCT01255800) found that the combinatorial therapy appears to be tolerable and showed signs of efficacy for HNSCC.

Abbreviations

HNSCC	head and neck squamous cell carcinoma
EGFR	epidermal growth factor receptor
HhP	Hedgehog signaling
Hh	Hedgehog

https://doi.org/10.1016/B978-0-12-821584-5.00017-1

EMT epithelial mesenchymal transition
SMO smoothened
GPCR G protein-coupled receptor
GLI glioma-associated oncogene
EGF epidermal growth factor
Vim vimentin
MTD maximum tolerated dose
RP2D recommended phase 2 dose
PFS progression-free survival

Conflict of interest

No potential conflicts of interest were disclosed.

Introduction

HNSCCs are a heterogeneous group of malignancies mainly derived from the mucosal epithelium in the oral cavity, pharynx, and larynx [1]. EGFR activation is common in HNSCC and is correlated with increased tumor growth and metastasis, poor prognosis, and resistance to chemotherapy and radiotherapy [2]. Several targeted therapies for EGFR have emerged in recent decades (e.g., gefitinib, erlotinib, cetuximab, panitumumab, and zalutumumab). Among them, cetuximab is the first to be approved for the treatment of HNSCC and is the most widely investigated. However, the initial response rate of cetuximab as monotherapy is only 10%–30%, suggesting that intrinsic resistance exists [3]. Moreover, patients who respond at the beginning display disease progression due to acquired resistance to cetuximab. The mechanism(s) of primary and acquired resistance remains elusive.

Hedgehog signaling plays a critical role in patterning and organ development during embryogenesis [4]. However, aberrant activation of the HhP in adults has been linked to carcinogenesis. To activate the HhP, Hh ligands first bind with transmembrane receptor patched (PTCH), releasing its inhibition on smoothened (SMO), a G protein-coupled receptor (GPCR)-like signal transducer [4]. Unblocked SMO triggers a downstream cascade that activates the transcription factor glioma-associated oncogene (GLI). Activated GLI proteins then translocate to the nucleus and mediate the transcription of Hh target genes. The HhP target genes include GLI1, a component of the Hh signal per se, and the GLI1 level is used as an indicator of HhP activity. Normally, the HhP is mainly quiescent in adult tissues but seems to be reactivated and responsible for tumorigenesis and tumor growth in many different tumor types [4,5]. Therefore, the HhP has been regarded as a potential target in cancer treatment.

The HhP and EGFR pathways have been reported to interact in numerous studies, both positively and negatively. For example, in keratinocytes, EGFR can signal via RAF/MEK/ERK to cooperate with GLI proteins to upregulate Hh target genes at the level of promoters [6]. In gastric cancer, epidermal growth factor (EGF) upregulates expression of target genes GLI1, BCL2, and PTCH1 [7]. However, it is converse in a human medulloblastoma cell line that EGFR signaling negatively regulates the Hh target genes PTCH and Hedgehog-interacting protein [8]. These dichotomous examples indicate that crosstalk between EGFR and HhP is context-dependent.

Both EGFR and HhP have been associated with epithelial-mesenchymal transition (EMT). EMT is a reversible developmental program that enables cells to suppress their epithelial features from changing to mesenchymal ones. In cancer, it has been demonstrated that EMT plays an important role in resistance to conventional and targeted therapies, including anti-EGFR therapy [9]. The process of EMT is characterized by changes in a series of markers: downregulation of epithelial-specific markers (e.g., cytokeratins and E-cadherin) and increase in mesenchymal markers (e.g., fibronectin, N-cadherin, Snail, Twist, ZEB1/2, and vimentin) [10]. In addition to the markers, EMT is also accompanied by changes in cell morphology.

In this chapter, researchers describe a previous study by Keysar et al. that uncovered the crosstalk between these two pathways and their roles in EGFR-dependent/independent EMT, migration/invasion, and anti-EGFR therapy susceptibility in HNSCC [11]. Their findings provided a rationale for co-targeting EGFR and HhP signaling in HNSCC patients. As expected, the combinatorial therapy achieved predominant efficacy in HNSCC patient-derived xenografts.

Crosstalk between EGFR and HhP in HNSCC

Since a previous study reported interplay between EGFR and HhP through MEK/ERK, Keysar et al. hypothesized that this interaction could contribute to EGFR resistance in HNSCC [11].

To test this hypothesis, they first inhibited or activated EGFR in different HNSCC cell lines. After inhibition of EGFR with erlotinib, the GLI1 level was upregulated in EGFR-dependent epithelial cells (HN11 and Tu-167 cell lines) but remained unchanged in EGFR-independent mesenchymal 584 cells. Conversely, following activation of EGFR with EGF, the GLI1 level was downregulated in HN11 and Tu-167 cells and remained unchanged in 584 cells. Here, GLI1 was an indicator of HhP activity. They further confirmed that erlotinib activated the GLI1 promoter, whereas EGF suppressed the promoter in both HN11 and Tu-167 cells via a GLI1 promoter-based luciferase assay. The 584 cells were not influenced by either drug. These results together indicated that EGFR activation decreases the HhP transcription factor GLI1 in EGFR-dependent HNSCC cells, contrary to a previous study showing that GLI1 levels were not modulated by EGFR signaling in human keratinocytes [6]. HN11 and Tu-167 cells chronically treated with 10 μM erlotinib also exhibited a high level of GLI1 and, to a lesser extent, GLI2.

To investigate resistance to erlotinib, the researchers next generated erlotinib-resistant HN11 and Tu-167 cells from chronic treatment with erlotinib. EMT has been reported to be related to drug resistance. The resulting resistant cells indeed exhibited significantly higher EMT-related genes, including ZEB1, SNAIL, TWIST, and VIM, and had higher c-FOS, an indicator of anti-EGFR insensitivity [12] but showed no GLI1 upregulation compared with parental cells. Flow cytometry also confirmed increased Vim-positive cells in erlotinib-resistant cells compared with parental cells, confirming EMT-like characteristics in erlotinib-resistant cells. The finding that resistant cells no longer overexpress GLI1 indicated that upregulation of GLI1 might only occur during the EMT process.

To further understand the relationship between the EGFR pathway, HhP, and EMT, the researchers treated HNSCC cells separately with HhP inhibitor IPI-926, erlotinib, or in combination, with or without addition of EGFR ligand EGF. Activation of EGFR with EGF (100 ng/mL) induced increased motility and spindle-like cellular morphology and upregulated the EMT genes ZEB1 and Vim, indicating an EMT-like state. These changes were only observed in three epithelial cell lines, HN11, Tu-167, and FaDu, but not 584 cells. This EGF-induced EMT-like state may be mediated by ZEB1 because siRNAs against ZEB1 significantly lowered the levels of VIM mRNA. As expected, the addition of erlotinib reversed EGF-induced EMT-related gene upregulation, whereas IPI-926 caused higher ZEB1 and VIM expression than EGF alone. These results together suggest that HhP inhibition might contribute to EGFR-driven EMT. Induction of ZEB1 and VIM by EGF was almost completely suppressed in both HN11 and Tu-167 erlotinib-resistant cells.

Having demonstrated that EGFR activation could induce an EMT-like state in EGFR-dependent cells, the invasion was then measured. Invasion was increased in EGF-treated cells but suppressed in erlotinib-treated cells. Consistent with previous findings, IPI-926 enhanced EGFR-driven invasion. However, proliferation was suppressed in invasive cells, and they were not more sensitive to EGFR inhibition than noninvasive cells.

The downstream pathways responsible for the EGFR-induced EMT-like state were clarified. The MEK/ERK inhibitor AZD6244, but not the PI3K/AKT inhibitor ZSTK474, blocked ZEB1 and VIM expression, indicating that MEK/ERK is primarily responsible for generating the mesenchymal phenotype.

Combination of cetuximab and IPI-926 in preclinical models

The above findings provide a rationale for cetuximab and IPI-926 combinatorial treatment in HNSCC because HhP inhibition could (1) enhance EGFR dependence and potentiate anti-EGFR therapy and (2) prevent EMT driven by EGFR inhibition. Cetuximab, IPI-926, and the combination were evaluated in a panel of patient-derived HNSCC xenografts. Cetuximab induced tumor shrinkage in four of six cases. However, no complete regression was observed in all four cases, and three regrew to baseline during the study. IPI-926 alone showed only modest efficacy in one case. The combination of cetuximab and IPI-926 caused complete regression in two of the four cetuximab-sensitive cases and significantly retarded regrowth in the other two cases.

To reveal the molecular events during treatment, markers relevant to EGFR signaling, HhP, and EMT were measured. Cetuximab and combination treatment upregulated the EMT-related genes ZEB2, TWIST1, and SNAIL in cetuximab-sensitive cases but not in cetuximab-resistant cases. The expression profile of EMT-related genes in cetuximab-sensitive xenografts was similar to that of the erlotinib-resistant HNSCC cell lines mentioned above. This resemblance suggested that after 1 week of treatment, EMT or selection for cells in EMT might have occurred. Components of EGFR signaling pathways, including pEGFR, pMAPK, and pAKT, were suppressed in all sensitive cases treated with cetuximab or in combination, whereas IPI-926 increased pMAPK in all four sensitive cases, pAKT in two cases, and pEGFR in one case. *E*-CAD was observed to decrease after IPI-926 treatment in cetuximab-sensitive cells, consistent with previous findings that HhP inhibition pushes cells into an EGFR-dependent EMT-like state.

Collectively, the Hedgehog and EGFR signaling pathways may negatively regulate each other. IPI-926 treatment sensitizes cetuximab therapy by pushing tumor cells further into an EGFR-dependent state.

Conclusion

The Hedgehog pathway has been reported to intersect with other signaling pathways in numerous papers, either positively or negatively. Keysar et al. first revealed crosstalk between the HhP and EGFR in HNSCC. These two pathways negatively regulate each other. On one hand, inhibition of Hedgehog could enhance EGFR signaling, further pushing the cells to an EGFR-driven EMT-like state. On the other hand, chronic blockade of EGFR upregulates GLI1, which might be the cause of EGFR-independent EMT and resistance to anti-EGFR therapy. Thus, the addition of IPI-926 was thought to have the potential to delay or block tumor recurrence by restoring EGFR dependence. Both preclinical results and a phase 1 clinical study showed signs of efficacy in HNSCC. Molecular events in tumor tissues also supported the above mechanisms, as expected. Collectively, IPI-926 might sensitize HNSCC cells to anti-EGFR therapy and shows promise for clinical translation as a combination with cetuximab.

References

[1] Johnson DE, Burtness B, Leemans CR, Lui VWY, Bauman JE, Grandis JR. Head and neck squamous cell carcinoma. Nat Rev Dis Primers 2020;6:92.

[2] Zimmermann M, Zouhair A, Azria D, Ozsahin M. The epidermal growth factor receptor (EGFR) in head and neck cancer: its role and treatment implications. Radiat Oncol 2006;1:11.

[3] Byeon HK, Ku M, Yang J. Beyond EGFR inhibition: multilateral combat strategies to stop the progression of head and neck cancer. Exp Mol Med 2019;51:1–14.

[4] Scales SJ, de Sauvage FJ. Mechanisms of Hedgehog pathway activation in cancer and implications for therapy. Trends Pharmacol Sci 2009;30:303–12.

[5] Gupta S, Takebe N, Lorusso P. Targeting the Hedgehog pathway in cancer. Ther Adv Med Oncol 2010;2:237–50.

[6] Kasper M, et al. Selective modulation of Hedgehog/GLI target gene expression by epidermal growth factor signaling in human keratinocytes. Mol Cell Biol 2006;26:6283–98.

[7] Seto M, et al. Regulation of the hedgehog signaling by the mitogen-activated protein kinase cascade in gastric cancer. Mol Carcinog 2009;48:703–12.

[8] Götschel F, et al. Synergism between Hedgehog-GLI and EGFR signaling in Hedgehog-responsive human medulloblastoma cells induces downregulation of canonical Hedgehog-target genes and stabilized expression of GLI1. PLoS One 2013;8:e65403.

[9] Tulchinsky E, Demidov O, Kriajevska M, Barlev NA, Imyanitov E. EMT: a mechanism for escape from EGFR-targeted therapy in lung cancer. Biochim Biophys Acta Rev Cancer 2019;1871:29–39.

[10] Ribatti D, Tamma R, Annese T. Epithelial-mesenchymal transition in cancer: a historical overview. Transl Oncol 2020;13, 100773.

[11] Keysar SB, et al. Hedgehog signaling alters reliance on EGF receptor signaling and mediates anti-EGFR therapeutic resistance in head and neck cancer. Cancer Res 2013;73:3381–92.

[12] Jimeno A, et al. C-fos assessment as a marker of anti-epidermal growth factor receptor effect. Cancer Res 2006;66:2385–90.

CHAPTER

27

Novel Toll-like receptor 9 agonist induces epidermal growth factor receptor (EGFR) inhibition and synergistic antitumor activity with EGFR inhibitors

Kewen Qian, Changhai Lei, and Shi Hu

Department of Biophysics, College of Basic Medical Sciences, Naval Medical University (Second Military Medical University), Shanghai, China

Abstract

Toll-like receptor 9 (TLR9) agonists can inhibit tumor cell proliferation, playing an antitumor effect. However, the specific mechanism of the effect is still unclear. This study demonstrated the effect of an immunomodified second-generation oligonucleotide (IMO), which is a TLR9 agonist, on the growth of GEO human colon cancer xenografts and an established derivative, GEO cetuximab-resistant derivatives colon cancer xenografts. IMO contains two structural domains. One is a 3′-3′-attached DNA domain, which makes IMO have two 5′-ends, so it can induce stronger immune responses after being recognized by TLR9. The other domain is an immunomodulatory domain. Considering that the occurrence and development of GEO are closely related to EGFR and that existing studies on the relationship between CpG and EGFR are still scarce, this research was based on the hypothesis that IMO can impair EGFR-related signaling pathways through both receptor targeting and immune regulation and enhance the inhibitory effect of EGFR inhibitors. The experiments tested the molecular levels of several growth factors related to cell proliferation, cell apoptosis, and angiogenesis with IMO treatment alone and in combination with the anti-EGFR monoclonal antibody cetuximab or the protein kinase inhibitor gefitinib. The results indicated that IMO effectively inhibits the growth of GEO human colon cancer xenografts by affecting EGFR-related signaling pathways. However, the inhibitory effect of IMO is relative; although the combination of IMO and gefitinib can enhance the therapeutic effect, it has no obvious inhibitory effect on EGFR in cetuximab-resistant GEO-CR cells. In addition, IMO can also have a synergistic effect on the treatment of GEO-CR through pathways unrelated to EGFR. It is interesting that the combination of IMO, cetuximab, and the chemical agent irinotecan can achieve a 90% inhibition rate in GEO colon cancer xenograft mouse models.

Novel Sensitizing Agents for Therapeutic Anti-EGFR Antibodies
https://doi.org/10.1016/B978-0-12-821584-5.00030-4

Abbreviations

TLR9 Toll-like receptor 9
IMO immunomodulatory oligonucleotides
ADCC antibody-dependent cell-mediated cytotoxicity
TGF-α transforming growth factor-α
NKs natural killer cells

Conflict of interest

No potential conflicts of interest were disclosed.

The ability of CpG DNA to affect tumor growth related to EGFR pathways

Unmethylated CpG oligonucleotides have a variety of immunomodulatory effects, such as activating macrophages, natural killer cells (NKs), and other innate immune cells, as well as promoting the release of various inflammatory cytokines, such as IL-6 and TNF-α [1,2]. The cell surface receptor of CpG is TLR9, which can then activate a series of signaling pathways intracellularly after recognizing CpG DNA [3,4]. CpG DNA has antitumor effects and can mediate antibody-dependent cell-mediated cytotoxicity (ADCC) effects [5], so it has certain research value in the anticancer adjuvant therapy. IMO contains 3′-3′-attached novel structures. Since TLR9 recognizes the DNA sequence at the 5′-end, compared to traditional CpG DNA, IMO can activate stronger TLR9-mediated immune responses [6,7]. Research on the effect of CpG DNA on EGFR-related signaling pathways is still insufficient. However, if the inhibitory effect of IMO on tumor cells is related to EGFR, it can effectively enhance the effectiveness of EGFR inhibitors.

IMO combined with cetuximab inhibits the expression and angiogenesis of signaling proteins and causes potent antitumor activity

Treatment with IMO or cetuximab alone, as well as the combination of these two drugs, was used to treat GEO colon cancer xenograft mice. It was found that IMO or cetuximab alone delayed tumor growth for a certain period of time. Compared with the untreated group, the tumor growth rate was delayed by 4 and 7 weeks. Then, the tumor recurred rapidly. However, when the two drugs were used in combination, tumor growth was significantly inhibited. Although a tendency of recurrence began to appear at the 11th week, the overall survival period was effectively prolonged. The median survival of GEO colon cancer xenograft mice treated with IMO or cetuximab alone was 9 and 13 weeks, respectively. All mice died during the experiment, while the combined administration group still retained an 80% survival rate at the end of the experiment.

Since then, researchers have tested the expression levels of several proteins closely related to tumor cell proliferation, cell apoptosis, and angiogenesis to further explore the specific mechanisms of the above antitumor effects [8]. According to the results of western blotting and immunoprecipitation, IMO can reduce the expression and phosphorylation of EGFR and its ligand transforming growth factor-α (TGF-α). In addition, two noteworthy molecules, bcl-2, a cellular protein that inhibits apoptosis, and VEGF are also inhibited, and the literature shows that CpG DNA can induce the expression of bcl-2 and VEGF on the surface of macrophages [9].

The in vitro effect of IMO is not obvious

It is worthy of further exploration that when GEO colon cancer cells were used for in vitro experiments to test the effect of IMO on tumor cell growth, no dose-dependent EGFR inhibitory effect was observed at different doses. According to the western blotting results, TLR9 expression was not detected on the surface of GEO colon cancer cells.

Cooperation of IMO and the EGFR tyrosine kinase inhibitor gefitinib has a combined effect

In addition to cetuximab, small-molecule protein kinase inhibitors are also used in targeted therapies for EGFR. The researchers tested the synergy of IMO on the small-molecule tyrosine kinase inhibitor gefitinib. Similar to cetuximab, the experiments used a single administration group of IMO or gefitinib and a combined administration group of the two drugs. Comparing the tumor volume with the no-treatment group, which reached a maximum volume of 2.0 cm^3 at the 8th week, IMO or gefitinib alone delayed the growth of the tumor as the maximum volume was reached at 12 and 14 weeks, respectively. However, when the two drugs were used in combination, the tumor volume did not reach the maximum volume until the end of the experiment. The downstream signaling pathways were also analyzed, and it was found that compared with gefitinib, IMO can inhibit the pMAPK, pAkt, and VEGF signaling pathways more strongly.

IMO has no inhibitory effect on the growth of cetuximab-resistant GEO-CR xenografts but still cooperates with cetuximab

To evaluate the dependence of IMO and cetuximab on EGFR signal sensitivity, the cetuximab-resistant GEO tumor xenograft derivative GEO-CR was used to construct a tumor xenograft mouse model [10]. The experiment still included the following groups: a control group, IMO treatment alone, cetuximab treatment alone, and combined administration of IMO and cetuximab. The results showed that IMO alone had no obvious effect on GEO-CR, while cetuximab alone had only a weak inhibitory effect. When the two drugs were used in combination, they exerted a significant synergistic inhibitory effect, as the tumor volume was still only 1.0 cm^3 after 6 weeks of treatment. These results suggest that IMO can also have a synergistic effect with cetuximab through EGFR-independent pathways.

Tumor xenografts can achieve eradication under administration of IMO in combination with cetuximab and irinotecan

After testing the synergistic effects of IMO on cetuximab and gefitinib using the two-drug combination protocol, the researchers further considered irinotecan, another commonly used drug for the treatment of colon cancer [11]. Compared with the general clinical dosing regimen that combines cetuximab and irinotecan [12], the therapeutic effect of IMO combined with cetuximab or irinotecan is more obvious. When cetuximab, irinotecan, and IMO were used together, they completely inhibited tumor growth in xenograft mice. According to existing experimental results, seven out of eight mice achieved complete elimination of tumor mass.

Discussion

In general, the study proved the following results. First, IMO has a certain antitumor effect and can inhibit the phosphorylation of EGFR, pMAPK, pAkt, cyclooxygenase-2, bcl-2, VEGF, and other signaling molecules and pathways related to tumor cell proliferation, apoptosis, and angiogenesis [8]. Second, IMO has a synergistic effect on the antitumor effects of EGFR inhibitors, such as the monoclonal antibody cetuximab and the small molecule protease inhibitor gefitinib. When the two drugs are used in combination, they can effectively delay tumor progression and improve the survival time of mice. In addition, IMO can also enhance the inhibitory effect of cetuximab on GEO-CR drug-resistant tumor cells through EGFR-independent pathways. Finally, when irinotecan, cetuximab, and IMO are used in combination, the occurrence of tumors is expected to be completely eradicated. However, most of these data came from the first experiment, and further verification and analysis are needed. Overall, this study suggests the potential value of IMO in EGFR targeted therapy.

References

[1] Kataoka T, et al. Antitumor activity of synthetic oligonucleotides with sequences from cDNA encoding proteins of Mycobacterium bovis BCG. Jpn J Cancer Res 1992;83:244–7.

[2] Tokunaga T, Yamamoto T, Yamamoto S. How BCG led to the discovery of immunostimulatory DNA. Jpn J Infect Dis 1999;52:1–11.

[3] Hemmi H, et al. A Toll-like receptor recognizes bacterial DNA. Nature 2000;408:740–5.

[4] Krieg AM. CpG motifs: the active ingredient in bacterial extracts? Nat Med 2003;9:831–5.

[5] Milas L, et al. CpG oligodeoxynucleotide enhances tumor response to radiation. Cancer Res 2004;64:5074–7.

[6] Kandimalla ER, Zhu F-G, Bhagat L, Yu D, Agrawal S. Toll-like receptor 9: modulation of recognition and cytokine induction by novel synthetic CpG DNAs. Biochem Soc Trans 2003;31:654–8.

[7] Agrawal S, Kandimalla ER. Intratumoural immunotherapy: activation of nucleic acid sensing pattern recognition receptors. Immuno-Oncol Technol 2019;3:15–23.

[8] Damiano V, et al. Novel Toll-like receptor 9 agonist induces epidermal growth factor receptor (EGFR) inhibition and synergistic antitumor activity with EGFR inhibitors. Clin Cancer Res 2006;12:577–83.

[9] Zheng M, Klinman DM, Gierynska M, Rouse BT. DNA containing CpG motifs induces angiogenesis. Proc Natl Acad Sci U S A 2002;99:8944–9.

[10] Fan Z, Masui H, Altas I, Mendelsohn J. Blockade of epidermal growth factor receptor function by bivalent and monovalent fragments of 225 anti-epidermal growth factor receptor monoclonal antibodies. Cancer Res 1993;53:4322–8.

[11] Cunningham D, et al. Cetuximab monotherapy and cetuximab plus irinotecan in irinotecan-refractory metastatic colorectal cancer. N Engl J Med 2004;351:337–45.

[12] Saltz LB, Meropol NJ, Loehrer Sr PJ, Needle MN, Kopit J, Mayer RJ. Phase II trial of cetuximab in patients with refractory colorectal cancer that expresses the epidermal growth factor receptor. J Clin Oncol Off J Am Soc Clin Oncol 2004;22:1201–8.

CHAPTER

28

Targeted cancer therapy: The future of drug combinations

Débora D. Meira[a], Matheus C. Casotti[a], Raquel F.R. Braga[a], Luiz C.G.S. Filho[a], Alessandra P. Guimarães[a], Camilly V. Campanharo[a], Daniel A. Duque[a], Débora G. Barbosa[a], Lorena M. Lopes[a], Victor N.G. Kohls[a], Lyvia N.R. Alves[a], Fernanda M. Garcia[a], Fernando L.H. Vieira[b], Vitor H. Almeida[c], Eldamaria V.W. Santos[a], Raquel S. Reis[a], and Iúri D. Louro[a]

[a]Núcleo de Genética Humana e Molecular, Departamento de Ciências Biológicas, Universidade Federal do Espírito Santo, Espírito Santo, Brazil [b]Representante da Universidade Federal do Espírito Santo no Comitê de Publicação do Projeto Elsa, Universidade Federal do Espírito Santo, Espírito Santo, Brazil [c]Instituto de Bioquímica Médica Leopoldo De Meis, Universidade Federal do Rio de Janeiro, Rio de Janeiro, Brazil

Abstract

Cancer is one of the most feared diseases in the world, being ranked as the second leading cause of death from disease. It is also observed that both chemotherapy and radiotherapy are very cytotoxic, and therefore, molecular targets are increasingly sought after and the development of new molecules with the ability to target tumor cells, sparing normal cells. Therefore, the main goal is to pursue more effective molecules with fewer side effects, greater patient adherence to treatment, and better quality of life. In the last decades, cancer research generated major progress in the understanding of the molecular basis of cancer, along with biotechnology advances, allowing for the development of new antineoplastic-targeted agents and a subsequent improvement in cancer treatment. Despite the progress, mechanisms of resistance to cancer therapy either inherited or acquired remain a hurdle, requiring new strategies to overcome them. EGFR (epidermal growth factor receptor) stands out as the main molecular target studied today, as it controls cell growth, differentiation, and death. EGFR, a transmembrane glycoprotein, belongs to the ErbB/HER family of receptors, which includes HER2 (ErbB2/neu), HER3 (ErbB3), and HER4 (ErbB4). Ligand binding leads to the formation of homo- or heterodimers between family members, facilitating receptor tail autophosphorylation. Phosphorylated receptors subsequently activate signaling pathways that regulate cell proliferation, survival, and transformation. EGFR inhibition by anti-EGFR monoclonal antibodies (mAbs) or tyrosine kinase inhibitors (TKIs)

Novel Sensitizing Agents for Therapeutic Anti-EGFR Antibodies

https://doi.org/10.1016/B978-0-12-821584-5.00032-8

represents a particularly successful molecular targeted therapy for tumors such as nonsmall-cell lung cancer and colorectal cancer. Anti-EGFR mAbs bind EGFR with higher affinity than the original ligands, preventing receptor activation. Moreover, they induce EGFR internalization and degradation, with consequent cell cycle arrest, inhibition of proliferation and angiogenesis, and promotion of in vitro and in vivo antibody-dependent cellular cytotoxicity (ADCC). Although exhibiting a plethora of antineoplastic mechanisms, numerous reports have described that several patients using EGFR inhibitors experience an initial clinical response followed by disease progression. Despite the benefits experienced by most patients bearing EGFR mutations, some of them will already present intrinsic resistance to EGFR-targeted therapy at diagnosis. Recently, several studies have shed light upon the mechanisms of acquired resistance to anti-EGFR mAbs and TKIs, and among them, the most important are the appearance of EGFR mutations, altered internalization mechanisms and downregulation of EGFR, mAbs inability to prevent the formation of ligand-induced heterodimers, *KRAS* mutations, and PTEN loss. These mechanisms culminate in a sustained activation of major intracellular signaling pathways controlled by MAPK and Akt, leading to persistent cell survival. Altogether, data suggest that altered signal transduction emerges as a major driving force in molecular target drug resistance, and therefore, one can expect that resistance could be overpowered by the combined use of specific inhibitors targeting such pathways in cancer cells. In this book chapter, we describe the molecular bases of cancer and the mechanism of monoclonal antibodies directed to the EGFR family, as well as their sensitizing agents capable of reversing tumor resistance. Therefore, future therapeutic approaches are likely to involve the combination of different antineoplastic-targeted agents.

Abbreviations

ADCC	antibody-dependent cellular cytotoxicity
ATP	adenosine triphosphate
AXL	AXL receptor tyrosine kinase
BCR	breakpoint cluster region protein
CA	clonogenic assay
CC	cervical cancer
c-Cbl	casita B-lineage lymphoma
CCV	clathrin-coated vesicle
CDC	complement-dependent cytotoxicity
CDKI	cyclin-dependent kinase inhibitor
CTLA-4	cytotoxic T-lymphocyte-associated protein 4
DNM1	dynamin-1
ECX	epirubicin plus capecitabine
EE	early endosome
EGF	epidermal growth factor
EGFR	epidermal growth factor receptor
Erb	erythroblastic oncogene
ERK 1/2	extracellular signal-regulated kinase 1/2
FOXO3A	forkhead box O3
GDP	guanosine diphosphate
GRB2	growth factor receptor bound 2
HGF	hepatocyte growth factor
HNSCC	head and neck squamous cell carcinomas
IGF-1R	insulin-like growth factor-1 receptor
IGF-BP	insulin-like growth factor-binding protein
IgG2	immunoglobulin G2
INCA	Brazilian National Cancer Institute
lncRNA	long noncoding ribonucleic acid
mAbs	monoclonal antibodies
MAPK	mitogen-activated protein kinase
MDGI	mammary-derived growth inhibitor

MTT	(3-(4,5-dimethylthiazol-2-yl)-2,5-diphenyltetrazolium bromide)
NK	natural killer cells
NSCLC	nonsmall-cell lung cancer
PBMC	peripheral blood mononuclear cells
PD-1	programmed cell death type 1
PDAC	pancreatic ductal adenocarcinoma
PDGFRA	platelet-derived growth factor A
PDK-1	phosphoinositide-dependent kinase-1
PD-L1	programmed death ligand 1
PI3K	phosphatidylinositol 3-kinase
PIK3CA	phosphatidylinositol-4,5-bisphosphate 3-kinase
PIP2	phosphatidylinositol 4,5-bisphosphate
PIP3	phosphatidylinositol (3,4,5)-trisphosphate
PTEN	phosphatase and tensin homolog
RTK	receptor tyrosine kinase
SF	surviving fraction
SFK	Src family of tyrosine kinases
Shc	Src homology and collagen
STAT	signal transducer and activator of transcription
T-DM1	ado-trastuzumab emtansine
TGF	transforming growth factor
TKI	tyrosine kinase inhibitor
TME	tumor microenvironment
TNFα	tumor necrosis factor α
VEGF	vascular endothelial growth factor
VEGFR	vascular endothelial growth factor receptor
v-erbB	viral erythroblastic oncogene B
WHO	World Health Organization

Conflict of interest

No potential conflicts of interest were disclosed.

Introduction

Currently, cancer is one of the most feared diseases in the world, being ranked as the second leading cause of death from disease [1]. According to the Brazilian Cancer Incidence Estimate by the National Cancer Institute (INCA), more than 600,000 new cases of cancer were estimated in 2020, being considered the second cause of death from disease in the country, following the same magnitude observed in the world [1]. Unfortunately, chemotherapy usually has side effects and high toxicity, and often times leads to multiple drug resistance. Cellular resistance to cytotoxic agents is the main reason why chemotherapy is not able to cure most cancers [2]. Furthermore, significant advances in surgery and radiotherapy are not expected. Given this scenario, the need for continued investments in the discovery of effective treatments for cancer control is clear. It is also clear that since the beginning of the development of antineoplastic agents, more efficient treatments are the ones that destroy tumor cells, not interfering with the growth of the patient's normal cells.

The main current molecular target is the epidermal growth factor receptor (EGFR) and human epidermal growth factor receptor 2 (HER2), because cell signaling through the ErbB/HER receptor family is closely related to the appearance of human tumors, being regulators of cellular functions, controlling several important pathways for cell proliferation, differentiation, and death [3,4]. Therefore, the EGFR has an important role in the development and progression of cancer, as its increased, mutated form or aberrant expression is frequent in several types of tumors and correlates with the aggressiveness of the disease [3]. All these factors are implicated in oncogenesis, and this is the rational for using EGFR inhibition as a potential target of anticancer therapies.

Over the past two decades, several EGFR-targeting agents have been developed, especially monoclonal antibodies (mAbs) and tyrosine kinase inhibitors (TKIs). Monoclonal antibodies target the extracellular portion of the EGFR (Fig. 1) and have multiple tumor inhibition mechanisms. In the first of them, mAbs block ligand-induced EGFR activation, with consequent cell cycle G1 arrest, inhibition of proliferation and induction of cell death by apoptosis. Furthermore, many mAbs interact with components of the immune system through the induction of antibody-dependent cellular cytotoxicity (antibody-dependent cell-mediated cytotoxicity, ADCC) and complement-dependent cytotoxicity. The third mechanism of antitumor action of mAbs lies in the inhibition of new blood vessel formation (angiogenesis), by inhibiting the expression of vascular endothelial growth factor, VEGF [4,5]. In addition to mAbs, another important class of anti-EGFR agents are inhibitors of receptor tyrosine kinase activity. They are small molecules that bind to EGF's intracytoplasmic portion (Fig. 1), on the ATP binding site, inhibiting receptor phosphorylation reactions [5]. One of the first compounds developed was PD153035, which has been intensively investigated in recent years. It reversibly inhibits EGFR autophosphorylation, blocking the proliferation and clonogenic capacity of several cell lines [6]. This class has other inhibitors that have been developed, such as the gefitinib, erlotinib, lapatinib, among others [4].

This book chapter aims to review the molecular basis of cancer, explaining how the genetic alterations that are capable of transforming normal cells into malignant occur, as well as

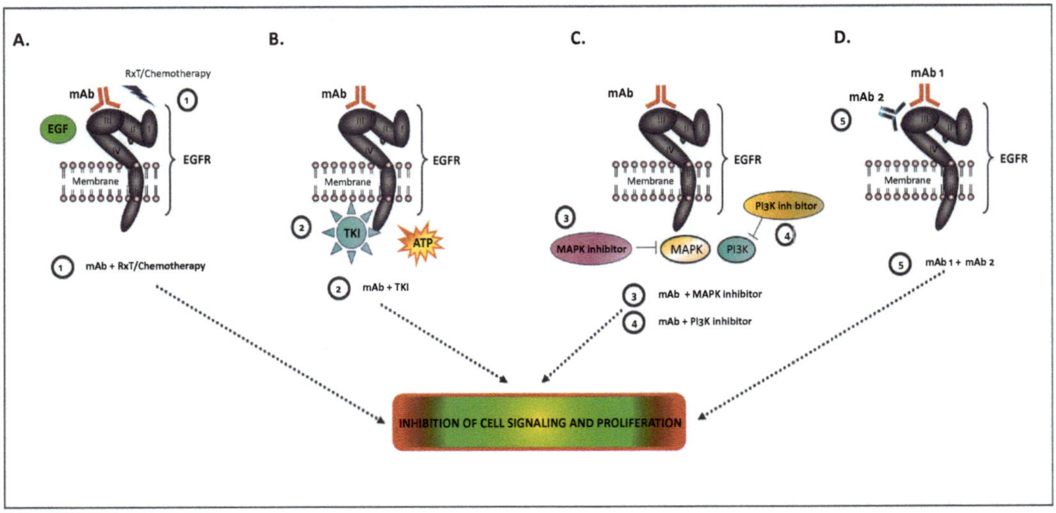

FIG. 1 Therapeutic mAbs: different scenarios and combinations to cancer treatment. Binding of mAbs to EGFR sensitizes cancer cells to (A) chemoradiation, (B) tyrosine kinase inhibitors, (C) PI3K and MAPK pathway inhibitors, and (D) other monoclonal antibodies leading to the inhibition of cell proliferation.

exploring the essential changes in cell physiology that are common to most tumors [7] emphasizing the role of endocytosis in tumor resistance. Furthermore, we describe the main targeted therapies with an emphasis on anti-EGFR monoclonal antibodies, as well as highlighting sensitizing agents/recent drugs that have the potential to reverse mAbs tumor resistance.

Molecular bases of cancer and growth factor pathway

Been considered one of the most feared diseases in the world, cancer was responsible for millions of deaths in recent years and, according to the World Health Organization (WHO), worldly estimates for the year 2030 indicate that about 11.5 million people may die from this disease [1]. Thus, it is increasingly necessary to have a clear understanding of all the molecular determinants involved in the formation of neoplasms, and, subsequently, to use this knowledge to develop and improve cancer treatment. It is known that the appearance of a tumor is associated with the occurrence of genetic alterations that progressively accumulate in a normal cell DNA, a process called carcinogenesis. Several lines of evidence suggest that carcinogenesis in humans is a multiple process, composed of several steps, which reflect the genetic alterations that control the progressive transformation of normal cells into malignant clones. Based on this hypothesis, the existence of essential changes in the physiology of a normal cell has been suggested that collectively allow for the emergence of neoplasms, which are shared by most human tumors. Such alterations are sustaining proliferative signaling, evading growth suppressors, avoiding immune destruction, deregulating cellular energetics, resisting cell death, genome instability and mutation, inducing angiogenesis, activating invasion and metastasis, tumor-promoting inflammation, and enabling replicative immortality [7]. Potential molecular targets in cancer are angiogenesis (VEGF and its receptor VEGFR, among others); extracellular matrix metalloproteinases involved in the tissue invasion process; growth factor receptors (EGFR, HER2, c-KIT, others); surface-specific receptors (CD20, among others); proteins involved in inflammation (cyclooxygenase 2—COX2); proteins involved in aberrant intracellular signaling (BCR-ABL, Ras, Akt, others); and additional therapeutic targeting of the hallmarks of cancer (specific inhibitors) are cyclin-dependent kinase inhibitors; immune-activating anti-CTLA4 mAb; telomerase inhibitors; inhibitors of HGF/c-Met; PARP inhibitors; proapoptotic BH3 mimetics; aerobic glycolysis inhibitors [4,7,8].

Among these new molecular markers in oncology, cell signaling through the ErbB/HER receptor family stands out, involved in the development of most tumors of epithelial origin, the best-known components being EGFR and HER2, both used as molecular targets by several new antineoplastic agents. EGFR (ErbB1) or epidermal growth factor receptor is a member of the ErbB growth factor family that includes HER2/neu (ErbB2), HER3 (ErbB3), and HER4 (ErbB4). The ErbB protein family has this name due to its homology with the erythroblastoma viral gene (v-erbB) and has several known ligands, among which the epidermal growth factor (EGF) stands out, transforming growth factor alpha (TGF-alpha), amphiregulin, heparin-binding EGF, epiregulin, betacellulin, epigene, and neuregulins 1, 2, 3, and 4 [3].These ligands are produced as glycosylated proteins, which are attached to the cytoplasmic membrane and subsequently cleaved by proteases, which release mature growth factors. It is important to point out that HER2 is an orphan receptor and has no known ligand so far [3]. ErbB receptors exist as monomers on the plasma surface and form homodimers (EGFR:EGFR) or heterodimers (EGFR:HER2; EGFR:HER3; EGFR:HER4) after associating with the ligand. However, HER2 is the preferred pair to dimerize with EGFR, suggesting that HER2 functions as a coreceptor. This characteristic of HER2 was explained by its conformation in the plasma membrane, which is very similar to the activated form of EGFR, and therefore, HER2 would always be in a configuration capable of dimerizing [3]. Moreover, heterodimers containing HER2 are more stable, with a lower rate of internalization and consequent degradation, and signal more potently than other dimers. Each of the ligands has specific binding properties that direct the formation of a particular dimeric combination [4,9,10].

EGFR is a 170 kDa protein with an extracytoplasmic domain (with four subdomains: I, II, III, and IV), a transmembrane region, and an intracytoplasmic domain with tyrosine kinase activity (Fig. 1). EGF binding, or other ligands, to EGF's extracytoplasmic domain leads to phosphorylation cascade responsible for transducing signals from the extracellular environment into the nucleus, where expression activation of several important cell proliferation genes occurs [3]. After dimerization, cross-activation of EGFR tyrosine kinase domain C-terminal region-specific residues takes place, by receptor trans and autophosphorylation. Activated tyrosine residues recruit adapter molecules (signal transducers) such as GRB2 (growth-factor-receptor-bound-2) and Shc (Src-homology-2-containing), which in turn activate Ras GTPase, and consequent activation of mitogen-activated protein kinase (MAPK) cascades. Ras is a GTPase that in its inactive form is bound to GDP (guanosine diphosphate). After activation, Ras undergoes an exchange of guanine nucleotides, binding to GTP (guanosine triphosphate) and activating Raf. The MAPK cascade is composed of Ras, Raf, MEK1/2, and ERK 1/2 (extracellular signal-regulated kinase 1/2) kinases and its activation leads to the transcription of genes that regulate cell proliferation, migration, invasion, and metastasis [11]. This is the main regulatory pathway for cell proliferation directly activated by EGFR and its exacerbated activation is essential for tumor development and progression [11,12]. Another direct target of EGFR is STAT5 (signal transducer and activator of transcription 5) [7]. In addition, EGFR C-terminal domain has a recognition site for the ubiquitin ligase Cbl, which is activated after EGF/EGFR binding, being responsible for receptor degradation and EGFR signaling termination [13].

EGFR does not have a PI3K (phosphatidylinositol 3-kinase) direct binding site, but activates the PI3K-Akt pathway through the Ras-PI3K-Akt pathway or through EGFR:HER3 heterodimers, since HER3 has a binding site for PI3K [3]. PI3K phosphorylates a membrane phospholipid (phosphatidylinositol biphosphate, PIP2), generating phosphatidylinositol triphosphate (PIP3). Akt has a domain that binds to PIP3, yielding its anchoring to the plasma membrane. Thus, Akt is phosphorylated/activated by PDK-1, inhibiting cell death pathways and positively modulating proliferation, cell cycle, and survival pathways. PTEN (phosphatase and tensin homolog) in its turn, inactivates this pathway due to its phosphatase ability that dephosphorylates PIP3 to PIP2. All these molecules together contribute to cell signaling and tumor development in an interconnected and well-organized way.

The dynamic and complex nature of signaling networks involved in cancer development and onset offers an excellent opportunity for pharmaceutical intervention. In accordance with this precept, in the last two decades, several new drugs have been developed that aim at inhibiting overexpressed or constitutively activated molecules in neoplastic cells, especially EGFR. The great advantage of these inhibitors is the preferential inhibition of EGFR in neoplastic cells, since they have a greater number of receptors (or the pathway is more activated) when compared to normal cells [4]. The accumulation of genetic alterations or changes in the expression of genes involved in cell growth, such as EGFR and HER2, contribute to the development of cancer. In this context, the high expression of EGFR is present in most tumors (bladder, breast, colorectal, glioma, lung, head and neck, ovary, esophagus, gastric, endometrial, and uterine cervix) and of HER2 in breast cancer, and receptor overexpression is associated with a worse prognosis [14]. It becomes clear, therefore, that EGFR and HER2 play an important role in the development and progression of cancer, as increased or aberrant expression of receptor/coreceptor (or its ligands) is frequent in several types of tumors and correlates with disease aggressiveness [15]. All these associated factors are implicated in oncogenesis, being the perfect rational for using EGFR inhibition as a potential target of antineoplastic therapies [4].

Cancer therapy and mechanisms underlying drug action/resistance

In the 1980s, Mendelsohn was the first researcher to propose the use of specific EGFR antagonists as antineoplastic therapy and his group produced two monoclonal antibodies of murine origin (mAb 528 and mAb 225), both with affinity for EGFR similar to EGF and TGF-alpha. These antibodies compete with natural ligands for EGFR binding, thus preventing its activation, in addition to promoting receptor internalization and degradation. These mAbs inhibited proliferation of breast, prostate, lung, bladder, pancreas, head and neck, and colon cancer cell lines both in vitro and in vivo [16,17]. In previous studies, our group has demonstrated that the combination of cetuximab with chemotherapy and radiotherapy in vitro (Fig. 1A) can be an effective strategy for the treatment of gynecological cancer [6]. We have also seen that the combination of monoclonal antibodies and TKI (Fig. 1B) or inhibitors of EGFR downstream pathways (Fig. 1C) or other monoclonal antibodies (Fig. 1D) has also been shown to be efficient in reversing the malignant phenotype of gynecological cancer cells [4,6,18]. Further ahead in this chapter, we will describe the mechanisms involved in resistance/sensitization to anti-EGFR monoclonal antibodies.

Reversing tumor resistance has been a challenge for modern oncology professionals, as many patients benefit from treatment with monoclonal antibodies and TKIs; however, after a certain time, antineoplastic cells become resistant to treatment. Furthermore, molecular determinants of treatment resistance and sensitivity are investigated. Wheeler et al. (2010) brilliantly described the main changes in resistance to anti-EGFR antibodies, which are (a) resistance to cetuximab resides in the overexpression of the EGFR ligand, TGF alpha; (b) EGFR overexpression was also involved in the development of acquired resistance; (c) failure in receptor ubiquitination and degradation, which is not removed from the plasma membrane; (d) activation of EGFR by Src family kinases and its increased expression; (e) EGFR heterodimerization with HER2 or HER3, with persistent activation of signaling via PI3K/Akt have also been reported; (f) EGFR translocation into the nucleus modulates repair enzymes activation and transcription factor interaction; (g) increased VEGF production, augmented angiogenesis, and therapy escape; (h) VEGFR1 contributes to cetuximab resistance; (i) mutations in both PTEN and RAS and constitutive activation of downstream EGFR pathways (j) mutations in *KRAS* cause this protein to remain GTP bound and consequently in constant activation, which allows for EGFR family members to remain always activated, regardless of ligand activation; (l) EGFRvIII, a truncated form of EGFR, which is constantly phosphorylated in a ligand-independent manner; (m) MDGI (breast-derived growth inhibitor) alters EGFR trafficking, leading to resistance to EGFR therapy [4,19].

As stated above, mAb cetuximab is indicated for metastatic colorectal cancer treatment in combination with irinotecan-based chemotherapy for EGFR-expressing tumors. However, there is no correlation between EGFR expression levels and cetuximab response in these patients. The main molecular marker associated with a worse clinical outcome for patients with metastatic colorectal cancer treated with cetuximab was *KRAS* mutation [19]. It is not very clear whether *B-RAF* and *PIK3CA* gene mutations confer the same level of resistance as mutations in *KRAS* [19]. Expression of another receptor tyrosine kinase, cMET, or hepatocyte growth factor receptor (HGFR), is also associated with resistance to cetuximab therapy [20]. Furthermore, in a model of cell lines and xenotransplantation, MEK (MAPK signaling pathway) inhibitors were able to inhibit in vitro and in vivo growth of *KRAS*-mutated cetuximab-resistant colorectal cancer cells, since this signaling pathway is activated in the presence of a gain-of-function mutation in *KRAS* [21]. It was also shown that head and neck cancer cells that are resistant to cetuximab have persistent activation of the Akt pathway and the combination with an inhibitor of the PI3K-Akt pathway sensitized the lineage to cetuximab with effective blockade of the Akt pathway [22]. It was also reported that dasatinib (tyrosine kinase inhibitor of Src family kinases) reversed resistance of mutant *KRAS* cetuximab-resistant colorectal cancer strains to both in vitro and in vivo (xenotransplantation model) with the inhibition of beta-catenin, MAPK, PI3K-Akt, and STAT pathways, and consequent inhibition of proliferation and induction of apoptosis in tumor lineages [23]. In conclusion, Weickhardt et al. reviewed strategies to overcome acquired resistance to EGFR family inhibitors by downstream targets of the Ras-MAPK and PI3K-Akt pathways, describing the "quadruple-negative" profile as those that do not have the four most important oncogenic alterations in the EGFR pathway as a possibly responsive profile to anti-EGFR monoclonal antibodies. These oncogenic alterations are loss of PTEN expression (the main negative regulator of the PI3K-Akt pathway) and mutation in the *BRAF*, *KRAS*, and *PIK3CA* genes. Therefore, these patients with a quadruple-negative profile and responders to cetuximab are those who express the PTEN protein and have wild-type *KRAS*, *BRAF*, and *PIK3CA* [4,24].

Wheler et al. established and characterized a model using high-throughput screening and examined the activation of proteins with tyrosine kinase activity in resistance models after chronic exposure to cetuximab. In this study, it was found that resistant cells exhibited increased EGFR expression secondary to receptor trafficking and degradation mechanisms. In addition, resistant cells exhibited strong activation of HER2, HER3, and c-MET (also known as hepatocyte growth factor receptor). High expression of EGFR promotes dimerization and transactivation of HER2 and HER3. Blocking EGFR and HER2 led to a loss of HER3 and PI3K/Akt pathway activity, suggesting that resistance to cetuximab is accompanied by dysregulation of receptor internalization/degradation and, in sequence, HER3-dependent EGFR activation [19]. Furthermore, Quesnelle and Grandis demonstrated that the inhibition of EGFR and HER2 is able to overcome an in vivo model (xenotransplantation) of acquired resistance to cetuximab [25]. De Roock et al. confirmed the negative effect of the *KRAS* mutation on cetuximab response, further demonstrating that *BRAF*, *NRAS*, and *PIK3CA* are also associated with a low response rate. They also suggest that cetuximab treatment should be accompanied by genotyping of *BRAF, NRAS,* and *PIK3CA* (exon 20) even in a wild-type *KRAS* population [26]. Moroni et al. proposed that response to anti-EGFR treatments is genetically based and suggest that patients should be selected for treatment based on EGFR copy number, where a high copy number is associated with response and sensitivity to cetuximab [4,27].

One of the most classic examples of monoclonal antibody in the literature today is trastuzumab, which is an anti-HER2 monoclonal antibody approved for the treatment of metastatic breast cancer with HER2 overexpression as: 1. monotherapy for patients who have received one or more chemotherapy treatments for metastatic disease and treatment failed; 2. in combination with paclitaxel for the treatment of patients who have not yet received chemotherapy for metastatic disease. Evidence suggests that resistance mechanisms to trastuzumab involve (a) trastuzumab not efficiently binding to HER2 in strains that overexpress EGFR family members; (b) RAC1 (a small GTPase of the Ras family) preventing the induction of degradation and internalization of HER2 by trastuzumab; (c) overexpression of EGFR family ligands; (d) *PIK3CA* activating mutations can uncouple Akt from regulation by upstream signaling pathways (via HER2:HER3 heterodimers). Furthermore, it was suggested that PTEN must be functional so that trastuzumab can inhibit Akt phosphorylation and loss of PTEN expression confers resistance to trastuzumab; (e) trastuzumab inhibits cell cycle progression by increasing p27 levels (an inhibitor of cyclin-dependent kinases and cell cycle arrest inducer) and p27 availability decrease is involved with trastuzumab resistance; (f) signaling via insulin-like growth factor-1 receptor and c-MET contributes to trastuzumab resistance through the activation of independent signaling or formation of HER2/IGF-1R heterodimers. These signaling pathways contribute to tumor resistance also due to a decrease in p27 levels [4,28].

Resistance to TKIs, unlike mAbs, usually includes target receptor mutations. However, resistance to lapatinib resides in alternative mechanisms, which include (a) increased expression of IGF-1R, with consequent heterodimerization with HER2, restoring PI3K pathway activation in resistant breast cancer lines; (b) activation of PI3K pathway through increased HER3 expression; c) de-repression of FOXO3A (Forkhead box O3), a transcription factor that increases signaling through the estrogen receptor and results in a phenotype no longer dependent on HER2, but dependent on hormonal signaling [29]. Corroborating these data, Garret and Arteaga (2011) described that resistance of HER2-positive neoplastic cells to lapatinib and trastuzumab resides in signaling by other tyrosine kinase-type receptors, increasing PI3K pathway activity, as well as the inhibition of trastuzumab binding by truncated forms of HER2 without the extracellular domain [4,30].

In previous studies, we have shown that mAbs matuzumab and cetuximab have different mechanisms of action and induction of death in cervical and vulval cancer cells by differently modulating these signaling pathways [4,6,18,31,32]. Our group demonstrated that for a mAb to be effective in inducing cancer cell death, it needs to negatively modulate the two main pathways (Akt and MAPK) and, if it fails to block one of the two pathways, it will be ineffective as an anticancer drug [18]. The monoclonal antibody (cetuximab) that was able to inhibit both pathways (Akt and MAPK) was shown to be an efficient mAb in inhibiting cell proliferation [6,18]. Furthermore, we have studied how to reverse tumor resistance by using specific inhibitors [6,18,31] (Fig. 1C), which has immediate clinical application [4]. Furthermore, it was possible to observe that the therapeutic association with cetuximab was able to circumvent the resistance and potentiate the effect of drugs such as cisplatin/RxT (Fig. 1A), trastuzumab and MAPK inhibitors (Fig. 1C), which could be used for the treatment of cervical cancer. In addition, cetuximab has also been shown to be effective in cancer therapy in combination with matuzumab (Fig. 1D), in addition to acting in EGFR/MAPK/Akt pathway inhibition, ADCC induction, and angiogenesis inhibition [6,18].

Therapeutic mAbs: Mechanisms of action and combinations

At the present time, thanks to the popularization of the EGFR target in the field of oncology, several anti-EGFR monoclonal antibodies have been registered for the treatment of human cancer [33]. However, tumor drug resistance immensely impairs the effectiveness of anti-EGFR monoclonal antibodies in overall patient survival. Thus, the combination of mAbs is an extremely important therapeutic strategy, as it enables the resensitization of cells resistant to single mAbs treatments [32]. We will describe in detail each of the main monoclonal antibodies that target the EGFR family (and some related targets) and, in Table 1, the synthesis of information (data on: Antibody; Brand name, Company; Target; Type; Action; Resistance; Sensibilization mechanism/pathway; Cancer type).

➢ Cetuximab (Erbitux, ImClone)

Cetuximab is a monoclonal antibody targeting epidermal growth factor (EGFR) approved for head and neck cancer and squamous cell carcinomas [34]. It acts on several mechanisms such as blocking the EGFR ligand, steric impediment of EGFR binding to its family members, inhibition of EGFR signaling cascade, and may cause G1 cell cycle arrest [33]. However, its response rate in monotherapies does not exceed 13% [34], showing the importance of association and sensitization studies [33]. Cetuximab is able to inhibit pro-angiogenic factors, which are often increased in cells with highly expressed EGFR; however, it also regulates immune response, which may affect the effectiveness of monotherapy, thus explaining the unpromising results, and the importance of developing research on these combinations [33]. Studies also relate acquired resistance by cetuximab to Akt potentiation, triggered by mutations in the EGFR extracellular domain [34,35]. In studies of acquired resistance to cetuximab, it is observed that the 2C4 antibody (inhibitor of HER2 heterodimerization) is capable of sensitizing resistant cells, thus proving to be a potential strategy to increase treatment efficacy [36]. As its resistance occurs through an EGFR-dependent mechanism [36], redirecting the mechanism of action through association with drugs such as gefitinib and erlotinib (TKIs that target EGFR) may be able to resensitize response to this drug [36]. When chronically exposed to cetuximab, the patient may experience c-Cbl-mediated internalization dysregulation (an E3 ubiquitin ligase to tyrosine 1045) and EGFR degradation. Sensitization therapies aim to modulate EGFR levels, which is able to bind and activate HER2 or HER3, thus maintaining MAPK and Akt pathways signaling in the presence of cetuximab, whereas the use of TKIs targeted to EGFR and HER2 blocks the transactivation of HER3 [36]. However, despite challenges related to mAb-only treatment, cetuximab may also present itself as a sensitizing agent in the cetuximab-lapatinib (TKI) combination, which results in downregulation of EGFR, thus being able to reverse gefitinib resistance in lung cancer cells with mutated EGFR [18].

TABLE 1 Monoclonal antibodies that target the EGFR family in oncology.

Antibody	Brand name (company)	Target	Type	Action	Resistance	Sensibilization mechanism/ pathway	Cancer type
Cetuximab	Erbitux (ImClone)	EGFR	Chimeric IgG1	Acts through the induction of internalization and degradation of the receptor, upregulation of p27kip1 and cell cycle stoppage at G1, apoptosis induction, angiogenesis inhibition, and induction of antibody-dependent cellular cytotoxicity (ADCC).	EGFR-, HER2-, and HER3-dependent mechanisms.	EGFR-directed TKIs, association to EGFR/HER2 dual tyrosine kinase inhibitor, inhibition to HER2 heterodimerization	Head and neck squamous cell carcinoma (HNSCC), colorectal cancer (CRC).
Panitumumab	Vectibix (Amgen)	EGFR	Human IgG2	Competes for the EGFR domain III with the ligand, possesses high affinity, preventing the binding of natural ligands.	Alteration in downstream molecules, activation of receptors (HER2 and MET), mutations in EGFR (extracellular domains).	PI3K/AKT/ mTOR, JAK/ STAT, PIK3CA/ PTEN, RAS/RAF	Metastatic CRC, solid tumors.
Nimotuzumab	Various brands	EGFR	–	Competes for the EGFR domain III with the ligand, has antiproliferative, antiangiogenic, and proapoptotic effects.	PIK3CA mutation and a RICTOR amplification.	PI3K/AKT/mTOR	HNSCC, pancreatic metastatic cancer, esophageal cancer, gastric cancer.

Continued

TABLE 1 Monoclonal antibodies that target the EGFR family in oncology—cont'd

Antibody	Brand name (company)	Target	Type	Action	Resistance	Sensibilization mechanism/ pathway	Cancer type
Necitumumab	Portrazza (Eli Lilly)	EGFR	Human IgG1	Acts in the EGFR extracellular domain III, blocking the binding to the EGFR ligand.	Unable to neutralize EGFRvIII and WT (wild-type) EGFR activation.	MAPK and PI3K/ AKT	Nonsmall-cell lung cancer (NSCLC) and solid tumors.
Matuzumab	Not developed commercially (Merck KGaA)	EGFR	Humanized IgG1	Binds to EGFR extracellular domain III.	Unable to inhibit the MAPK pathway in A431 lineages.	PI3K/AKT	Pancreatic cancer, HNSCC.
Trastuzumab	Herceptin (Genentech)	HER2	Humanized IgG1k	Internalization and downregulation of HER2 from cell surface, inhibition of PI3K/Akt pathway, cell cycle stoppage at G1, inhibition of the angiogenesis, and ADCC induction.	Steric hindrance by extracellular matrix (ECM) proteins, signaling increase of the growth factor receptor similar to insulin-I, activation of bypass receptors, constitutive activation of PI3K/ AKT pathway, or impaired ADCC response.	PI3K/AKT	HER2-overexpressing breast cancer, HER2-overexpressing metastatic gastric or adenocarcinoma of the gastroesophageal junction.
Pertuzumab	Perjeta (Genentech)	HER2	Humanized IgG1	Inhibition of signaling processes (downstream) through blockage of HER2 dimerization with other HER family members and ADCC induction.	PTEN or PI3K mutation, suppression of miRNA150 in ovarian cancer.	PI3K/AKT	HER2+ metastatic breast cancer and ovarian cancer.

Monoclonal antibodies clinically used in oncology (except matuzumab). Information about mAbs (action; resistance; sensibilization mechanism/pathway; cancer type) was extracted from articles listed in the references. Only major clinical indications are indicated. ADCC, antibody-dependent cellular cytotoxicity; CRC, colorectal cancer; EGFR, epidermal growth factor receptor; HNSCC, head and neck squamous cell carcinoma; NSCLC, nonsmall-cell lung cancer.

➢ Matuzumab (not developed commercially, Merck KGaA)

Matuzumab, also called EMD 72000, is a humanized anti-EGFR IgG1 mAb that was initially studied for pancreatic cancer treatment as a single drug at a dose between 400 and 1600 mg, taken weekly [37]. This mAb acts by binding to EGFR with high affinity and inducing ADCC in head and neck squamous cell carcinoma lines [18]. However, although matuzumab effectively binds to EGFR and prevents its phosphorylation, it has been shown to fail to inhibit the MAPK pathway; therefore, cetuximab is more efficient in inhibiting A431 cell proliferation [18]. Our group has shown that this mAb had no effect on cell proliferation, as there was no downregulation of EGFR, although matuzumab inhibited receptor autophosphorylation [31]. These preclinical findings might explain the overall unsuccessful results obtained in phase I and II studies testing matuzumab. No evidence of clinical activity was observed when matuzumab was administered as monotherapy in patients with epithelial ovarian cancer and, phase II studies showed that matuzumab combined with epirubicin, cisplatin, and capecitabine (ECX), or pemetrexed, does not increase response or survival of patients with advanced esophagic-gastric and NSCLC cancers, respectively. Moreover, it was reported that Takeda Pharmaceutical Company Limited discontinued matuzumab development based on the negative clinical findings to date [31]. We have shown that matuzumab alone did not inhibit cervical cancer cell proliferation and combined with chemoradiation did not enhance cytotoxic effects on gynecological cancer cell lines. Despite inhibiting autophosphorylation, matuzumab was not able to induce EGFR downregulation and persistent activation of downstream signaling pathways was observed (matuzumab failed to inhibit Akt and ERK 1/2 phosphorylation elicited by EGF, although matuzumab inhibits EGFR and HER2 phosphorylation). Accordingly, we analyzed the activation of EGFR downstream targets to determine partners involved in the signaling pathway elicited by EGF in matuzumab-treated cells. In this setting, PI3K/Akt pathway inhibition, unlike MAPK inhibition, sensitized gynecological cancer cells to matuzumab treatment in vitro. These results reinforced the paradigm that several signal transduction pathways control tumor growth and contribute to resistance and blockade of Akt signaling, which is a determinant factor to overcome resistance to matuzumab [31]. Based on these findings, we proposed a hypothesis that explains one possible mechanism of ineffectiveness of matuzumab and how to overcome it. Matuzumab, differently from cetuximab, was not able to induce EGFR downregulation, with persistent signaling and gynecological cancer cell proliferation. Although the combination of matuzumab with chemoradiation or a MAPK pathway inhibitor did not trigger benefits over single treatments, we observed that targeting PI3K, in combination with matuzumab (Fig. 1C), markedly reduced A431 and Caski cell survival, highlighting the importance of PI3K/Akt pathway [31].

➢ Combination of cetuximab (Erbitux) + matuzumab

The combination of anti-EGFR antibodies promotes the creation of complexes and, consequently, favors downregulation of EGFR, which inhibits tumor growth [18]. Studies carried out in A431 cells point to clinical potential in the combination of cetuximab and matuzumab, justified by the increase in negative regulation and inhibition of EGFR and HER2 signaling resulting from concomitant therapy with different mAbs. It was demonstrated that treatment with anti-EGFR mAbs could enhance receptor downregulation and reduce cell proliferation. Furthermore, cetuximab and matuzumab can bind simultaneously to different regions in EGFR domain III (Fig. 1D), and when the combination of cetuximab with matuzumab

(50 μg/mL each) was tested in clonogenic assay, they synergistically reduced the number and size of A431 colonies. Likewise, we demonstrated a strong reduction in ERK1/2 and Akt phosphorylation when mAbs were combined, with no changes in total protein. As EGFR phosphorylation activates the MAPK and Akt pathways, their simultaneous blockage may explain the synergistic effects obtained by this combination, which indicates the possibility of shared and complementary mechanisms [18].

- Panitumumab (Vectibix , Amgen)

Panitumumab (Vectibix), a humanized monoclonal antibody to immunoglobulin G2 (IgG2), was first approved in 2006 for the treatment of metastatic colorectal cancer [33,38]. It competes for EGFR extracellular domain III and performs a high-affinity interaction, preventing binding of natural ligands. In a study by Trivedi et al., panitumumab was compared to cetuximab in head and neck cancer and a similar degree of EGFR inhibition was found between them; however, panitumumab was less effective in mediating antitumor immune mechanisms [39]. The Fc region of the IgG2 structure of panitumumab has low affinity for the CD16 receptor on natural killer (NK) cells and, therefore, is unable to induce antibody-dependent cell-mediated cytotoxicity [40].

- Nimotuzumab (various brands)

Nimotuzumab is a humanized IgG1 monoclonal antibody that specifically binds to the extracellular domain of the EGFR and inhibits the downstream signaling cascade and has an intermediate affinity against EGFR. This mAb activates natural killer cells and induces dendritic cell and tumor antigen-specific CD8+ T cell maturation [33]. Its pharmacodynamic properties are unique and allow for a long-lasting low-toxicity treatment [41]. In a study by Palomo et al., a positive clinical outcome was found for a cancer patient treated with a combination therapy of anti-EGFR mAb nimotuzumab with anti-NGcGM3 therapy (NGcGM3/VSSP vaccine) [42].

- Trastuzumab (Herceptin , Genentech)

The first antibody approved for the treatment of HER2-positive metastatic breast cancer was trastuzumab (Herceptin), a humanized anti-HER2 monoclonal antibody [43]. Trastuzumab has several mechanisms of action to inhibit tumor growth in vitro and in vivo, acting through internalization and downregulation of HER2, induction of apoptosis through the inhibition of the PI3K/Akt pathway, upregulation of cyclin-dependent kinase inhibitor (CDKI) p27kip1 causing cell cycle arrest in G1, inhibition of angiogenesis, and antibody-dependent cell-mediated cytotoxicity (ADCC) [38,43]. Due to its success, it is, nowadays, a first-line drug for the treatment of breast cancer. However, while clinical results are encouraging, approximately half of HER2-positive cancers are primarily resistant to trastuzumab or become resistant during long-term treatment [43].

- Pertuzumab (Perjeta , Genentech)

Pertuzumab, a humanized monoclonal antibody, acts to block HER2 coreceptor role [43,44], inhibiting downstream mitogenic signaling processes, showing itself to be a potential combination in trastuzumab sensitization, bypassing possible resistance mechanisms.

➤ Combination of trastuzumab + pertuzumab

Increased effectiveness of two different mAbs combination with the same target in relation to their administration alone is due to each antibody particular mechanism of action. Combination of trastuzumab and pertuzumab, for example, shows downregulation of HER2 by trastuzumab and concomitant inhibition of its dimerization by pertuzumab, resulting in an efficient therapy [43]. Although there are difficulties with respect to long-term tolerability, the efficacy of the triple combination lapatinib-trastuzumab-pertuzumab may be even higher than the double combination described above [43].

➤ Ado-Trastuzumab emtansine (Kadcyla, Genentech)

Ado-trastuzumab emtansine or T-DM1 (ado-trastuzumab emtansine; Kadcyla, Genentech Inc.) is a 2013-approved antibody-drug conjugate formed by combining maytansinoid DM1, a highly potent microtubule-destabilizing agent, with trastuzumab [38,44]. A phase III study demonstrated increased overall survival compared with lapatinib/capecitabine in patients with HER2-positive metastatic breast cancer previously treated with trastuzumab and a taxane, which led to the approval of this drug. In addition, randomized controlled trials in patients with advanced metastatic breast cancer demonstrate that Kadcyla is a highly effective treatment with a very favorable safety profile [38,44].

➤ Bevacizumab (Avastin, Genentech)

Bevacizumab was the first angiogenesis inhibitor approved for the treatment of nonsmall-cell lung cancer patients (NSCLC) [45]. This humanized mAb targets vascular endothelial growth factor (VEGF) [45], through binding of mAb to all circulating soluble VEGF-A isoforms, avoiding the interaction between VEGF-A and VEGFR, consequently inhibiting pathways that promote vascularization [46]. Studies in colorectal cancer indicate a significant interference between VEGF and EGFR signaling pathways [47]; however, mAbs combinations targeting these pathways have not shown efficacy, unlike the association with tyrosine kinase inhibitors (TKI), which showed interesting results, even in resistant cancers, similar to the association of bevacizumab with erlotinib, for example [47]. Erlotinib can potentiate the activity of bevacizumab independently of *RAS*, appearing to attenuate the excessive accumulation of intracellular EGFR resulting from exposure to bevacizumab [47].

➤ Ramucirumab (Cyramza, Eli Lilly)

Ramucirumab is a human mAb that acts in angiogenesis, playing a significant role in tumor survival [45,48]. This mAb targets VEGFR-2 extracellular domain. To date, there is no clinical trial or even clear guidance on which groups of patients can benefit from a combination of EGFR-VEGF-targeted treatments, although both are intrinsically related to anticancer therapy [48].

➤ Necitumumab (Portrazza, Eli Lilly)

Necitumumab is a humanized monoclonal antibody to immunoglobulin G1 (IgG1), which acts on EGFR extracellular domain III, blocking EGFR ligand binding. It is currently used for nonsmall-cell lung cancer treatment (NSCLC) [18]. In 2015, the use of necitumumab combined with two chemotherapy drugs (gemcitabine and cisplatin) was approved, as it provides a survival advantage in patients with NSCLC [38].

➢ pan-HER

Bearing in mind the frequent development of acquired resistance to single HER family antibody treatments, Jacobsen et al. developed an alternative therapeutic protocol model called pan-HER. This strategy consists of a mixture of three pairs of mAbs in a balanced ratio targeting non-overlapping epitopes, which simultaneously inhibit EGFR, HER2, and HER3, avoiding compensatory mechanisms of tumor development through HER family members [49,50]. Pan-HER represents an innovative scheme to overcome primary and acquired resistance due to tumor heterogeneity and plasticity in relation to family dependence on HER [50]. The combination of three mAbs pairs composing pan-HER leads to potent growth inhibition across an extremely broad spectrum of tumor cell lines in xenograft models, including cell lines with acquired resistance to therapeutic antibodies and expressions of mutant forms of *TP53* and *KRAS* [49,50]. In addition, pan-HER is highly effective even in the presence of ligands from the HER family, demonstrating that it is efficient to overcome acquired resistance due to increased production of ligands [50]. It also prevented the compensatory upregulation of HER, often associated with responses to single receptor inhibitors, in addition to inhibiting proliferation of cells that acquired, in vitro, resistance to the inhibitory mechanisms of cetuximab, trastuzumab, or pertuzumab [49,50]. In order to assess the efficacy of pan-HER under conditions of acquired resistance, cells resistant to cetuximab, trastuzumab, or pertuzumab were generated and tested in animals for sensitivity to pan-HER therapy in several difficult-to-treat cancer models, including xenografts from patients with ovarian, colorectal, lung, and pancreatic cancer in patient-derived xenograft (PDX) models [49,50]. As a result, pan-HER overcame the effects of cetuximab and trastuzumab, and the combination of trastuzumab and pertuzumab was not able to reverse cell resistance to any of the single antibodies, reaffirming the need to target more than one HER family member in order to overcome acquired resistance. Furthermore, in certain models, the inhibitory action of pan-HER was not only sustained, but remained after the withdrawal of mAbs combination [49]. Despite being able to overcome patient's resistance and increase antitumor effects, this strategy challenges the issue of co-application toxicity of six different antibodies, since cutaneous, gastrointestinal, or other adverse effects will limit the clinical application of pan-HER [49]. A plausible solution to overcome pan-HER toxicity would be to test less complex mixtures, from two to four mAbs, which seems to be less toxic and with satisfactory results. This technique is supported by pancreatic cancer studies in which combinations including anti-HER2 and anti-EGFR antibodies were employed [49].

Future expectations regarding the PD-1/PD-L1 pathways

Recent studies combining therapies targeting programmed death receptor-1 (PD-1) and its ligand, PD-L1, have had EGFR as one of the most targeted pathways [51]. Specifically for the treatment of nonsmall-cell lung cancer, research development to identify biomarkers that direct combinations of this target with other mAbs has become necessary. Briefly, PD-1/PD-L1 blocking mechanisms are able to prevent T cells downregulation during antitumor immunity effector phase; therefore, drugs can stimulate an effector T cell response [45]. It has been suggested that anti-CTLA-4 mAbs may increase PD-1/PD-L1 activity blocking [45]. This combination of checkpoint inhibitors enabled the development of several studies, and this therapy is often used to construct additional treatments [51]. We would like to point out some promising PD-1/PD-L1 mAbs, such as avelumab (Bavencio), aimed at PD-L1; nivolumab (Opdivo, anti-PD-1 mAb); durvalumab (Imfinzi), a PD-L1 inhibitor indicated for second-line therapy for patients with advanced urothelial cancer [45], acting as an immunotherapeutic agent that blocks the interaction between programmed cell death ligand 1 (PD-L1) and its receivers [52]. Atezolizumab (Tecentriq) is a humanized monoclonal antibody directed to the co-stimulatory molecule B7 and the programmed death receptor-1 (PD-1) [45,53] whose tests have been promising in the treatment of patients with NSCLC [45]. Studies comparing the efficacy of atezolizumab monotherapy, atezolizumab-bevacizumab combination, and sunitinib therapy pointed to a superior efficacy of the combination of these mAbs compared to sunitinib [53].

➢ Other mAbs combinations

The potential observed with respect to mAbs mixtures/associations in increasing EGFR degradation points to an important therapeutic strategy for certain cancers, even if more aggressive [54]. Modotuximab is an example of these mixtures, as is Sym004, a new 1:1 mixture of two chimeric monoclonal antibodies (992 and 1024) that target non-overlapping EGFR epitopes and was targeted to patients with head and neck squamous cell carcinoma [55], inducing EGFR internalization in the cytoplasmic compartment, resulting in receptor inactivation [56]. Another potential mix in research is MM151, which brings together three different monoclonal IgG1 antibodies [57]. The combination of different monoclonal antibodies directed at different cellular targets is so important that Morrocco et al. described more than 200 clinical studies in development to treat different types of cancer (melanoma, NSCLC, CRC, breast, ovarian, pancreatic, gastric, prostate, bladder, thyroid, uterine cancer, central nervous system malignancies, HNSCC, esophageal adenocarcinoma, hematological malignancies, glioblastoma, advanced or metastatic malignancies, advanced solid tumors, ovarian, fallopian tube, or primary peritoneal cancer, chronic lymphocytic leukemia, acute myeloid leukemia, gastroesophageal junction, small-cell lung cancer, hepatocellular carcinoma). The combinations are as varied as possible, such as: atezolizumab + ado-trastuzumab emtansine (targets PD-L1 + HER2); atezolizumab + obinutuzumab + ibrutinib (targets PD-L1 + CD20); atezolizumab + pertuzumab + trastuzumab (targets PDL-1 + HER2 + HER2); avelumab + utomilumab + rituximab (targets PD-L1 + CD137 + CD20); cetuximab + avelumab (targets EGFR + PD-L1); cetuximab + pembrolizumab (targets EGFR + PD-1); durvalumab + rituximab (targets PDL-1 + CD20); ipilimumab + nivolumab (targets CTLA-4 + PD-1); ipilimumab + nivolumab + lirilumab (targets CTLA-4 + PD-1 + KIR); nivolumab + nimotuzumab (targets PD-1 + EGFR); tremelimumab + durvalumab + olaparib (targets CTLA-4 + PD-L1) [38].

A different way of classifying tumor resistance

Cancers are difficult to treat due to their heterogeneity, and many of them exhibit dysregulation of the EGFR family. Therefore, monoclonal antibodies are drugs for targeted cancer therapy and have significantly improved cancer treatment. Since the first monoclonal antibody approval, rituximab, for hematological malignancies, almost 30 additional mAbs have been approved in oncology. Furthermore, the introduction of trastuzumab has clearly improved the outcome of patients with breast cancer, even though treatment response is modest. Part of mAbs therapy failure is due to cancer complexity and various resistance mechanisms associated with targeted therapies, as only a few patients have a lasting response to currently available treatments. Resistance can be intrinsic or acquired: in intrinsic mechanisms, antibody is not effective, even if the antigen is present in tumor cells. In contrast, in acquired resistance, tumor cells show an initial sensitivity to treatment, but after a variable period, they stop responding to it [38].

Mechanisms that confer resistance to treatment with anti-EGFR antibodies are diverse and not all of them are completely elucidated, but some are well described and can be divided accordingly [33]: (1) pre-target, which includes drug resistance factors prior to binding of monoclonal antibody to EGFR receptor; (2) on-target, resistance factors to monoclonal antibody binding to EGFR; (3) post-target, resistance mechanisms downstream of anti-EGFR monoclonal antibody binding to EGFR; (4) off-target, resistance mechanisms through other signaling molecules or receptors after monoclonal antibody binding to EGFR.

In pre-target, the tumor microenvironment (TME) is associated with resistance to monoclonal antibody therapy. Studies have shown that the accumulation of TME components, such as hyaluronic acid (HA), can form a physical barrier capable of interfering with access of anti-EGFR antibodies and immune cells to target tumor cells, thus promoting resistance [58]. Furthermore, the cytosolic protein MDGI (mammary-derived growth inhibitor) is involved in EGFR intracellular accumulation by altering receptor traffic and its plasma membrane organization, and as a result, therapy with cetuximab becomes ineffective [59].

In on-target, it was described by Wheeler et al. [19] that EGFR receptor overexpression is an important mechanism of acquired resistance to cetuximab; however, clinical studies have not confirmed the relationship between EGFR expression level and receptor inhibitor therapy. Thus, the approach of analyzing EGFR levels serves as a biomarker to verify cetuximab treatment response likelihood. In addition, EGF and TNFα ligand overexpression competes with cetuximab for EGFR receptor binding, preventing mAb adhesion, and contributing to acquisition of acquired resistance to cetuximab, since the inhibition of ligand binding (generated by the antibody) prevents the activation of the tyrosine kinase domain and prevents downstream activation of pro-survival pathways. Mutation of the gene encoding domain III (408–468) of the EGFR also affects cetuximab receptor binding, as does panitumumab, which competitively binds to EGFR domain III (386–391), and mutations that alter the binding site are also associated with resistance [29]. Ligand binding to EGFR promotes receptor heterodimerization, leading to transphosphorylation of cytoplasmic tail tyrosine residues. Thus, overexpression of HER family ligands is also associated with resistance to cetuximab, through EGFR heterodimerization with HER2 or HER3, with persistent activation of signaling via PI3K/Akt. As with cetuximab, heterodimerization of HER2 and HER3 is a well-known mechanism of resistance to trastuzumab, where the ligand activates HER3 which dimerizes and phosphorylates HER2 even when linked to the monoclonal antibody. Moreover, HER2 expression reduction can impair trastuzumab binding and internalization, causing treatment resistance. The increased association between EGFR and ubiquitin ligase CBL demonstrated an increase in ubiquitination and EGFR downregulation, that is, failure in receptor ubiquitination and degradation, which is not removed from the plasma membrane, leading to resistance [29]. EGFR also functions as a transcription factor in the nucleus, increasing cell proliferation, thus playing an important role in acquired resistance to cetuximab. EGFR translocation to the nucleus modulates the activation of repair enzymes and interacts with transcription factors, through Src family kinases (SFKs), disrupting G1/Scycle progression. Interestingly, SFK activity loss leads to the loss of nuclear EGFR, thus increasing membrane EGFR and resensitizing for cetuximab. Even though there is no knowledge of a point mutation that may be related to resistance to cetuximab and panitumumab, some clinical studies analyzing the EGFRvIII variant, a truncated form of EGFR, suggest that this protein contributes to cetuximab resistance, as this variant remains constantly phosphorylated in a ligand-independent manner [29].

In post-target, after ligand activation of HER family receptors, dimerization and activation stimulate many complex and highly regulated intracellular signaling pathways, including RAS/RAF/MEK/ERK (MAPK) and PI3K/ Akt axes, promoting tumor cell proliferation, survival, invasion, and angiogenesis. Consequently, mutations in these pathways lead to signaling disturbances, such as resistance to mAb target therapy. It was described by Meira et al. that a PI3K-Akt pathway inhibitor induced matuzumab tumor cell sensitivity, reversing MAPK pathway resistance, showing how these pathways are closely related and involved in tumor resistance [31]. *PIK3CA* gene, which encodes the catalytic subunit of PI3K and occasionally mutates, confers acquired resistance to anti-EGFR therapies as it also alters downstream signaling pathways. On the other hand, PTEN, a molecule responsible for the downregulation of the PI3K/Akt signaling cascade, acts by converting PIP-3 back into PIP-2, thus interrupting signaling, so mutations that generate loss or inactivation of *PTEN* cause resistance. *KRAS* and *NRAS* mutations or *KRAS* amplification causes constitutive binding to GTP and, remaining constitutively active, regardless of ligand presence. *KRAS* is an attractive target, as it accounts for 40% of total resistance cases. Other important mutations in this pathway include *BRAF* gain of function, leading to increased kinase activity and exacerbated activation of the MAPK pathway and cetuximab/panitumumab resistance [60]. Activation of EGFR by Src family kinases and its increased expression [19] and overexpression of NCAPG (non-SMC condensin I complex subunit G) cause resistance to trastuzumab through the activation of SRC/STAT3 signaling pathway, as demonstrated by Jiang et al. . In addition, a high activity between SFKs and EGFR contributes to acquired resistance to cetuximab, therefore blocking SFK activity with dasatinib could resensitize tumors to cetuximab.

In off-target, part of the resistance mechanisms is related to other receptors, and it is important to highlight the tyrosine kinase type that, like the HER family receptors, can activate the same downstream pathways and signaling cascades; thus, dysregulation of such molecules generates resistance to targeted anti-EGFR therapies. Overexpression of IGF-1R, or downregulation of IGFBP-3 and IGFBP-4 (insulin-like growth factor-binding proteins 3 and 4), which negatively regulate binding of IGF to IGF-1R, may lead to maintenance of IGF-1R-induced PI3K/Akt signaling, conferring resistance to cetuximab. Amplification of *MET* (*MET* proto-oncogene, receptor tyrosine kinase) that accumulates on the cell surface and stimulates ErbB3 phosphorylation, which in turn activates downstream PI3K/Akt signaling axis, as well as the overexpression of HGF (hepatocyte growth factor, MET ligand), is associated with acquired resistance to cetuximab or panitumumab [61]. Overexpression of

VEGFR1 (soluble vascular endothelial growth factor receptor-1) confers resistance to cetuximab, as well as increased VEGF production, leading to increased angiogenesis and escape from therapy. AXL (AXL receptor tyrosine kinase), when overexpressed, leads to its aberrant activation that confers refractoriness or an insufficient response to cetuximab, activation of EGFR and MAPK pathway signaling. PDGFRA (platelet-derived growth factor receptor alpha) and FGFR1 (fibroblast growth factor receptor 1) are causes of resistance when overexpressed [62].

Mechanisms other than overexpression of resistance-associated receptors are also causes of resistance, such as long non-coding RNA (lncRNA), causes resistance to anti-EGFR monoclonal antibodies, long-chain lncRNA MIR100HG produced miR-100 and miR-125b, synergistically inhibits five negative modulators of the Wnt/β-catenin signaling pathway, resulting in increased Wnt signaling, and supporting resistance to cetuximab [63]. P120-catenin (p120ctn), a tumor suppressor, negative expression confers resistance to cetuximab, especially if associated with EGFR overexpression [64]. PD-1 is a cell surface molecule with two ligands, PD-L1 and PD-L2. Pembrolizumab (Keytruda) is a humanized immunoglobulin G4 monoclonal antibody that targets PD-1 receptor, being an effective first- and second-line treatment option for patients with advanced NSCLC. Some studies demonstrated that expression of mutant EGFR in bronchial epithelial cells induced PD-L1 expression, causing treatment resistance [65]. Patients with EGFR mutations showed positive responses to the association of anti-PD-1 therapy and target therapy against the EGFR receptor, probably because of the downregulation of PD-L1 induced by the inhibition generated by erlotinib, which proved to be safe and viable in combination therapy [66].

Knowledge about resistance mechanisms is helping design new clinical trials phases I-IV to provide adaptive therapeutic interventions for patients with refractory or recurrent cancers through the inhibition of alternative pathways and sensitization of resistant tumors. Some of these trials have had favorable results and are now available for clinical application, being able to offer individualized treatments for specific patients.

Does endocytosis/EGFR downregulation explain resistance to mAbs?

Cancer cell development is linked to overexpression of genes involved in cell growth, such as growth factor receptors. Therefore, tumor cells tend to have a greater amount of growth factor receptors in their cell membranes compared to healthy cells, especially EGFR and HER2, yielding them more sensitive to growth factors and more prone to disordered cell growth [67]. This feature is interesting for targeted therapies, since EGFR-specific antibodies have a greater activity on tumor, reducing treatment side effects [44,68,69].

Anti-EGFR antibodies bind more avidly to the receptor than the natural ligand, obstructing tumorigenesis signaling [49]. However, overexpression of mutated EGFR monomers increases dimerization frequency, triggering endocytosis and signaling reactions independent of a ligand [67]. Through RTKs recycling, EGFRs will later return to the cell membrane and continue to signal in an excessive and disorderly manner, increasing cell proliferation, angiogenesis, and inhibiting apoptosis, contributing to tumor growth. Therapeutic antibodies can inhibit growth or kill cancer cells by two mechanisms: immune-mediated (e.g., ADCC, antibody-dependent cellular cytotoxicity) and interference with tumorigenesis (e.g., triggering apoptosis, inhibiting cell proliferation, or blocking angiogenesis) [38]. EGFR number on cell membrane and its influence on resistance/sensitivity to monoclonal antibodies vary according to the mechanism analyzed. When evaluating antibody-dependent cellular cytotoxicity (ADCC), therapies that use anti-EGFR mAbs aim to induce immune responses by ADCC; therefore, these therapies use the greater amount of EGFR in tumor cell membrane. At high EGFR endocytosis, mAb ability to bind receptors is impaired, leading to less efficient treatments [70]. High EGFR dimers formation and constancy in receptor endocytosis is a resistance cause to anti-EGFR mAbs therapies. This can be seen in cases where EGFR mutations lead to a reduced capacity for endocytosis, in which scenario treatments with cytotoxic anti-EGFR mAbs are more effective [67,70,71]. In addition, tumor cell resistance was reduced with the use of dynamin blockers [70], a GTPase responsible for endocytosis in eukaryotic cells. Endocytosis starts with membrane invagination involving a clathrin coat vesicle (CCV), of about 0.2 μm in diameter [72]. CCVs can select receptors destined for cell entry, the first step in receptor's downregulation process. Small endocytic vesicles may lack clathrin coatings, and larger vesicles called "macropinosomes" are responsible for macropinocytosis [67]. Clathrin-mediated EGFR endocytosis is involved in gefitinib refractoriness in wild-type EGFR nonsmall-cell lung cancer (NSCLC) cell lines. Adequate endocytic uptake of signaling receptors are regulators of cellular behavior, including growth, development, and differentiation [73–75]. The abnormal expression and dysregulation of intracellular traffic machinery induce cancer development, such as NSCLC. EGFR endocytosis terminates activated signaling or recycles back to cell surface to continue signaling. EGFR internalization affects cell growth and survival through interactions with various adapters or signaling molecules. EGFR can be located in organelles, nucleus, and trigger the transcription of survival proteins, yielding development and maintenance of cancer [76,77]. Macropinocytosis has a role in the elimination mechanism induced by RAS in ductal adenocarcinoma of the pancreas (PDAC) [78]. Macropinosomes fuse with lysosomes, leading to proteolytic degradation and released amino acids meet cell's metabolic needs [79]. Elimination and hydrolysis of extracellular protein via macropinocytosis have been found to support *KRAS*-driven cell proliferation in a medium lacking essential amino acids [80,81]. Continued proliferation of cells in PDAC places a great demand on the molecular components, for which synthetic metabolic pathways are upregulated [82,83] or macromolecules are reused, especially useful in a nutrient-restricted tumor microenvironment. The decision to recommit to synthesis or elimination appears to be, at least in part, governed by oncogenic signaling [84]. Constitutive activation of the PI3K-Akt–mTOR pathway induces nucleotide and fatty acid synthesis, as well as protein translation [85], while mutant *RAS* induces both autophagy and macropinocytosis [78,79,86].

By providing continuity over the endocytosis process itself, internalized endocytic vesicles fuse with a population of small vesicles and tubules called early endosomes (EEs). These structures have a certain acidity, which facilitates dissociation of many ligands from their receptors. Newly released ligands accumulate in the lumen of EEs, are transferred to late endosomes and finally to lysosomes for degradation. Such receivers show two destinations. First, they can be recycled back to the plasma membrane by returning directly from the EEs or passing through a population of organelles called "recycling endosomes." The recycling kinetics vary between two routes, with the direct return being much faster, forming a substantial intracellular pool of recycling receptors. Or alternatively, especially in the case of activated RTKs, recycling can be made relatively inefficient with receptors (with or without bound ligand) being transferred to late endosomes and lysosomes for degradation ("downregulation") [67]. Endocytic organelles are functionally organized to allow for receptor/ligand recycling or degradation selection. Most signaling events occur in the plasma membrane, but endocytic organelles may have important roles in signal transduction, in addition to simply providing a mechanism for RTK reduction [67]. RTKs pathway may be different when coupled to monoclonal antibodies. Ligand-induced endocytosis and RTK degradation can attenuate growth-promoting signals. Cooperative effects of mAb combinations, in bivalent versions of anti-ErbB mAbs without Fc, inhibit tumorigenic growth, negatively regulating RTKs. mAb-driven RTK internalization pathway is distinct from ligand-induced endocytosis. Cooperative EGFR effects reveal that synergy requires binding to distinct epitopes and cooperative action of mAbs combinations includes the generation of much larger receptor-mAb complexes. mAb-RTK complexes endocytosis rates proportional to their size. Unlike the rapid EGF-induced endocytosis, mAb-induced homodimers undergo relatively slow endocytosis and weak ubiquitylation. The synergistic downregulation of RTK by mAbs combinations is a mechanism of tumor cell sensitization with a relevant role in cancer treatment [87].

Our research group demonstrated that endocytosis and receptor degradation induced by anti-EGFR mAbs culminate in the inactivation of growth factor receptors and suppression of downstream signaling pathways, reducing the proliferative/survival potential of cancer cells. As anti-EGFR mAb cetuximab efficiently induces EGFR degradation and decreases cell survival, it was used as a positive control to investigate if matuzumab could induce EGFR downregulation. A431 and Caski cells were treated with either matuzumab or cetuximab (100 μg/mL) for 24 h, and as expected, 24-h treatment with cetuximab induced a robust reduction of 50% and 70% in EGFR protein content in A431 and Caski cells, respectively. Treatment with matuzumab alone failed to induce EGFR downregulation after 24 h of treatment that could explain the sustained cell proliferation and survival [31]. Our group observed that the combination of matuzumab and cetuximab triggers synergistic cell proliferation inhibition effects (Fig. 1D) and EGFR downstream signaling pathways, suggesting for a potential role in cancer therapy. We also observed that the combination of cetuximab and matuzumab induced EGFR degradation, as the amount of total EGFR was greatly reduced after the combined treatment. We also observed a strong reduction in ERK1/2 and Akt phosphorylation when mAbs were combined. EGFR phosphorylation activates MAPK and Akt pathways. Their simultaneous blockage may explain the synergistic effects obtained by this combination. We demonstrated that matuzumab's inability to inhibit A431 cell proliferation reflects the complexity of cells that heavily depend on the MAPK pathway, possibly developing resistance to molecular target therapy, which could be overcome by the combination of anti-EGFR mAbs or by different strategies that target the EGFR pathway [18]. We highlighted that impaired EGFR downregulation might be the possible biological mechanism responsible for matuzumab's inefficacy in vitro in gynecological cancer cell lines and that combination with cetuximab was able to revert resistance [31]. We observed that matuzumab alone effectively induced ADCC in A431 cells [18], confirming that ADCC induction and inhibition of cell proliferation and EGFR degradation are different mechanisms triggered by monoclonal antibodies. Tested gynecological cancer cell lines harbor genetic modifications that overactivate the EGFR pathway: A431 has *EGFR* gene amplified (30 copies) and Caski cells harbor a *PIK3CA* exon 9 activating mutation (E545K), while C33A has a *PTEN* mutation. These genetic changes turn the EGFR pathway constitutively active. Altered signaling differs among these cell lines and may differentially respond to PI3K/ Akt pathway modulation. EGF-elicited signal transduction is not the only mechanism mediated by anti-EGFR mAbs, since these molecules can also induce ADCC and, in primary cervical cancer cell lines obtained from cervical biopsies, ADCC induction was dependent on EGFR expression. Matuzumab effectively induced ADCC in A431 and Caski cells, while no ADCC was observed in the C33A cell line, reinforcing that the induction of ADCC depends on a certain level of EGFR cell surface expression [31]. Our studies suggest differences between cetuximab and matuzumab antiproliferative effects and their combinatorial ability to synergistically inhibit proliferation and downstream signaling pathways. We have also demonstrated that cetuximab combined with trastuzumab synergistically reduces cell proliferation and activation of downstream signaling pathways (with concomitant reduction of MAPK and Akt phosphorylation) in cervical cancer cells, corroborating that monoclonal antibodies combination is an efficient strategy for the cancer treatment [6].

Exosomes, membranous vesicles 30–150 nm in diameter, constitutively released by almost all cell types play important roles in intercellular communication through the transfer of intercellular charges, such as miRNAs, mRNAs, and proteins [88–90]. Exosomes derived from stromal cells can affect the therapeutic response and induce drug resistance in cancer cells [91,92]. Several studies have described that exosomes released by drug-resistant tumor cells can transmit the drug-resistant phenotype to drug-sensitive cells, mainly by conferring drug efflux pumps and miRNAs [93,94]. This has been recognized as a new drug resistance "spread" mechanism. H1975 cell-derived exosomes from EGFR-TKI-resistant cells can induce gefitinib resistance in PC9-sensitive cells through the activation of the PI3K/Akt signaling pathway, because PI3K/Akt is a major EGFR signaling pathway and abnormal activation of the PI3K/Akt signaling pathway is one of the most relevant mechanisms of acquired resistance to targeted therapies in patients with EGFR-mutated NSCLC [93]. RTK endocytosis and its recycling to membrane may be an explanation for cetuximab in conjunction with Akt blockers, which would lead to interference in the signaling cascade involved in endosome formation [95,96], being more effective against tumor cells compared to cetuximab alone or combined with chemoradiotherapy [31], corroborating that high endocytic frequency of EGFR proteins in the cell membrane may be a key factor in resistance of tumor cells to treatments with anti-EGFR mAbs [70].

We believe that the correct combination of anti-EGFR antibodies and their potent ability to induce degradation of anti-EGFR receptors can be an efficient and intelligent strategy to sensitize cancer cells. Because cancer treatment is very complex, new strategies that use the combination of monoclonal antibodies with sensitizing agents, such as TKIs and Akt inhibitors (Fig. 1), are valuable, as they can help overcome resistance to current treatments, blocking of downstream EGFR pathways, such as the Akt/mTOR pathway, being essential to block tumor proliferation and survival.

To visualize the interaction of all proteins related to resistance and sensitivity to the new anti-EGFR therapeutic strategies addressed in this book chapter, we created two protein–protein interaction (PPIN) networks, which can be visualized in Fig. 2A and B, respectively (PPIN associating resistance proteins to anti-EGFR mAbs and PPIN associating sensitivity proteins to anti-EGFR mAbs). ERBB2, ERBB3, ERBB4, MAP2K1, NRAS, PIK3CA, VEGFA, DNM1, and PTEN stand out as central points in the PPIN of resistance to anti-EGFR mAbs, and stand out as central points in the PPIN of sensitivity to anti-EGFR mAbs, mainly AKT1, VEGFA, and SRC. If we look at the major proteins involved in both resistance and sensitivity, we observe that VEGFA is present in both networks, suggesting that its inhibition should be used in combinations targeting the VEGF/VEGFR pathway. Also appearing in both resistance and sensitivity network is the PI3K/AKT pathway, and, as we have already demonstrated in our studies [4,31], this is a promising target in oncology. Note in Fig. 2A that other members of the EGFR family are involved in resistance to anti-EGFR mAbs (ERBB2, ERBB3, and ERBB4), which suggests that the combination of monoclonal antibodies to different members of the EGFR family is also an interesting strategy, as described by other authors [87]. Different forms of receptor heterodimerization control cell proliferation in a subtle way through endocytosis [75,87]. We believe it is very important to invest in this class of drugs (anti-ERBB2, anti-ERBB3, and anti-ERBB4 antibodies) that can modulate endocytosis, as this process is closely related to tumor resistance. From these protein–protein interaction networks created considering relevant articles in the field of oncology, we suggest that these molecular targets should be studied together to understand how to reverse tumor resistance and how to better apply and combine novel sensitizing agents for therapeutic anti-EGFR antibodies.

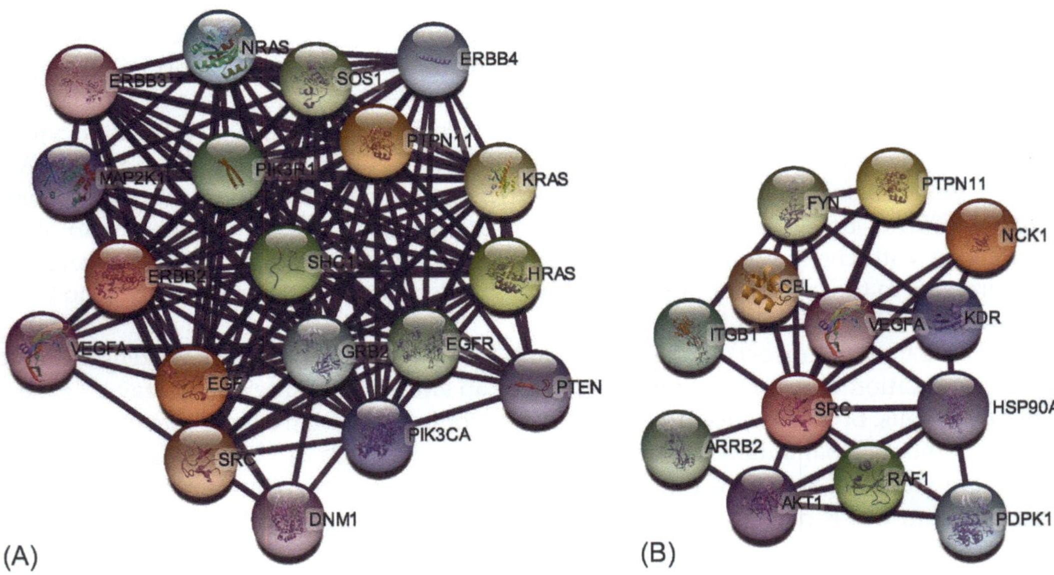

FIG. 2 Protein–protein interaction networks related to resistance and sensitivity to monoclonal antibodies. Predicted networks have non-directional edges that represent significance percentile of the physical interaction between proteins with a role in (A) monoclonal antibody resistance and (B) monoclonal antibody sensitivity. The yFiles Organic layout algorithm was used to predict network organizational structure. Networks (A) and (B) were created in Cytoscape by accessing the STRING database via stringApp.

Conclusion and future tendencies

After three decades of research in oncology, in which relatively unspecific treatments were used, we are living new paradigms, in which several therapeutic options have been offered to cancer patients. Although drugs with a specific molecular target, especially monoclonal antibodies, are very promising and efficient in the destruction of tumor cells, they also produce side effects worthy of medical attention, such as diarrhea, skin rash, and anaphylaxis.

Despite this, they are already being used in the clinic, with promising results and specific subgroups of patients who have specific molecular targets and no resistance markers can benefited of this new class of drugs together with surgery, chemotherapy, and/or radiotherapy. We know that correct selection of patients is of paramount importance for the antineoplastic therapy to work, and the development of new laboratory/biotechnological strategies to target it is expected. Currently, most researchers are directing antineoplastic therapy based on the combination of several drugs with a targeted target and not just a single target in combination with chemotherapy/radiotherapy. This is because the tumor has several escape and compensation mechanisms, in addition to complex signaling pathways making treatment difficult.

New inhibitors and sensitizing agents are not so specific for single targets, but with multiple specificity that target certain families of membrane receptors and kinases involved in the signal transduction and oncogenesis. Once EGFR is inhibited, tumor cell compensates by signaling HER2, HER3, cMET, IGF-1R, and other receptors, a phenomenon known as oncogenic shift. *KRAS*, *BRAF*, and *PIK3CA* mutations are intrinsic resistance mechanisms, which are quite different from the acquired resistance discussed previously.

We sought to clarify the mechanism of action of some monoclonal antibodies and their sensitizing agents that are under clinical study and those already used by cancer patients. In addition, we recalled a little of the molecular bases of cancer, as well as the molecular determinants of sensitivity/resistance that are necessary to understand how these new drugs work in the cellular environment. We believe that the future of technological innovation in oncology lies in the combination of targeted molecular immunotherapy (monoclonal antibodies) and its sensitizing agents with chemotherapy/radiotherapy for the best cancer treatment, ensuring patient's quality of life. Investing in therapeutic strategies that stimulate the degradation of EGFR-type receptors in rational combinations according to the resistance mechanism presented by the patient will be of paramount importance for the evolution of current cancer treatments. Our preclinical studies indicated that Akt pathway and EGFR may not be completely responsible, but cooperate in the resistance of cancer cells to mAbs, suggesting a rationale for clinical strategies directed to patients displaying a resistant profile to anti-EGFR therapies. We developed two protein–protein interaction (PPIN) networks, associating resistance and sensitivity proteins to anti-EGFR mAbs. The right time to use combination, which combination, dose, and when to undergo neoadjuvant treatment are questions that remain unanswered. Future therapeutic approaches are likely to involve the combination of different antineoplastic-targeted agents, including monoclonal antibodies combinations.

References

[1] INSTITUTO. NACIONAL DE CÂNCER JOSÉ ALENCAR GOMES DA SILVA (Brazil), Rio de Janeiro; 2021. Available at: http://www.inca.gov.br/. Access: July 2021.

[2] Rodrigues VHSL, Cunha Jr GF. Mecanismos múltiplos de resistência à quimioterapia. In: JRQ G, editor. Manual de Oncologia. 3. ed. São Paulo: BBS Editora; 2008. p. 199–204.

[3] Citri A, Yarden Y. EGF-ERBB signaling: towards the systems level. Nat Rev Mol CellBiol 2006;7:505–16.

[4] De Almeida VH, Ferreira CG, Meira DD. Novos fármacos com alvo molecular específico em oncologia. Revista da Sociedade Brasileira de Cancerologia 2011;XIII (48):183–90.

[5] Bergers G, Hanahan D, Coussens LM. Angiogenesis and apoptosis are cellular parameters of neoplastic progression in transgenic mouse models of tumorigenesis. Int J Dev Biol 1988;42:995–1002.

[6] Meira DD, et al. Combination of cetuximab with chemoradiation, trastuzumab or MAPK inhibitors. Br J Cancer 2009;101:782–91.

[7] Hanahan D, Weinberg RA. Hallmarks of cancer: the next generation. Cell 2011;144:646–74.

[8] Hoff PM, Katz A, Zogbi Y. Drogas com alvo molecular específico. In: Ferreira CGE, Rocha JC, editors. Oncologia Molecular. 2. ed. Atheneu: São Paulo; 2010. p. 407–13.

[9] Lenferink AE, Pinkas-Kramarski R, Van de Poll ML, et al. Differential endocytic routing of homo- and heterodimeric ErbB tyrosine kinases confers signaling superiority to receptor heterodimers. EMBO J 1998;17:3385–97.

[10] Tzahar E, Waterman H, Chen X, et al. A hierarchical network of interreceptor interactions determines signal transduction by Neu differentiation factor/neuregulin and epidermal growth factor. Mol Cell Biol 1996;16:5276–87.

[11] Katz M, Amit I, Yarden Y. Regulation of MAPKs by growth factors and receptor tyrosine kinases. BiochimBiophys Acta 2007;1773:1161–76.

[12] Dhillon AS, Hagan S, Rath O, et al. MAP kinase signaling pathways in cancer. Oncogene 2007;26:3279–90.

[13] Levkowitz G, Waterman H, Ettenberg SA, et al. Ubiquitin ligase activity and tyrosine phosphorylation underlie suppression of growth factor signaling by c-Cbl/Sli-1. Mol Cell 1999;4:1029–40.

[14] Kersemaekers AM, Fleuren GJ, Kenter GG, et al. Oncogene alterations in carcinomas of the uterine cervix: overexpression of the epidermal growth factor receptor is associated with poor prognosis. Clin Cancer Res 1999;5:577–86.

[15] Ozanne B, Richards CS, Hendler F, et al. Over-expression of the EGF receptor is a hallmark of squamous cell carcinomas. J Pathol 1986;149:9–14.

[16] Zafir-Lavie I, Michaeli Y, Reiter Y. Novel antibodies as anticancer agents. Oncogene 2007;26:3714–33.

[17] Mendelsohn J, Baselga J. The EGF receptor family as targets for cancer therapy. Oncogene 2000;19:6550–65.

[18] Meira DD, et al. Different antiproliferative effects of matuzumab and cetuximab in A431 cells are associated with persistent activity of the MAPK pathway. Eur J Cancer 2009;45:1265–73.

[19] Wheeler DL, Dunn EF, Harari PM. Understanding resistance to EGFR inhibitors- impact on future treatment strategies. Nat Rev Clin Oncol 2010;9:493–507.

[20] Inno A, et al. Is There a Role for IGF1R and c-MET Pathways in Resistance to Cetuximab in Metastatic Colorectal Cancer? Clin Colorectal Cancer 2011;4:325–32.

[21] Yoon J, Koo KH, Choi KY. MEK1/2 inhibitors AS703026 and AZD6244 may be potential therapies for KRAS mutated colorectal cancer that is resistant to EGFR monoclonal antibody therapy. Cancer Res 2011;71 (2):445–53.

[22] Rebucci M, et al. Mechanisms underlying resistance to cetuximab in the HNSCC cell line: role of AKT inhibition in bypassing this resistance. Int J Oncol 2011;38(1):189–200.

[23] Dunn EF, et al. Dasatinib sensitizes KRAS mutant colorectal tumors to cetuximab. Oncogene 2010;30(5):561–74.

[24] Weickhardt AJ, Tebbutt NC, Mariadason JM. Strategies for overcoming inherent and acquired resistance to EGFR inhibitors by targeting downstream effectors in the RAS/PI3K pathway. Curr Cancer Drug Targets 2010;10(8):824–33.

[25] Quesnelle KM, Grandis JR. Dual Kinase Inhibition of EGFR and HER2 Overcomes Resistance to Cetuximab in a Novel In Vivo Model of Acquired Cetuximab Resistance. Clin Cancer Res 2011;17(18):5935–44.

[26] De Roock W, et al. Effects of KRAS, BRAF, NRAS, and PIK3CA mutations on the efficacy of cetuximab plus chemotherapy in chemotherapy-refractory metastatic colorectal cancer: a retrospective consortium analysis. Lancet Oncol 2010;11(8):753–62.

[27] Moroni M, et al. Gene copy number for epidermal growth factor receptor (EGFR) and clinical response to antiEGFR treatment in colorectal cancer: a cohort study. Lancet Oncol 2005;6(5):279–86.

[28] Tim J, Kruser TJ, Wheeler DL. Mechanisms of resistance to HER family targeting antibodies. Exp Cell Res 2010;316(7):1083–100.

[29] Chen FL, Xia W, Spector NL. Acquired resistance to small molecule ErbB2 tyrosine kinase inhibitors. Clin Cancer Res 2008;14(21):6730–4.

[30] Garrett JT, Arteaga CL. Resistance to HER2-directed antibodies and tyrosine kinase inhibitors: mechanisms and clinical implications. Cancer Biol Ther 2011;11(9):793–800.

[31] Meira DD, Almeida VH, Mororó JS, Caetano MS, Nóbrega IP, Batista D, et al. Efficient Blockade of Akt signaling is a determinant factor to overcome resistance to Matuzumab. Mol Cancer 2011;10(1):151–8.

[32] Arndt J, Meira DD. Terapias alvo atualmente utilizadas para o tratamento do câncer de pulmão. Revista da Sociedade Brasileira de Cancerologia 2012;XIV(49):30–6.

[33] Cai W, Zeng L, Wang LF, et al. The latest battles between EGFR monoclonal antibodies and resistant tumor cells. Front Oncol 2020;10.

[34] Picon H, Guddati AK. Mechanisms of resistance in head and neck cancer. Am J Cancer Res 2020;10(9):2742–51.

[35] Nair S, Trummell HQ, Rajbhandari R, Thudi NK, Nozell SE, Warram JM, Willey CD, Yang ES, Placzek WJ, Bonner JA, Bredel M. Novas mutações de ectodomínio de EGFR associadas com ativação independente de ligante e resistência a cetuximabe na cabeça e câncer de pescoço. PLoS One 2020;15, e0229077.

[36] Wheeler DL, Huang S, Kruser TJ, et al. Mechanisms of acquired resistance to cetuximab: role of HER (ErbB) family members. Oncogene 2008;27:3944–56.

[37] Kapoor D, Bhatt S, Kumar M, et al. Ligands for Targeted Drug Delivery and Applications. Basic Fundamentals of Drug Delivery; 2019. p. 307–42.

[38] Marrocco I, Romaniello D, Yarden Y. Cancer Immunotherapy: The Dawn of Antibody Cocktails. Methods Mol Biol 2019;1904:11–51. https://doi.org/10.1007/978-1-4939-8958-4_2. 30539465.

[39] Trivedi S, Srivastava RM, Concha-Benavente F, Ferrone S, Garcia-Bates TM, Li J, et al. Anti-EGFR targeted monoclonal antibody isotype influences antitumor cellular immunity in head and neck cancer patients. Clin Cancer Res 2016;22:5229–37. https://doi.org/10.1158/1078-0432.CCR-15-2971.

[40] García-Foncillas J, Sunakawa Y, Aderka D, Wainberg Z, Ronga P, Witzler P, Stintzing S. Distinguishing Features of Cetuximab and Panitumumab in Colorectal Cancer and Other Solid Tumors. Front Oncol 2019;20 (9):849. https://doi.org/10.3389/fonc.2019.00849. PMID: 31616627; PMCID: PMC6763619.

[41] Mazorra Z, Chao L, Lavastida A, Sanchez B, Ramos M, Iznaga N, Crombet T. Nimotuzumab: beyond the EGFR signaling cascade inhibition. Semin Oncol 2018;45(1–2):18–26. https://doi.org/10.1053/j.seminoncol.2018.04.008. Epub 2018 May 2. PMID: 30318080.

[42] Palomo AG, Medinilla AL, Segatori V, Barroso MDC, Blanco R, Gabri MR, Pérez AC, Monzón KL. Synergistic potentiation of the anti-metastatic effect of anti EGFR mAb by its combination with immunotherapies targeting the ganglioside NGcGM3. Oncotarget 2018;9(35):24069–80. https://doi.org/10.18632/oncotarget.25290. PMID: 29844873; PMCID: PMC5963610.

[43] Tóth G, Szöőr Á, Simon L, Yarden Y, Szöllősi J, Vereb G. The combination of trastuzumab and pertuzumab administered at approved doses may delay development of trastuzumab resistance by additively enhancing antibody-dependent cell-mediated cytotoxicity. MAbs 2016;8(7):1361–70. https://doi.org/10.1080/19420862.2016.1204503. Epub 2016 Jul 5. PMID: 27380003; PMCID: PMC5058622.

[44] Burgess AW, Henis YI, Hynes NE, Jovin T, Levitzki A, Pinkas-Kramarski R, Yarden Y. EGF receptor family: twisting targets for improved cancer therapies. Growth Factors 2014;32(2):74–81. https://doi.org/10.3109/08977194.2014.896355. Epub 2014 Mar 18. PMID: 24641597.

[45] Bonomi PD, Gandara D, Hirsch FR, et al. Predictive biomarkers for response to EGFR-directed monoclonal antibodies for advanced squamous cell lung cancer. Annals of Oncology 2018;29(8):1701–9.

[46] Garcia J, Hurwitz HI, Sandler AB, et al. Bevacizumab (Avastin) in cancer treatment: A review of 15 years of clinical experience and future outlook. Cancer Treat Rev 2020;86, 102017.

[47] Mésange P, Bouygues A, Ferrand N, Sabbah M, Escargueil AE, Savina A, Chibaudel B, Tournigand C, André T, de Gramont A, Larsen AK. Combinations of Bevacizumab and Erlotinib Show Activity in Colorectal Cancer Independent of RAS Status. Clin Cancer Res 2018;24(11):2548–58. https://doi.org/10.1158/1078-0432.CCR-17-3187. Epub 2018 Feb 28. PMID: 29490990.

[48] Bignucolo A, De Mattia E, Cecchin E, et al. Pharmacogenomics of Targeted Agents for Personalization of Colorectal Cancer Treatment. Int J Mol Sci 2017;18(7):1522.

[49] Yarden Y, Sela M. *Cancer* Immunotherapy: More Is (Much) Better. Clin Cancer Res 2015;21(18):4030–2. https://doi.org/10.1158/1078-0432.CCR-15-0996. Epub 2015 May 27. PMID: 26019173.

[50] Jacobsen HJ, Poulsen TT, Dahlman A, et al. Pan-HER, an antibody mixture simultaneously targeting EGFR, HER2 and HER3 effectively overcomes tumor heterogeneity and plasticity. Clin Cancer Res 2015;21:4110–22.

[51] Johnson CB, Win SY. Combination therapy with PD-1/PD-L1 blockade: An overview of ongoing clinical trials. Oncoimmunology 2018;7(4), e1408744. https://doi.org/10.1080/2162402X.2017.1408744. PMID: 29632719; PMCID: PMC5889289.

[52] Ferris RL, Haddad R, Even C, et al. Durvalumab with or without tremelimumab in patients with recurrent or metastatic head and neck squamous cell carcinoma: EAGLE, a randomized, open-label phase III study. Ann Oncol 2020;31(7):942–50.

[53] Mcdermott DF, Huseni MA, Atkins MB, et al. Clinical activity and molecular correlates of response to atezolizumab alone or in combination with bevacizumab versus sunitinib in renal cell carcinoma. Nat Med 2018;24(6):749–57.

[54] Ferraro DA, Gaborit N, Maron R, et al. Inhibition of triple-negative breast cancer models by combinations of antibodies to EGFR. Proc Natl Acad Sci U S A 2013;110(5):1815–20.

[55] Machiels J, Specenier P, Krauss J, et al. A proof of concept trial of the anti-EGFR antibody mixture Sym004 in patients with squamous cell carcinoma of the head and neck. Cancer Chemother Pharmacol 2015;76(1):13–20.

[56] Napolitano S, Martini G, Martinelli E, et al. Therapeutic efficacy of SYM004, a mixture of two anti-EGFR antibodies in human colorectal cancer with acquired resistance to cetuximab and MET activation. Oncotarget 2017;8 (40):67592–67,604.

[57] Napolitano S, Martini G, Martinelli E, et al. Antitumor efficacy of triple monoclonal antibody inhibition of epidermal growth factor receptor (EGFR) with MM151 in EGFR-dependent and in cetuximab-resistant human colorectal cancer cells. Oncotarget 2017;8(47):82773–82,783.

[58] Singha NC, et al. Tumor-associated hyaluronan limits efficacy of monoclonal antibody therapy. Molecular Cancer Therapeutics 2015;14(2):523–32.

[59] Nevo J, et al. Mammary-derived growth inhibitor alters traffic of EGFR and induces a novel form of cetuximab resistance. Clin Cancer Res 2009;15(21):6570–81.

[60] Kasper S, et al. Molecular dissection of effector mechanisms of RAS-mediated resistance to anti-EGFR antibody therapy. Oncotarget 2017;8(28):45898–45,917.

[61] Bardelli A, et al. Amplification of the MET receptor drives resistance to anti-EGFR therapies in colorectal cancer. Cancer Discov 2013;3(6):658–73.

[62] Brand TM, et al. AXL mediates resistance to cetuximab therapy. Cancer Res 2014;74(18):5152–64.

[63] Lu Y, et al. LncRNA MIR100HG-derived miR-100 and miR-125b mediate cetuximab resistance via Wnt/β-catenin signaling. Nat Med 2017;23(11):1331–41.

[64] Landmesser ME, et al. Loss of p120ctn causes EGFR-targeted therapy resistance and failure. PLoS ONE 2020;15 (10).

[65] Akbay EA, et al. Activation of the PD-1 pathway contributes to immune escape in EGFR-driven lung tumors. Cancer Discov 2013;3(12):1355–63.

[66] Yang JCH, et al. Pembrolizumab in Combination With Erlotinib or Gefitinib as First-Line Therapy for Advanced NSCLC With Sensitizing EGFR Mutation. J Thorac Oncol 2019;14(3):553–9.

[67] Mellman I, Yosef Y. Endocytosis and cancer. Cold Spring Harb Persp Biol 2013;5(12). https://doi.org/10.1101/cshperspect.a016949. PMCID: PMC3839607. PMID: 24296170.

[68] Lee JM, et al. Novel strategy for a bispecific antibody: induction of dual target internalization and degradation. Oncogene 2016;35(34):4437–46.

[69] Volk AL, et al. Bispecific Antibody Molecule Inhibits Tumor Cell Proliferation More Efficiently Than the Two-Molecule Combination. Drugs in R&D 2021;21(2):157–68.

[70] Chew HY, et al. Endocytosis inhibition in humans to improve responses to ADCC-mediating antibodies. Cell 2020;180(5):895–914.

[71] Joseph SR, et al. An ex vivo human tumor assay shows distinct patterns of EGFR trafficking in squamous cell carcinoma correlating to therapeutic outcomes. J Invest Dermatol 2019;139(1):213–23.

[72] Brodsky FM. Diversity of clathrin function: New tricks for an old protein. Annu Rev Cell Dev Biol 2012;28:309–36.
[73] Evans PR, Owen DJ. Endocytosis and vesicle trafficking. CurrOpin Struct Biol 2002;12(6):814–21.
[74] Maldonado-Baez L, Wendland B. Endocytic adaptors: recruiters, coordinators and regulators. Trends Cell Biol 2006;16(10):505–13.
[75] Roepstorff K, Grovdal L, Grandal M, Lerdrup M, van Deurs B. Endocytic downregulation of ErbB receptors: mechanisms and relevance in cancer. Histochem Cell Biol 2008;129(5):563–78.
[76] Kim B, et al. Clathrin-mediated EGFR endocytosis as a potential therapeutic strategy for overcoming primary resistance of EGFR TKI in wild-type EGFR non-small cell lung cancer. Cancer Med 2020;10(1):372–85. https://doi.org/10.1002/cam4.3635. PMCID: PMC7826488, PMID: 33314735.
[77] Jayashankar V, Edinger AL. Macropinocytosis confers resistance to therapies targeting cancer anabolism. Nat Commun 2020;11(1121). https://doi.org/10.1038/s41467-020-14928-3. PMCID: PMC7048872. PMID: 32111826.
[78] Commisso C, Davidson SM, Soydaner-Azeloglu RG, Parker SJ, Kamphorst JJ, Hackett S, Grabocka E, Nofal M, Drebin JA, Thompson CB. Macropinocytosis of protein is an amino acid supply route in Ras-transformed cells. Nature 2013;497:633–7.
[79] Michalopoulou E, et al. Macropinocytosis renders a subset of pancreatic tumor cells resistant to mTOR inhibition. Cell Rep 2020;30(8):2729–42. https://doi.org/10.1016/j.celrep.2020.01.080. . PMCID: PMC7043007, PMID: 32101748.
[80] Kamphorst JJ, et al. Human pancreatic cancer tumors are nutrient poor and tumor cells actively scavenge extracellular protein. Cancer Res 2015;75:544–53. https://doi.org/10.1158/0008-5472.CAN-14-2211.
[81] Palm W, et al. The utilization of extracellular proteins as nutrients is suppressed by mTORC1. Cell 2015;162:259–70. https://doi.org/10.1016/j.cell.2015.06.017.
[82] Elstrom RL, Bauer DE, Buzzai M, Karnauskas R, Harris MH, Plas DR, Zhuang H, Cinalli RM, Alavi A, Rudin CM, Thompson CB. Akt stimulates aerobic glycolysis in cancer cells. Cancer Res 2004;64:3892–9.
[83] Engelman JA, Luo J, Cantley LC. The evolution of phosphatidylinositol 3-kinases as regulators of growth and metabolism. Nat Rev Genet 2006;7:606–19.
[84] Pavlova NN, Thompson CB. The emerging hallmarks of cancer metabolism. Cell Metab 2016;23:27–47.
[85] Howell JJ, Ricoult SJ, Ben-Sahra I, Manning BD. A growing role for mTOR in promoting anabolic metabolism. Biochem Soc Trans 2013;41:906–12.
[86] Karsli-Uzunbas G, Guo JY, Price S, Teng X, Laddha SV, Khor S, Kalaany NY, Jacks T, Chan CS, Rabinowitz JD, White E. Autophagy is required for glucose homeostasis and lung tumor maintenance. Cancer Discov 2014;4:914–27.
[87] Friedman, et al. Synergistic down-regulation of receptor tyrosine kinases by combinations of mAbs: Implications for cancer immunotherapy. Proc Natl Acad Sci USA 2005;102(6):1915–20. https://doi.org/10.1073/pnas.0409610102. PMCID: PMC548578. PMID: 15684082.
[88] Colombo M, Raposo G, Thery C. Biogenesis, secretion, and intercellular interactions of exosomes and other extracellular vesicles. Annu Rev Cell Dev Biol 2014;30:255–89.
[89] Ela S, Mager I, Breakefield XO, Wood MJ. Extracellular vesicles: biology and emerging therapeutic opportunities. Nat Rev Drug Discov 2013;12:347–57.
[90] Thery C. Exosomes: secreted vesicles and intercellular communications. F1000 Biol Rep 2011;3:15.
[91] Au Yeung CL, Co NN, Tsuruga T, et al. Exosomal transfer of stroma-derived miR21 confers paclitaxel resistance in ovarian cancer cells through targeting APAF1. Nat Commun 2016;7:11150.
[92] Wang J, Hendrix A, Hernot S, et al. Bone marrow stromal cell-derived exosomes as communicators in drug resistance in multiple myeloma cells. Blood 2014;124:555–66.
[93] Qu L, Ding J, Chen C, et al. Exosome-transmitted lncARSR promotes sunitinib resistance in renal cancer by acting as a competing endogenous RNA. Cancer Cell 2016;29:653–68.
[94] Wei F, Ma C, Zhou T, et al. Exosomes derived from gemcitabine-resistant cells transfer malignant phenotypic traits via delivery of miRNA-222-3p. Mol Cancer 2017;16:132.
[95] Chew CL, et al. In vivo role of INPP4B in tumor and metastasis suppression through regulation of PI3K–AKT signaling at endosomes. Cancer Dis 2015;5(7):740–51.
[96] Sugiyama MG, Gregory DF, Costin NA. Akt-ing up just about everywhere: compartment-specific Akt activation and function in receptor tyrosine kinase signaling. Front Cell Dev Biol 2019;7:70.

Index

Note: Page numbers followed by *f* indicate figures and *t* indicate tables.

A

AAV
  antibodies production, 44
  vector, 42–44
ADAM protease, 14
ADCC. *See* Antibody-dependent cell-mediated cytotoxicity (ADCC); Antibody-dependent cellular cytotoxicity (ADCC)
Adenocarcinoma, pancreatic ductal, 208–209
Adenosine triphosphate (ATP), 6
Ado-trastuzumab emtansine, 203
Afatinib, 3–4*t*, 122
  tumor growth inhibitory effect of, 120
AFM21 CD16A, 8–9
AKT, 15*f*
  activation, 20–21, 115
  HGF rescue occurs via MET-dependent activation of, 116
  pathway, 21, 56
    inhibition, 56
    signaling, 21
  phosphorylation, 57, 59
ALDH$^+$ cell population
  CT16 enhances the radiosensitivity of cancer cells and reduces, 86–87
  enrichment of, 85
Amivantamab, 3–4*t*
Amplification
  HER2 gene, 139
  of *MET*, 207–208
  MET activation and ERBB2, 37
Angiogenesis, 18–19
  IMO combined with cetuximab inhibits expression and, 186
Antagonists, EGFR, 168–170
Antiautophagic peptides
  modulate the effect of anti-EGFR therapy on tumor cells, 176
Antibody-dependent cell-mediated cytotoxicity (ADCC), 64, 186
Antibody-dependent cellular cytotoxicity (ADCC), 5, 106–107, 192
Antibody-receptor binding, EGFR ECD mutations affet, 30–31
Anti-EGFR
  treatments
    autophagy and response of cancer cells to, 175–176
    targeting autophagy solve the problem of resistance to, 175
  tyrosine kinase inhibitors, 6–7
Anti-HER3 antibody
  neutralizing, 65
    mAb33, 67
Anti-HER-targeted therapy, 56
Anti-pan-RAS antibody, 147
Antisense oligonucleotides, 9–10
Antistem cell effect of CT16, 86
Antitumor activity, 87
  of CT16, 86
  monoclonal antibodies, 192
  signaling proteins and causes potent, 186
  of ZD6474, 127
Antitumor effect, of EGFR inhibitors, 126
Apoptosis, cell proliferation and induction of, 36–37
Autophagosome, 174–175
Autophagy, 174, 176
  facilitates tumor resistance to targeted biological therapies, 174–175
  inhibitors, 175
  regulatory signaling, 174
Autophosphorylation, HER2, 57

B

BAY 86-9766, 139–140
BCR-ABL mutations, 32
BDR4, 93
BET. *See* Bromodomain and extra terminal (BET)
Bevacizumab, 120, 203
Biological therapies, autophagy facilitates tumor resistance to targeted, 174–175
Bioluminescent resonance energy transfer (BRET), 30
Bispecific antibodies, 108
BRAF mutations
  mutated HT29 cells, 134
  as predictor of response, 18
  V600E mutation, 77
BRD4. *See* Bromo-containing domain protein 4 (BRD4)

Breast cancer, 48–49
cells, NEDD4 expression, 168
BRET. *See* Bioluminescent resonance energy transfer (BRET)
Bromo-containing domain protein 4 (BRD4), 92
expression of, 95–96
gene silencing, 94
inhibition, 92
siRNA, 93
Bromodomain and extra terminal (BET)
blocking prevents the acquisition of cetuximab resistance, 95
family members, pharmacological inhibition, 93–94
inhibition, 95
protein family, 92
BxPC3, 51
model, 51
cell-burdened nude mouse model, 58

C

CACO2 cells, 162–163
CAGE, 176
CAL33 CTXR clone, 92–93
Calu-3 cells, 51
Cancer. *See also specific types*
breast, 48–49
cells, NEDD4 expression, 168
colorectal (*see* Colorectal cancer (CRC))
and growth factor pathway, molecular bases, 193–195
metastatic (*see* Metastatic colorectal cancer (mCRC))
nonsmall (*see* Nonsmall cell lung cancer (NSCLC))
potential molecular targets in, 193
therapy and mechanisms, 195–198
Cancer cells
development, 208
inhibitory effect of Pan-HER on, 51
RAS activation in, 146
Cancer stem cells (CSCs), 84, 156–157
EMT and, 85
Notch2/3 reduces number of, 87
in PDX models, 88
stimulation of, 156–157
Carboxy-terminal fragments (611-CTF), HER2, 78
Carcinogenesis, 193
Carcinogenic signaling pathway, targeting, 120
Caski cells, 210–211
Catumaxomab (Removab), 56
CCV. *See* Clathrin coat vesicle (CCV)
CDK9 activation, 92
Cell
culture models, 20
metabolism, disordered, 162
proliferation and induction of apoptosis, 36–37
surface receptor of CpG, 186
Cellular resistance, to cytotoxic agents, 191
Cetuximab (Erbitux), 3–4*t*, 5, 16, 18–19, 36–37, 43, 64, 120, 142, 146, 152, 180, 198–201, 199–200*t*
in clinic, 16–17
cytotoxicity and radiosensitizing potential of, 153
efficacy of, 164
ERBB2 signaling overcomes resistance to, 79
HGF rescued apoptosis of DiFi cells induced by, 116
HNSCC model of, 94
induced radiosensitization, JNK2 depletion enhances, 153
inhibitory effect of on CRC proliferation, 163
and IPI-926 in preclinical models, 182
mAb, 196
mAb33 and, 66
MEHD7945A inhibits growth of two tumor cell lines, 64
receptor tyrosine kinase (RTK) overexpression mediates resistance to, 95
resistant ReHCC827 cells, 59
SYM004 compared with, 36
therapeutic effects of, 30
ZD6474 and, 127–128
Cetuximab-KRAS-esiRNA complex, in vivo antitumor effect of, 134
Cetuximab-protamine-KRAS-esiRNA complex, effect of, 134
Cetuximab resistance
BET blocking prevents acquisition of, 95
head and neck squamous cell carcinoma (HNSCC) model of acquired, 92–93
HNSCC model of acquired, 92–93
resistant ReHCC827 cells, 59
sensitivity, 146
in vitro, pharmacological inhibition of BET family members, 93–94
in xenotransplantation (PDX) models, 92
Cetuximab-resistant (CR)
cells
cell line G465E-XL, 31
cell line HCC827 CR, 76–77
CRC cells, 36–37
ERBB2 amplification in, 76–77
GEO-CR xenografts, 187
phosphorylated in, 78
ReHCC827 cells, 59
SLC1A5 as overexpressed in cancer tissues, 162
patients, EGFR in, 79–80
tumor cells caused by K-RAS mutation, 147
xenograft models, 78
Cetuximab-sensitive CRC cell lines, 114
Cetuximab-treated FaDu cells, 154
Chemotherapy, 156–157
Clathrin coat vesicle (CCV), 208–209
Clinical Laboratory Improvement Amendment (CLIA), 18

cMet
receptors, 168
signaling pathway, 106–108
tyrosine kinase, 9
Colorectal cancer (CRC), 4–5, 17–18, 21, 100, 138. *See also* Metastatic colorectal cancer (mCRC)
cells
from cetuximab-induced G1 arrest, 115
inhibition of SLC1A5 induces DNA damage, 163–164
EGFR and, 36
malignant transformation of, 114
monoclonal antibodies in, drug resistance of targeted, 42
proliferation, inhibitory effect of cetuximab on, 163
SLC1A5 as overexpressed in cancer tissues of cetuximab-resistant, 162
tumor xenograft models of, 37–39
Colorectal cancer cell lines (DLD-1), 146
cell syngeneic colorectal cancer cell line, 148
cetuximab-sensitive, 114
Combination therapy, 141
CpG
cell surface receptor of, 186
DNA, 186
to affect tumor growth related to EGFR pathways, 186
oligonucleotides, 186
CRC. *See* Colorectal cancer (CRC)
CrossMab method, 56
CSCs. *See* Cancer stem cells (CSCs)
c-Src activity, 100
CT16
antistem cell effect of, 86
antitumor activity of, 86
enhances radiosensitivity of cancer cells and reduces $ALDH^+$ cell populations, 86–87
reduce proliferation of tumor cells, 87
C-terminal fragment, of HER2, 78
611-CTF, 78
CTXR HNSCC models, 93
Cyclooxygenase-2 (COX-2), 127
Cytoplasmic tyrosine kinase domain (TKD), 14
Cytotoxic agents, cellular resistance to, 191
Cytotoxicity and radiosensitizing potential of cetuximab, 153

D

Dasatinib, 100
ddPCR. *See* Droplet digital polymerase chain reaction (ddPCR)
Degradation
of HER3 proteasome, 169
PTEN, 15*f*, 21
Depatuxizumab, 8
DIFI, 114
DIRAS. *See* Distinct subgroup of the Ras family (DIRAS)
Disordered cell metabolism, 162
Distinct subgroup of the Ras family (DIRAS), 176
D-MUT tumors, 147
xenograft, 147
DNA repair mechanism, 72–73
DNA synthesis, 121
Double-strand breaks (DSB), 72
Downstream expression, 31
Droplet digital polymerase chain reaction (ddPCR), 31
Drug resistance, 140
acquired, 49
HER3 as essential regulatory target for acquired, 70–71
through tumor xenografts and cells, 71
Drug-resistant tumor, 106
cell lines, 18–19
cells, 158–159
DSB. *See* Double-strand breaks (DSB)
Duct carcinoma, 49
Duligotuzumab, 8
DVD-IG CRTB6, 56
inhibitory effect of, 57
monotherapy, 58
D-WT cells, proliferation of, 147

E

Early endosomes (EEs), 209–210
E-cadherin, 20–21
EEs. *See* Early endosomes (EEs)
EGFR. *See* Epidermal growth factor receptor (EGFR)
EGFR tyrosine kinase inhibitors (EGFR-TKIs), 174
EMD 72000. *See* Matuzumab
EMT. *See* Epithelial-mesenchymal transition (EMT)
Endocytic organelles, 209–210
Endocytosis, 208–212
Enzymatically produced (e)siRNA, 133
Epidermal growth factor (EGF), 30, 48, 180
heparin-binding, 21
kinase inhibitor, 65
Epidermal growth factor receptor (EGFR), 2, 14
activation of, 14, 18, 126, 180
antagonists, 168–170
with antibody mixture, 52
and autophagy regulatory signaling, 174
biology, 14–16
blockers, 84
blocking antibodies, 15*f*
in cetuximab-resistant patients, 79–80
coexpressed in genotypically diverse panel of cetuximab-sensitive CRC cell lines, 114
combined activation of EGFR and MET, 115
constitutive activation of EGFR downstream effector molecules, 21

Epidermal growth factor receptor (EGFR) *(Continued)*
and CRC, 36
C-terminal domain, 194
degradation, 164
ubiquitin–proteasome pathway-mediated, 163
dependent intracellular signaling, 37
disorders of EGFR internalization and degradation, 19
downregulation, 208–212
dual blockade of, 85–87
ECD mutants, 31
ECD mutations
in CRC patients, 30–31
MM-151 reduced frequency of, 32
epidermal-associated, 15–16
epitopes of, 36
expression and corresponding carcinogenic effects, 169
four-in-one antibodies effectively inhibit, 57
gene copy number (GCN) as a predictor of response, 17
HER3 dimerization, 70
and HhP in HNSCC, 181–182
inhibition, 181
phosphoproteome modifications after cetuximab-mediated, 153
inhibitors, 20–21, 65, 70, 84
antitumor effect of, 126
and irradiation, 84–85
primary resistance of, 138–139
resistant cells, 86
responsive cell lines, 85
tolerance, 64
in vitro, Notch2/3 inhibits resistance to, 85–86
in vivo, Notch2/3 inhibits resistance to, 86
inhibitory effect of Pan-HER on, 52
intracellular tyrosine kinase domain of, 36
to JNK2, JIP-4 connects, 154
kinase activation, 14
ligand, 48–49
precursor, 14
state, 17–18
monoclonal antibodies target, 199–200*t*
mutations, 17
Novel EGFR targeting strategies, 7–10
nuclear transport of, 164
overall survival (OR) prolongation in patients with, 88
overexpression of, 3
Pan-HER targets, 50–51
pathways, CpG DNA to affect tumor growth related to, 186
phosphorylation, 8, 57
and signaling, 152
positive tumors, 76
role in human cancer, 16
signal expression of SCC cell lines, 152
signaling, activation, 84
signaling pathway, 14–16, 174
downstream, 64, 164
signal transduction of, 120
structure with extracellular domain, 2*f*
subcellular localization of, 20
targeted antibodies
function, 16
mechanisms of resistance, 15*f*, 17–19
angiogenesis, 18–19
targeted drugs, clinical prospects, 156
targeted therapy, 176, 180
targeting, 3–7
TKIs against, 107
T790M substitution of, 3–7
tyrosine kinase domain (TKD) of, 16, 36
tyrosine kinase inhibitor gefitinib, 187
variant III (EGFRvIII), 4, 7–9
Epithelial cells, transform into mesenchymal cells, 20–21
Epithelial-mesenchymal transition (EMT), 20–21, 157, 180–181
and stem cell-like properties in NSCLC cells, 84–85
Epithelial to mesenchymal transition (EMT), 84
and CSCs, 85
ERbB
and HER receptor regulation, 70
proteins, 78, 193–194
receptors, 193–194
tyrosine kinases, 76
ERBB2, 76
amplification
in HCC827 CR, 76–77
mediates cetuximab resistance through Erk1/2 signaling, 76–77
resistance to cetuximab without, 77–78
dual inhibition of, 79–80
signaling
heregulin activates, 77–78
resistance to cetuximab, 79
therapeutic antibodies against, 79
ERK
activity, 147
Erk1/2 signaling, ERBB2 amplification mediates cetuximab resistance through, 76–77
pathway, efficacy, 147
phosphorylation of, 139–140
Erlotinib, 3–4*t*, 6, 88
resistant tumors, L858R/T790M, 108
Estrogen receptor, 156
Exosomes, 211
Extracellular domain (ECD) mutations, 30
Extracellular matrix (ECM), HNSCC cells growth in vivo, 152
Extracellular vesicles (EVs) regulation, 175–176

F

FACS. *See* Fluorescence-activated cell sorting (FACS)
FaDu cells
  cetuximab-treated, 154
  radiosensitization effect on, 153
FAK. *See* Focal adhesion kinase (FAK)
FL518, 57, 59
Fluorescence-activated cell sorting (FACS), 84–85
Focal adhesion kinase (FAK), 101, 168
  inhibitors of, 168–169
  and paxillin, 101
FOLFOX and targeted agents, 100
Forkhead box O3 (FOXO3A), 197
Four-in-one antibodies
  antitumor activity both in vitro and in vivo, 57–58
  design and characterization of, 56–57
  effectively inhibit EGFR, HER2, HER3, and VEGFR2, 57
  interfere with HER/MET crosstalk, 58
  as overcome resistance to HER therapy, 59
Furmonertinib, 3–4*t*
Futuximab (Sym-004), 9

G

Gefitinib, 3–4*t*, 6, 88
  EGFR tyrosine kinase inhibitor, 187
Gene silencing
  BRD4, 94
  by siRNA, 132
Genes with tensin homologs, 156
Gene transfer, 44
Genome-wide copy analyses, 76–77
Genome-wide expression profiling, 77
GEO cell lines, clones of, 141
GEO-CR xenografts, cetuximab-resistant, 187
GLI. *See* Glioma-associated oncogene (GLI)
GLI1 level, 181
Glioma-associated oncogene (GLI), 180
GLI1 promoter, 181
  luciferase assay, 181
Glucose transporter-1 (GLUT1), 162
Glutamine transporter, 162, 164
GPNA, inhibition of SLC1A5, 162–163
G protein-coupled receptor (GPCR)-like signal transducer, 180
Growth factor receptor. *See also* Epidermal growth factor receptor (EGFR)
  hepatocyte, 196
  human epidermal (*see* Human epidermal growth factor receptor (HER))
Guanosine triphosphate (GTP), 132

H

HB-EGF. *See* Heparin-binding EGF (HB-EGF)
HCC827 cells, NSCLC cell line, 76–77
HCC827 CR, ERBB2 amplification in, 76–77
HCC1954 (breast cancer cell line), xenografts of, 58–59
HCT116 tumors, KRAS mRNA levels in, 134
Head and neck squamous cell carcinoma (HNSCC), 16–17, 43, 64, 180
  cells growth in vivo, 3-dimensional (3D) extracellular matrix (ECM) study, 152
  clinical cohort of, 95–96
  EGFR and HhP in, 181–182
  model, 19–20
    of acquired cetuximab resistance, 92–93
    of cetuximab, 94
    PDX, 95
  secondary resistance to cetuximab in, 92
Hedgehog pathway, 183
Hedgehog signaling, 180
Heparin-binding EGF (HB-EGF), 21
Hepatocyte growth factor (HGF), 114
  induces resistance to cetuximab through MET activation, 115
  rescued apoptosis of DiFi cells induced by cetuximab, 116
  rescue occurs via MET-dependent activation of AKT and MAPK, 116
  rescues CRC cells from cetuximab-induced G1 arrest, 115
Hepatocyte growth factor receptor (HGFR), 196
HER. *See* Human epidermal growth factor receptor (HER)
HER2, 48–50, 193–194
  autophosphorylation, 57
  carboxy-terminal fragments (611-CTF), 78
  C-terminal fragment of, 78
  four-in-one antibodies effectively inhibit, 57
  gene, 38
    amplification, 139
HER3, 48–50
  downstream signals, 64
  as essential regulatory target for acquired drug resistance, 70–71
  four-in-one antibodies effectively inhibit, 57
  ligand neuromodulation proteins, 50
  overexpression, 95
  phosphorylation, 57–58
  proteasome, degradation, 169
  signaling, 65
    pathways, 64
  upregulation of, 56, 65–66
HER4, 48
Heregulin, 139
  activates ERBB2 signaling, 77–78
  upregulation, 76
HER2-HER3 dimer, 50
HER/MET crosstalk, four-in-one antibodies interfere with, 58

HER3-PI3K-AKT, 64
Heterodimers, 193–194
Heterogeneity, 49, 52
HGF. *See* Hepatocyte growth factor (HGF)
Hh ligands, 180
HhP, 180
  and EGFR in HNSCC, 181–182
HNSCC. *See* Head and neck squamous cell carcinoma (HNSCC)
HOV. *See* Human papilloma virus (HOV)
HSC6 cells, 94
HT29 cells, BRAF-mutated, 134
HT-29 xenograft, 121
Human epidermal growth factor receptor (HER), 48, 56
  family receptors, 51
  increased expression of HER family growth factors, 21
  Pan-HER (*see* Pan-HER)
  signaling as critical for resistance to HER-targeted therapy, 58–59
  targeted therapy, MET/HER signaling as critical for resistance to, 58–59
  therapy, four-in-one antibodies as overcome resistance to, 59
Human papilloma virus (HOV), 95–96
Hyperphosphorylation
  of JNK2, 154
  of Rac-MEK4/7-JNK-c-Jun signals, 153

I

IBD. *See* Inflammatory bowel disease (IBD)
Icotinib, 3–4*t*
IGF-1 receptors, 9
Immunomodulatory effects, 186
Immunomodulatory oligonucleotides (IMO)
  in combination with cetuximab and irinotecan, 187–188
  combined with cetuximab inhibits expression and angiogenesis, 186
  cooperation of, 187
  in vitro effect of, 187
Inflammatory bowel disease (IBD) and KRAS-amplified mCRC, 102
Inhibitors
  autophagy, 175
  EGF kinase, 65
  epidermal growth factor receptor (EGFR), 20–21, 65, 70, 84
    antitumor effect of, 126
    and irradiation, 84–85
    primary resistance of, 138–139
    resistant cells, 86
    responsive cell lines, 85
    tolerance, 64
    in vitro, Notch2/3 inhibits resistance to, 85–86
    in vivo, Notch2/3 inhibits resistance to, 86
  of focal adhesion kinase (FAK), 168–169
  kinase, 65
  MEK, 141–142, 196
  Met, 59–60
  miR-123, 175
  small-molecule kinase, 32
  tyrosine kinase (*see* Tyrosine kinase inhibitors (TKIs))
Integrin-linked kinase (ILK) activation, 20–21
Integrins, 168
Intracellular RAS-RAF-MEK-MAPK signaling pathway, 139
Intracellular signal transduction, oncogenic signaling pathway, 120–121
In vitro
  four-in-one antibodies as antitumor activity, 57–58
  Notch2/3 inhibits resistance to EGFR inhibitors, 85–86
  SLC1A5 promotes proliferation of CRC both, 162–163
  validation of KRAS-siRNA-anti-EGFR antibody complexes, 133
In vivo
  antitumor effect of cetuximab-KRAS-esiRNA complex, 134
  experiments, 43–44
  four-in-one antibodies as antitumor activity, 57–58
  Notch2/3 inhibits resistance to EGFR inhibitors, 86
  SLC1A5 promotes proliferation of CRC both, 162–163
IPI-926 in preclinical models, cetuximab and, 182
Irradiation, EGFR inhibitors and, 84–85

J

JIP-4 connects EGFR to JNK2, 154
JNJ-61186372 antibody, 106–109
JNK2
  depletion enhances cetuximab-induced radiosensitization, 153
  hyperphosphorylation of, 154
JQ1 sensitized drug-resistant cell lines, to cetuximab, 94

K

Keyhole edge limpet hemocyanin (KLH), 7–8
Kinase inhibitor, EGF, 65
Knobs-into-holes (KiH) method, 56
KRAS
  gene, 138–139
  gene mutation, 6
  inhibitors, 132
  mRNA levels in HCT116 tumors, 134
  mutant, 132
    GRO, 114
  mutation, 146
    cetuximab-resistant tumor cells caused by, 147

as predictor of response, 18
RAS amplification and, 101–102
role of, 101
protein, 132
silencing and tumor inhibitory activity, 133
KRAS-siRNA-anti-EGFR antibody complexes
generation and characterization of, 132–133
in vitro validation of, 133

L

Lapatinib, 3–4*t*
LAPTM4B. *See* Lysosomal-associated protein transmembrane 4 beta (LAPTM4B)
Ligand–receptor binding, EGFR ECD mutations affet, 30–31
Ligands, 193–194
EGFR, 48–49
epidermal growth factor receptor (EGFR), 48–49
precursor, 14
state, 17–18
Hh, 180
induced activation in signaling pathway, 107
induced EGFR activation, 192
neuromodulation proteins, 50
LIM1215 cells, inhibit growth of, 31
lncRNAs. *See* Long noncoding RNAs (lncRNAs)
Long noncoding RNAs (lncRNAs), 175–176
L858R/T790M erlotinib-resistant tumors, 108
LY3164530, 8–9
Lysosomal-associated protein transmembrane 4 beta (LAPTM4B), 175–176
Lysosomes, 174–175

M

mAbs. *See* Monoclonal antibodies (mAbs)
Macroautophagy, 174
Macrophages, 186
Macropinocytosis, 208–209
Macropinosomes, 208–209
Malignant transformation of CRC, 114
Mammary-derived growth inhibitor (MDGI), 20
transfected cells, 20
MAPK, 139
activation, 140–141
augmenting activation of, 115
HGF rescue occurs via MET-dependent activation of, 116
signaling pathway, 153
Matuzumab, 199–200*t*, 201
Maximum tolerated dose (MTD)-targeted inhibition, 139–140
mCRC. *See* Metastatic colorectal cancer (mCRC)
MDA-MB-231 cells, 58
MDGI. *See* Mammary-derived growth inhibitor (MDGI)
MEHD7945A, 57, 64, 70–71
inhibits radiation-induced survival and overcomes cross-resistance to radiation, 65
inhibits growth of
human tumor xenografts, 65
two tumor cell lines, 64
mechanism of, 71
in radiation-damaged cells, 72–73
Mesenchymal cells, epithelial cells transform into, 20–21
Mesenchymal epithelial transition (MET), 38
activation, 59–60, 114, 141–142
and ERBB2 amplification, 37
HGF induces resistance to cetuximab through, 115
amplification, 59–60, 114
coexpressed in a genotypically diverse panel of cetuximab-sensitive CRC cell lines, 114
combined activation of EGFR and MET, 115
expression
in CRC microenvironment, 114
downregulation, 116
inhibitors, 139, 141–142, 196
development, 147–148
pathway, 115
inhibition of, 115
phosphorylation of, 58
signaling as critical for resistance to HER-targeted therapy, 58–59
MET. *See* Mesenchymal epithelial transition (MET)
Metastasis of OSCC cells, 114
Metastatic colorectal cancer (mCRC), 43
cell lines, 21
IBD and KRAS-amplified, 102
overall survival (OS) of mCRC patients, 100
phase IB study, 100
RAS amplification and KRAS mutation, 101–102
role of KRAS mutation, 101
safety and efficacy, 100
time to treatment failure (TTF), 101
treatment, combination synergistic chemotherapy, 100
miR-123 inhibitors, 175
miRNA (micro-RNA) regulation, 175–176
Mitogen activated protein kinase (MAPK) cascades, 194
MM-151, 30
inhibit growth of
cetuximab-resistant clones in spontaneous models, 31
LIM1215 cells overexpressing EGFR ECD mutants, 31
reduced frequency of EGFR-ECD allele mutations, 32
Molecular heterogeneity, 42

Molecular markers, 196
in oncology, 193–194
Monoclonal antibodies (mAbs), 5, 36, 39, 70, 138, 197–198
antitumor action of, 192
block ligand-induced EGFR activation, 192
cetuximab, 196
combinations, 205
in CRC drug resistance of targeted, 42
drug resistance of targeted, 43–44
mAb33, 65
and cetuximab, 66
osimertinib with, 65–66
mAb-protamine-KRASesiRNA complexes, 134
target EGFR, 199–200*t*
therapeutic, 192*f*
and TKIs mechanistic differences and activity comparison between, 121
Monotherapy, ZD6474 as, 127–128
Morphogenesis, 114
MTJ1G, 158–159
MTT test, 163
Multiantibody mix therapy, 9
Multigene mutations, 138–139
Multiple cancer cell lines, synergistic antitumor effect against, 127–128
Multiple myeloma (MM) cells, 147–148
Multitargeted therapy strategy, 158
Multivesicular bodies (MVBs), 175–176
Mutation
BCR-ABL, 32
BRAF
mutated HT29 cells, 134
predictor of response, 18
as predictor of response, 18
V600E, 77
V600E mutation, 77
EGFR ECD mutations
in CRC patients, 30–31
MM-151 reduced frequency of, 32
KRAS, 146
cetuximab-resistant tumor cells caused by, 147
as predictor of response, 18
RAS amplification and, 101–102
role of, 101
multigene, 138–139
MVBs. *See* Multivesicular bodies (MVBs)

N
Nanoantibodies, 9
NanoLuc-EGFR, 30
Naphthoquinone-derived molecules, 9
National Cancer Institute (INCA), 191
Natural killer cells (NKs), 186
NDRG1 (N-myc downstream regulatory gene 1), upregulation of, 169–170
Necitumumab, 3–4*t*, 6, 199–200*t*, 203–204
NEDD4 expression, in breast cancer cells, 168
Nimotuzumab, 3–4*t*, 5, 199–200*t*, 202
NKs. *See* Natural killer cells (NKs)
Nonreceptor tyrosine kinase PYK2, 168
Nonsmall cell lung cancer (NSCLC), 43, 106, 174
cell line HCC827, 76–77
cells, 21
cell lines, 84–85
EMT and stem cell-like properties in, 84–85
potential therapeutic efficacy of CT16 in, 87
models, 19–20
xenograft model, 8–9
Nonsmall cell lung carcinoma (NSCLC), 64
Notch antagonist tarextumab, 158–159
Notch receptors, 159
Notch2/3 receptors, ALDH$^+$ cell population on, 85
Notch signaling pathway, 157
NSCLC. *See* Nonsmall cell lung carcinoma (NSCLC)
Nuclear localization signal domain (NLS), 14

O
Oligonucleotides
antisense, 9–10
CpG, 186
Olmutinib, 3–4*t*
Oncogenic shift, 19–20
Oncogenic signaling pathway, intracellular signal transduction, 120–121
Oncology, molecular markers in, 193–194
Oral squamous cell carcinoma (OSCC) cells, metastasis, 114
Osimertinib, 3–4*t*, 174–175
inhibits growth of human tumor xenografts, 66
with mAb33, 65–66
Overall survival (OR) prolongation, in patients with EGFR, 88

P
Pancreatic ductal adenocarcinoma (PDAC), 208–209
proliferation of cells in, 208–209
Pan-HER, 50, 204
activity, 51
antiproliferative effect of, 50
induced ADCC levels, 51–52
inhibitory action of, 204
inhibitory effect of, 51
on EGFR, 52
targets EGFR, 50–51
treated BxPC3 or Calu-3 tumors, 51
Panitumumab (Vectibix), 3–4*t*, 5, 16, 36, 199–200*t*, 202
FOLFOX4, 101
therapeutic effects of, 30
PAN-SFK inhibitor, 20
Patient-derived xenograft (PDX), 158

CRC cell model, 31
models, 65–66, 204
CSCs, 88
of HNSCC patients, 95
PDX-7157 tumors, 95
Patient's metastatic lesion, 66
Paxillin, focal adhesion kinase (FAK) and, 101
PD-1. *See* Programmed death receptor-1 (PD-1)
PD98059, 147
PDAC. *See* Pancreatic ductal adenocarcinoma (PDAC)
PD-1/PDL1 pathways, 204–205
PDX. *See* Patient-derived xenograft (PDX)
PECAPJ49 CTXR, 92–93
Peripheral tissues, 162
Pertuzumab, 48, 199–200*t*, 202–203
Phosphatase and tensin homolog (PTEN), 194
degradation, 15*f*, 21
expression, 21
PTEN-PIK3CA-AKT, 138
Phosphatidylinositol triphosphate (PIP3), 194
Phosphoproteome
array technique, 153
modifications after cetuximab-mediated EGFR inhibition, 153
Phosphorylation
AKT, 57, 59
EGFR, 8, 57, 152
ERK, 139–140
HER3, 57–58
of mesenchymal epithelial transition (MET), 58
Met inhibitors, 59–60
RNA polymerase, 92
Phosphotyrosine-binding protein, 14
PI3K (phosphatidylinositol 3-kinase)
Akt pathway, 146, 196
binding site, 194
inhibition, 92
phosphorylates, 194
PIK3CA gene, 207
PIP3. *See* Phosphatidylinositol triphosphate (PIP3)
Plasma membrane, 120
Pn vitro experiment, 33
Polymerase phosphorylation, RNA, 92
Programmed death receptor-1 (PD-1), 204–205, 208
Proliferation, 146–147
of CRC, 162–163
of D-WT cells, 147
Protamine, 132–133
Protein–protein interaction (PPIN) networks, 211–212, 212*f*
Proteins
ERBB, 78
kinase, RAF, 139
KRAS, 132
modification, 14
Proteomic analysis, 169
PTEN. *See* Phosphatase and tensin homolog (PTEN)
PYK2, 168
and FA, 171
inhibitors of, 168–169
PYK2-NDRG1-NEDD4, 170–171

R

Rac-MEK4/7-JNK-c-Jun signals, hyperphosphorylation of, 153
Radiation
damaged cells, mechanism of MEHD7945A, 72–73
MEHD7945A inhibits radiation-induced survival and overcomes cross-resistance to, 65
X-ray, 153
Radiosensitization, effect on FaDu cells, 153
Radiotherapy, 156–157
tumor recurrence after, 87
RAF, protein kinase, 139
Ramucirumab, 203
RAS
activation in cancer cells, 146
activity, 147
amplification and KRAS mutation, 101–102
RAS-RAF-MEK-MAPK, 138–140
pathway, 140
signaling system, 141–142
activation, 141–142
R451C, 31
Receptor tyrosine kinase (RTK), 14, 32, 66, 171
activation of, 92–93
blocking BET inhibit expression of, 94
expression of, 92, 95–96
overexpression mediates resistance to cetuximab, 95
signal, 120
signaling pathway, 42
Recycling endosomes, 209–210
Redox homeostasis, 162
ReHCC827 cells, 59
cetuximab resistant, 59
Resistant cells, 196–197
Rindopepimut immunotherapy, 7–8
RNA
interference techniques, 9–10
polymerase phosphorylation, 92
RNAi, functional, 43–44
RTK. *See* Receptor tyrosine kinase (RTK)

S

SCC cell lines, EGFR signal expression of, 152
SCC47 cells, 94
SCCHN. *See* Squamous cell carcinoma of head and neck (SCCHN)
Sensitivity, cetuximab resistance, 146
Sensitization therapies, 198–201

Serine–threonine kinase, 18
SFKs. *See* Src family kinases (SFKs)
Signaling networks, dynamic and complex nature of, 194–195
Signaling pathway, 43–45
  EGFR downstream, 164
  intracellular RAS-RAF-MEK-MAPK, 139
  ligand-induced activation in, 107
  RTK, 42
  targeting carcinogenic, 120
Signaling proteins, and causes potent antitumor activity, 186
Signaling receptors, adequate endocytic uptake of, 208–209
Signal transducer, G protein-coupled receptor (GPCR)-like, 180
Signal transduction, 70–71, 115
  intracellular, 120–121
  pathways, 3
  in tumor cells, 120
siRNA
  BRD4, 93
  drugs, limitation, 133
  gene silencing by, 132
  targeting, codelivery, 132
SLC1A5, 162
  downregulation of, 163
  inhibition of, 163
    GPNA, 162–163
    induces DNA damage in CRC cells, 163–164
  overexpressed in cancer tissues of cetuximab-resistant CRC patients, 162
  promotes proliferation of CRC both in vitro and in vivo, 162–163
  role of, 162
  targeting, 163
  ubiquitin–proteasome pathway-mediated EGFR degradation, 163
Small interfering RNAs (siRNAs), 132
Small-molecule inhibitors of receptor, tyrosine kinases (TKIs), 48
Small-molecule kinase inhibitors, 32
Sodium-dependent transporter, 162
Soft agar colony formation assays, 127–128
Sorafenib, 141
S-phase cells, 162–163
SPR. *See* Surface plasmon resonance (SPR)
Squamous cell carcinoma of head and neck (SCCHN), 67
Src family kinases (SFKs), 14–15, 94, 206–207
Stem-like cells, triple-negative breast cancer (TNBC), 158
Student's t-test, 121
Surface plasmon resonance (SPR), 57, 85
SW480 cells, 162–163
SYM004, 36–37
  compared with cetuximab, 36
Synergistic antitumor effect, against multiple cancer cell lines, 127–128
Synergistic effect, 49

T

Targeted agents, FOLFOX and, 100
Targeting carcinogenic signaling pathways, 120
Tensin homologs, genes with, 156
Tetravalent antibodies, 57
Therapeutic antibodies, 208–209
Therapeutic mAbs, mechanisms of action and combinations, 198–204
TICs. *See* Tumor-initiating cells (TICs)
TKIs. *See* Tyrosine kinase inhibitors (TKIs)
TLR9, 186
TNBC. *See* Triple-negative breast cancer (TNBC)
Transformation, 147
Transforming growth factor-α (TGF-α), 127, 186
Transmembrane tyrosine receptors, 36
Transporter, sodium-dependent, 162
Trastuzumab, 199–200*t*, 202
  treatment, effect of, 50–51
Triple-negative breast cancer (TNBC), 156, 168–169
  cell line function research, 169
  stem-like cells, 158
Tu-167 cells, 181
Tumor
  drug-resistant, 106
  growth, 58
    inhibitory effect of afatinib, 120
  recurrence after radiotherapy, 87
  resistance, 195–196, 205–208
  xenograft, 187–188
    and cells, drug resistance through, 71
    models of human colorectal cancer, 37–39
    osimertinib inhibits growth, 66
Tumor cells, 16–18, 20–21, 49
  A431, 14
  antiautophagic peptides modulate effect of anti-EGFR therapy on, 176
  CT16 reduce proliferation of, 87
  drug-resistant tumor cell lines, 18–19
  resist EGFR blockade, 15*f*
  signal transduction in, 120
Tumorigenesis, 14, 76
Tumor-initiating cells (TICs), 87, 156–157
Tumor microenvironment (TME), 206
Tyrosine kinase domain (TKD)
  cytoplasmic, 14
  of EGFR, 16, 36
Tyrosine kinase inhibitors (TKIs), 2–6, 9, 56, 141, 192, 192*f*
  anti-EGFR, 6–7
    therapeutic effect, 7
  drugs, 32

against EGFR, 107
first-generation EGFR, 106
and mAbs, mechanistic differences and activity comparison between, 121
Tyrosine kinases, 48
ERBB receptor, 76
of HER family, 48–49
Tyrosine residues, 194

U

Ubiquitin–proteasome pathway-mediated, EGFR degradation, 163
Unc-51-like kinase (ULK), 174

V

Vandetanib. *See* ZD6474
Vargatef, 122
Vector immunoprophylaxis (VIP)
cetu/VIP Dalo, 44–45
cross design and effect presentation of, 44–45
ficl/VIP Dalo, 44–45
phased therapeutic strategy of antibody transformation in vivo based on, 42–43
VEGF, 18–19, 126
inhibitor, 18–19
signal transduction of, 120
VEGFR, inhibition of, 126–127
VEGFR1, overexpression of, 207–208
VEGFR2, 126
four-in-one antibodies effectively inhibit, 57
Vimentin expression levels, 20–21
Vim-positive cells, 181
VIP. *See* Vector immunoprophylaxis (VIP)

W

Wild-type CRC, 42
Wild-type receptor (pNanoLuc-EGFR WT), 30–31

X

Xenograft
models
cetuximab-resistant (CR), 78
Xenografts
D-MUT tumors, 147
GEO-CR, 187
HCC1954 (breast cancer cell line), 58–59
HT-29, 121
inhibits growth of human tumor, 66
model
cetuximab-resistant, 78
NSCLC, 8–9
patient-derived, 95
patient-derived (*see* Patient-derived xenograft (PDX))
tumor, 187–188
xenograft models of CRC, 37–39
Xenotransplantation (PDX) models, cetuximab resistance in, 92
X-ray radiation, 153

Z

ZD6474, 3–4*t*, 126
antitumor activity of, 127
and cetuximab, 127–128
efficiently inhibited growth of EGFR-resistant tumors, 126–127
as monotherapy, 127–128

FSC
www.fsc.org
MIX
Papier | Fördert
gute Waldnutzung
FSC® C083411

Kolibri 360 GmbH
Ferdinand-Jühlke-Straße 7
99095 Erfurt, Deutschland
produktsicherheit@kolibri360.de